Etiology of Hodgkin's Disease

NATO ASI Series
Advanced Science Institutes Series

A series presenting the results of activities sponsored by the NATO Science Committee, which aims at the dissemination of advanced scientific and technological knowledge, with a view to strengthening links between scientific communities.

The series is published by an international board of publishers in conjunction with the NATO Scientific Affairs Division

A	Life Sciences	Plenum Publishing Corporation
B	Physics	New York and London
C	Mathematical	Kluwer Academic Publishers
	and Physical Sciences	Dordrecht, Boston, and London
D	Behavioral and Social Sciences	
E	Applied Sciences	
F	Computer and Systems Sciences	Springer-Verlag
G	Ecological Sciences	Berlin, Heidelberg, New York, London,
H	Cell Biology	Paris, Tokyo, Hong Kong, and Barcelona
I	Global Environmental Change	

PARTNERSHIP SUB-SERIES

1. Disarmament Technologies	Kluwer Academic Publishers
2. Environment	Springer-Verlag
3. High Technology	Kluwer Academic Publishers
4. Science and Technology Policy	Kluwer Academic Publishers
5. Computer Networking	Kluwer Academic Publishers

The Partnership Sub-Series incorporates activities undertaken in collaboration with NATO's Cooperation Partners, the countries of the CIS and Central and Eastern Europe, in Priority Areas of concern to those countries.

Recent Volumes in this Series:

Volume 278 — Obesity Treatment: Establishing Goals, Improving Outcomes, and Reviewing the Research Agenda
edited by David B. Allison and F. Xavier Pi-Sunyer

Volume 279 — Organization of the Early Vertebrate Embryo
edited by Nikolas Zagris, Anne Marie Duprat, and Antony Durston

Volume 280 — Etiology of Hodgkin's Disease
edited by Ruth F. Jarrett

Series A: Life Sciences

Etiology of Hodgkin's Disease

Edited by

Ruth F. Jarrett

University of Glasgow
Glasgow, Scotland

Plenum Press
New York and London
Published in cooperation with NATO Scientific Affairs Division

Proceedings of a NATO Advanced Research Workshop on
the Aetiology of Hodgkin's Disease,
held May 2–5, 1994,
at Loch Lomond, Scotland

NATO-PCO-DATA BASE

The electronic index to the NATO ASI Series provides full bibliographical references (with keywords and/or abstracts) to about 50,000 contributions from international scientists published in all sections of the NATO ASI Series. Access to the NATO-PCO-DATA BASE is possible in two ways:

—via online FILE 128 (NATO-PCO-DATA BASE) hosted by ESRIN, Via Galileo Galilei, I-00044 Frascati, Italy

—via CD-ROM "NATO Science and Technology Disk" with user-friendly retrieval software in English, French, and German (©WTV GmbH and DATAWARE Technologies, Inc. 1989). The CD-ROM also contains the AGARD Aerospace Database.

The CD-ROM can be ordered through any member of the Board of Publishers or through NATO-PCO, Overijse, Belgium.

ISBN-13: 978-1-4613-8005-4   e-ISBN-13: 978-1-4613-0339-8
DOI: 10.1007/978-1-4613-0339-8

© 1995 Plenum Press, New York
Softcover reprint of the hardcover 1st editon 1995
A Division of Plenum Publishing Corporation
233 Spring Street, New York, N. Y. 10013

10 9 8 7 6 5 4 3 2 1

PREFACE

This volume reports the proceedings of a NATO Advanced Workshop held at Cameron House Hotel, Loch Lomond, Scotland, from May 2 - 5, 1994. The major impetus for this workshop was the realisation, over the past 7 years, that the Epstein-Barr virus is associated with a proportion of cases of Hodgkin's disease and is likely to play an aetiological role. There were four main aims of the workshop: first, to discuss the recent findings in relation to Epstein-Barr virus and the aetiology of Hodgkin's disease; second, to relate these data to the epidemiology of Hodgkin's disease; third, to discuss other potential aetiological factors and finally, to discuss future directions for research into Hodgkin's disease.

Leading experts in the field have contributed chapters to this volume. There is some overlap among chapters, particularly regarding Epstein-Barr virus, thereby allowing different groups to express views on similar topics. Perhaps, however, the most surprising feature of the workshop was the lack of controversy regarding the role of Epstein-Barr virus in Hodgkin's disease, an association that was treated with great scepticism at the beginning of the decade.

The first three chapters, by Alexander, Taylor *et al.*, and Levine *et al.*, discuss the epidemiology of Hodgkin's disease with particular attention to clustering and genetic susceptibility. These chapters represent the first attempt to bring together epidemiological and molecular studies in Hodgkin's disease.

Young and Niedobitek provide a lucid overview of the biology of the Epstein-Barr virus and recent data relating to involvement of this virus in lymphomas. Data relating to Hodgkin's disease and Epstein-Barr virus are summarised by Jarrett, and Jiwa *et al.* describe the correlation between the presence of Epstein-Barr virus and other features of the biology of Hodgkin's disease, such as the expression of cellular antigens.

The next chapter by Jarrett and Armstrong describes the molecular epidemiology of EBV-associated Hodgkin's disease and the possible involvement of other viruses in Epstein-Barr virus-negative cases. Hamilton-Dutoit provides a clear review of Hodgkin's disease in the context of HIV infection, and Levine *et al.* discuss the data relating to the possible involvement of human herpesvirus-6.

Attention then turns to the biology of Hodgkin's disease and the Reed-Sternberg cell. Dürkop *et al.* and Gruss and Dower provide up-to-date reviews on the CD30 molecule and its ligand. Trümper *et al.* then summarise their innovative studies on single, micromanipulated Reed-Sternberg cells, including experiments performed at the DNA and RNA level. Poppema and Visser focus on the nature of the reactive component of Hodgkin's lesions and also discuss intriguing new findings relating to expression of HLA.

The difficulty of propagating Reed-Sternberg cells is described in chapters by Kapp *et al.* and Krajewski *et al.* Data relating to Hodgkin's-derived cell lines are summarised. Recent data on the development of potential animal models, particularly the SCID mouse system, are described.

The final chapters by Weiss and Piris *et al.* review the literature and present new data relating to the involvement of oncogenes and tumour-suppressor gene in Hodgkin's disease.

The participants wish to thank NATO for its sponsorship of this meeting, which was greatly appreciated.

ACKNOWLEDGMENT

I am indebted to Bette Gibb, without whom the workshop and this book would not have come to fruition.

CONTENTS

EPIDEMIOLOGY: OVERVIEW AND SPATIAL CLUSTERING

Freda E Alexander

Department of Public Health Sciences, The University of Edinburgh, Medical School, Teviot Place, Edinburgh, EH8 9AG, UK

INTRODUCTION

The first section of this chapter provides an overview of the epidemiology of Hodgkin's disease (HD), much of which is well known. Particular attention will be paid to the possible involvement of infectious agents in the aetiology and here attention becomes focused on young people, especially young adults. The second part considers clustering of HD in much greater depth. It is a basic belief of the author that clustering should not be considered in isolation, but forms an integral part of the investigation of infectious agents and HD. For this reason there is a single concluding paragraph covering the two sections.

OVERVIEW

The two-disease hypothesis

One of the most striking features of the descriptive epidemiology of HD is the bimodal age-incidence curve with its prominent peak in young adults (20-29 years) in developed countries. The second peak in older people (aged over 60 years) remains evident in most (Glaser and Swartz, 1990; Parkin *et al.*, 1992) but not all (McKinney *et al.*, 1989) registries. This was one of the factors which prompted MacMahon (1966) to suggest that HD encompasses two separate diseases, with an "infectious aetiology" applying to younger cases whilst the causes of older cases might be similar to those of non-Hodgkin's lymphoma (NHL). A specific hypothesis for HD in young adults, the late-host-response model (Fig.1), has been formulated (Gutensohn and Cole, 1977) and will be discussed below.

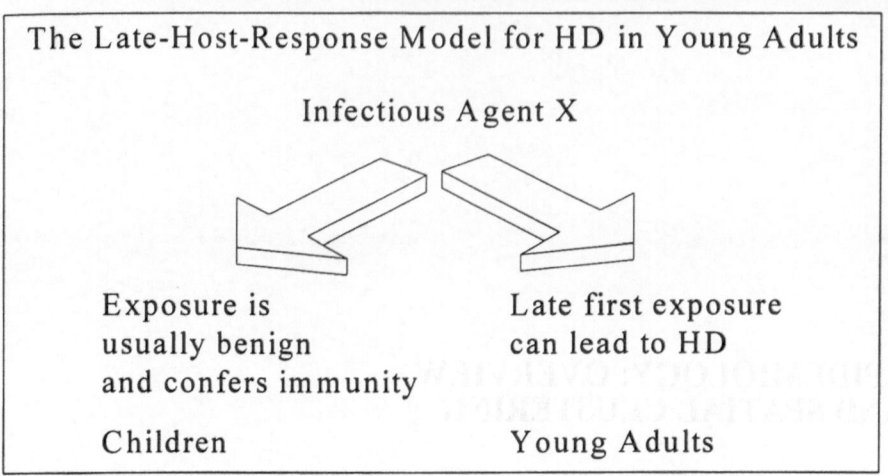

Fig. 1:   The late-host-response model.

Since the two-disease hypothesis was raised by MacMahon (1966), supporting epidemiological evidence has accumulated. The sex ratio and the Rye-type distribution are both age-specific (Glaser, 1986; Alexander *et al.,* 1991); in general, male cases predominate but rates for females in the young adult peak in developed countries are essentially the same as those for males. Most cases in young adults, especially in women, are of nodular sclerosing (HDNS) type, and the age-incidence curve for HDNS disease has a single mode (Fig. 2) .

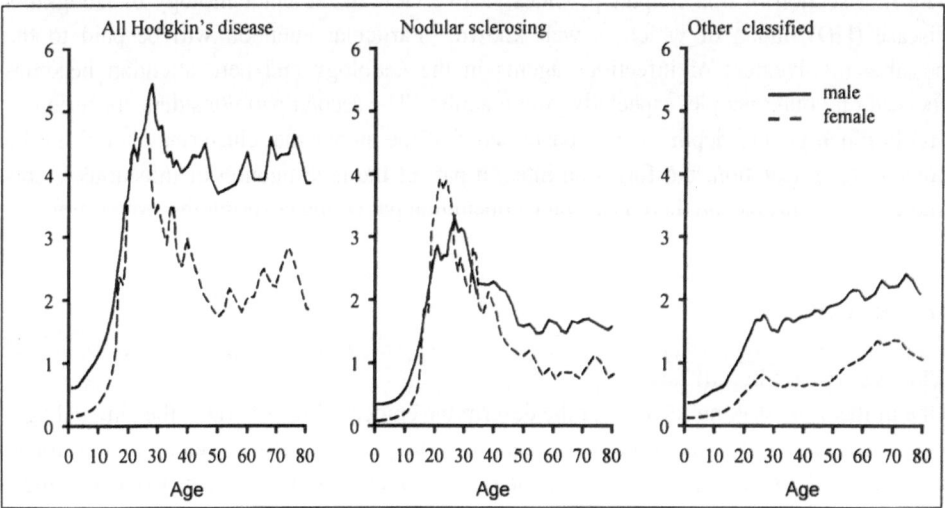

Fig. 2:   Age-incidence of Hodgkin's disease.

A recent international study of time trends (Hartge *et al.,* 1994) shows rising rates for young adults and a recent decline in rates in older people. These trends are evident, primarily, in the US data but appear also in other "industrialised areas such as Quebec, Birmingham and Sweden".

Ecological (Alexander *et al.,* 1991; Gutensohn and Cole, 1977) and case-control (Gutensohn, 1982) studies show that area and personal socio-economic and demographic characteristics associated with risk of HD in young and older people are different. Areas with higher socio-economic status had higher rates of HD in young people but lower rates in older (50 years and more) people, with both the trends and the interaction with age being statistically significant (Alexander *et al.,* 1991). Compared with controls, young cases come from smaller families and are more likely to have lived in a single family unit (Gutensohn, 1982); these patterns were reversed for older subjects ($\geq$55 years).

For HD in young people (or young adults) these results provide the basic evidence for an infectious aetiology. Supportive findings discussed below often apply to young rather than older cases and provide confirmation of the two disease hypothesis. Amongst these we must include the results of analyses of spatial clustering which will be discussed in the second half of the chapter.

Finally, the proportions of tumours which contain Epstein-Barr virus (EBV) genomes within the Reed-Sternberg cells (EBV-RS+) show a striking difference according to age; the proportion in young adults is low compared to that in older people (those 50 years and older). This question will be considered in more detail in later chapters but is relevant here as a final piece of evidence for the two-disease hypothesis. It is intriguing and somewhat disconcerting that these data provide support for viral involvement in the pathogenesis of HD which is strongest outside the young adult age group.

In the remainder of the overview attention will be focused on HD in young people since this aspect of epidemiology is most relevant to the general aims of the workshop. As far as older people are concerned, MacMahon's hypothesis of aetiological similarity with NHLs is supported in part by reports of common risk factors (e.g., herbicides, pesticides, immune suppression) but there are some notable differences. In particular, HD has not shared the rapid increase with time which has been seen for NHL (Hartge *et al.,* 1994) and the frequency of EBV involvement in HD is much greater than in NHL.

Children and young adults: the polio model
The discussion above is consistent with two alternative interpretations of the epidemiological distinction between HD in children and young adults:
 i) The polio model (Fig. 3): there is a single common infectious agent which is associated with HD in both children and young adults. Late first exposure conveys a higher risk of HD and is typical of HD in young adults in developed countries. Earlier first exposure, especially when associated with host factors which diminish the immune response (e.g., malnutrition), is a major cause of HD in children.
 ii) Distinct aetiologies: the alternative proposes that the aetiological agent which is active under the late-host-response model for HD in young adults has no relevance to childhood HD

Arguments supporting the polio model include considerations of parsimony and simplicity; additional support comes from the continuity of the age-incidence curves from the pattern typical of developing countries, with a peak in childhood, to that of developed countries

3

with the young adult peak and including a recognised intermediate stage (Stalsberg, 1973; Merk *et al.*, 1990). One of the most powerful arguments has come from the demonstration by Correa and O'Conor (1971) of a strong inverse correlation in selected international data (from Doll *et al.*, 1970) between incidence rates of HD in children and young adults. However, a recent analyses of the total computerised database (MacFarlane *et al.*, 1995) has failed to confirm this association. Registries not incorporated in the previous analysis included Baltic states and countries of central and eastern Europe where high rates in young adults were not accompanied by very low rates in children. These new data support the Distinct Aetiology model.

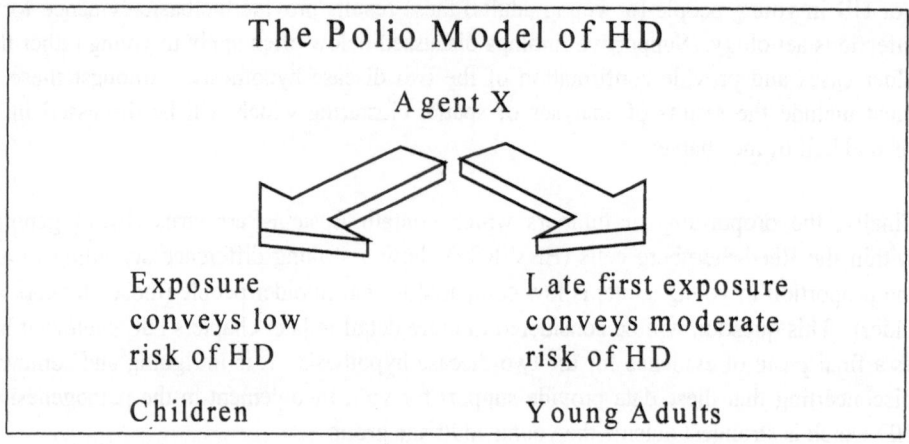

Fig. 3:   The polio model of Hodgkin's disease.

Analytical studies tend to suggest that the same risk factors apply to HD in children and young adults but the direction of the association is different (Gutensohn and Shapiro, 1982). Insofar as the risk factors are proxies for age at first exposure to the unknown agent X, these results are supportive of the polio model.

The strongest argument against the polio model is the demonstration that childhood HD is much more likely to be EBV-RS+ than HD in young adults (see Jarrett and Armstrong, this volume). These findings indicate that the polio model is unlikely to hold with EBV as agent X.

The aetiology of HD in children in developed countries is poorly understood because of the rarity of this disease.

The late-host-response model for Hodgkin's disease in young adults
The hypothesis is that late first exposure to a common infectious agent X can, in the presence of suitable host factors, lead to HD after a lengthy and variable latent period. Epidemiology can contribute to understanding and confirmation of both parts of this hypothesis.

Reference has already been made to both ecological and analytical epidemiological studies which show that HD in young adults is associated with factors which are proxies for late exposure to common infectious agents; these include family structure (higher risk for members of small families), position within the family (higher risk for first-born) and childhood socio-economic factors (higher risk for higher parental social status and longer parental education). College entrants who had earlier reported fewer childhood infections experienced increased frequency of HD (Paffenberger *et al.*, 1977).

Higher risk of HD has also been reported in follow-up of cohorts of cases of infectious mononucleosis (IM) (e.g., Munoz *et al.*, 1978). Since IM is known to be caused by late first exposure to EBV this is often interpreted as evidence that late exposure to EBV can cause HD. The findings are, however, consistent with the more general interpretation that the type of person and the kind of lifestyle which leads to late first exposure to EBV is associated with HD; in this form the findings support the late-host-response model but give few pointers to the aetiological agent. Up to now, almost no studies have tested for EBV-association in cases of HD arising following a history of IM; one study (Jehn *et al.*, 1981) was negative but the techniques involved are no longer state-of-the-art and this is a high priority research topic.

Some epidemiological studies have shown direct links between illness associated with other herpesviruses and HD in young adults: these include herpes simplex (Serraino *et al.*, 1991) and herpes zoster (Serraino *et al.*, 1991; McKinney *et al.*, 1990).

Seroepidemiological studies have consistently associated high antibody titres to human herpesvirus-6 (HHV-6) and EBV-VCA and EBV-EA with HD (Clark *et al.*, 1990; Henle and Henle, 1973; Evans and Gutensohn, 1984; Levine *et al.*, 1992). For the EBV antigens, the evidence includes prospective data from blood samples collected several years in advance of the HD diagnoses (Mueller *et al.*, 1989). These results are not restricted to young adults and the biological significance of age-specific effects is unclear. As for the IM cohorts, two interpretations are possible: the findings may indicate a direct causal effect of exposure to EBV and/or HHV-6 or point to a less direct association with HD.

Positive results from analyses of spatial clustering in this age group provide further support for the hypothesis. These analyses study residence at diagnosis which serves as proxy for location at the (unknown) time of exposure; since the latent period for HD is believed to be both lengthy (years rather than days) and variable, migration would be expected to dilute the effect. Further dilution is predicted since we have a hypothesis under which risk of HD in exposed people is low. It follows that case clustering is not essential for the late-host-response model; that positive results have been obtained may point to population microepidemics of the aetiological agent and to the importance of high-titre exposure as has been postulated for childhood leukaemia (Kinlen *et al.*, 1990).

The most critical host factor is clearly age. That immune suppression is likely to play some role may be deduced from the high incidence of HD in persons seropositive for HIV (see Hamilton-Dutoit, this volume). Inherited susceptibility is also important (Mack *et al.*, 1995; see Taylor *et al.*, this volume) but its effect has been difficult to quantify since it is quite rare to find pairs of HD cases who are first degree blood relatives. No systematic register of more diffused HD families has been assembled and environmental factors must also explain some of the observed familial aggregation (Dorkin, 1987).

Attention has recently been drawn to the possibility that hormonal factors, especially female reproductive history, may be key host factors. Evidence supporting the hypothesis that increasing parity reduces risk of HD in women (possibly through an effect on the immune system) has been reviewed by Glaser (1994), and finds further support from a large Norwegian record linkage study involving 1.3 million women (Kravdal and Hansen, 1993).

CLUSTERING

Background

The study of clusters and clustering of cases of HD has a long and controversial history (Grufferman, 1977; Smith, 1978; Alexander, 1990). For HD, the start of cluster reports came later than, for example, leukaemia. The reports involved larger numbers and were somewhat more persuasive than those for other conditions but shared with them the essential anecdotal status. The best known example is Albany High School in New York State (USA). Initially, four cases were observed and subsequent enquiries amongst relatives and friends found nine further cases linked to the original four over a 20-year period (Vianna et al., 1971). This was striking but no control data are available and statistical evaluation is impossible.

Problems inherent in investigations of (chronic) disease clusters include the impossibility of formal statistical analyses of anecdotal reports, the lack of precise definitions of a "cluster", and until recently, lack of appropriate statistical methodology. The rationale of using these investigations to elucidate aetiology lies in the belief that exposures will aggregate at times and places where cases cluster; their utility has been questioned (Rothman, 1987) but HD appears to be one disease for which the results integrate in a timely fashion with other research (Alexander and Cuzick, 1992).

Formal statistical analyses

The first attempt at formal statistical analyses involved follow-up studies of schools classified by the presence or absence of HD in a 5-year "baseline" period; the occurrence of HD in a subsequent time period was investigated. In the first application (Vianna and Polan, 1973) ten cases were identified in 1965-1969 in eight schools, which had had cases between 1960 and 1964, but there were no cases in 16 matched "control" schools. The differences here were statistically highly significant but the results were not confirmed elsewhere (Grufferman et al., 1979; Zack et al., 1977) and may be attributed to an unknown extent to ascertainment bias (Pike et al., 1974; Grufferman and Delzell, 1984).

Space-time interaction testing (Knox, 1964) is a simple but statistically valid method of determining if there is a tendency for cases to occur together simultaneously in space and time. It was designed for infectious diseases with short latent periods. There have been numerous applications to HD (reviewed in Alexander, 1990). These have involved over 4,000 cases and taken definitions of spatial proximity from 30 days to 2 years. The results have been equivocal and applications to age/diagnostic sub-groups as well as several interpretations of "proximity" have led to insuperable problems of multiple testing. It has rarely been applied following the demonstration (Chen et al., 1984) that this method has low statistical power when latent periods are long and variable (as is predicted by the late-host-response model for HD).

Spatial clustering

Good statistical methodologies for the investigation of spatial clustering of disease in human populations have become available relatively recently. There are, in general, two types of methods. In the first, counts of cases in suitable disjoint areas are compared with those which would arise for a random distribution. The only application of this method to HD has shown significant clustering in young people in Scotland (Table 1) (Urquhart, 1989).

The alternative methodology involves examination of distances between cases after adjustment has been made for the variability of the population-at-risk. Related methodologies in the UK (Alexander *et al.*, 1989; Alexander, 1990) and the USA (Glaser, 1990) have found significant spatial clustering; the second of these revealed pronounced geographical clustering of HD through most of the San Francisco Bay area and this was concentrated in young people.

Table 1: Hodgkin's disease, aged 0-44, Scotland 1975-1984. Distribution of 739 cases between 739 areas of approximately equal population size.

Number of cases	Expected number of areas	Observed number of areas
0	292	272.1
1	261	271.8
2	115	135.8
3	46	45.2
4	18	11.3
5	5	2.3
6+	2	0.4

$\chi^2 = 15.96$, df = 5, p <0.01

Typically, distance-based methodologies can identify "clusters" of cases; Fig. 4 shows "clustered" cases which are centres of small case aggregations and which can be grouped into clusters, when clustered cases are nearest neighbours of one another.

It is not adequately appreciated that clusters can and do occur by chance. If the case distribution were random then 8-9% of all cases would be classified as clustered according to the definition used here. The test for spatial clustering involves examination of the overall percentage of clustered cases; in one analysis 13% of cases of HD in young people (p<0.05), but only 9% of cases in persons over 50 years of age, were clustered (Alexander *et al.*, 1991a). This provides further evidence for the two-disease hypothesis. At present it is not known if HD in children shows spatial clustering. For young people, altogether, a higher percentage (29%) of clustered cases has been reported for isolated areas (Alexander *et al.*, 1991b); this had been predicted for a disease where risk was increased by exposure to microepidemics of a common infectious agent (James, 1990).

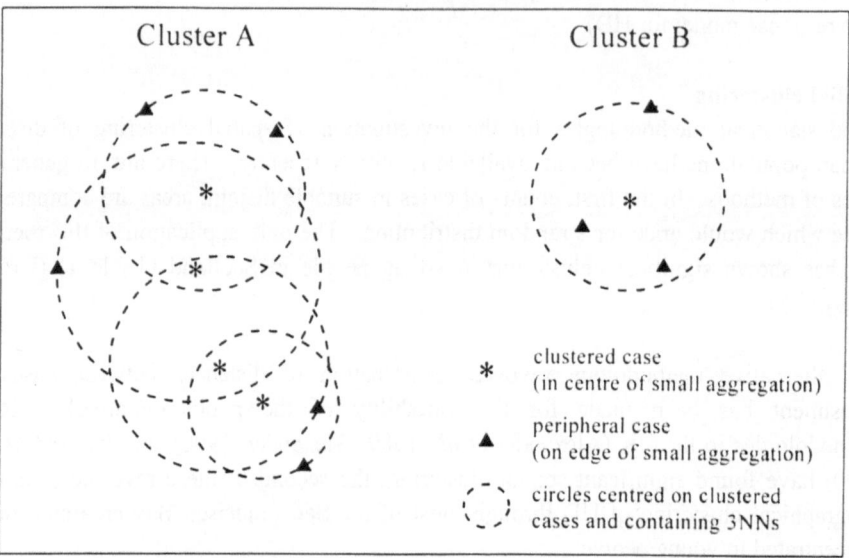

Fig. 4:   Clustering of Hodgkin's disease.

Clustering at earlier time periods

Case-control study designs are available for the investigation of past linkage of HD cases in school, workplace or by geographical proximity (e.g., Pike and Smith, 1974), and here, overlap in time is normally required. A review of studies of this type, reported up to 1989, concluded that there was little evidence of workplace linkage (Alexander, 1990). No studies suggested substantial risk for school or school-age contact but 4/6 reported relative risks in the range 1.2 - 1.9. Positive and negative results appear to be separated by their eligibility criteria; negative results (e.g., Smith *et al.,* 1977) were not restricted to subjects whose schooling had been in the study region and this could lead to substantial dilution of any effect. A recent study (Ross and Davis, 1990) has reported significant aggregation of childhood and teenage residences of HD cases but, surprisingly, the evidence is restricted to older cases (diagnosed over the age of 40 years). This study considers HD cases diagnosed over a relatively short time period, 1974-1982. Negative results, especially for young people, may be explained by this time window being too short to include all cases arising in young people who had lived together during their childhood and adolescence.

Overall, these data offer limited support for the hypothesis that HD cases in young adults have had common exposures during late childhood and adolescence (Davis, 1986; Grufferman, 1977). The exposures could involve non-infectious agents (Grufferman, 1977) but, after considering the overall epidemiological and virological results, the most plausible interpretation is that the clustering points to susceptibility of adolescents to microepidemics of agent X.

Control data are not necessary for analyses of geographical proximity using complete residential histories; a new and powerful statistical methodology (Riise *et al.,* 1991) tests for space-time clustering at each year of age within birth cohorts. This has been applied to multiple sclerosis for which a similar aetiological hypothesis applies and correspondingly equivocal results have been obtained from existing analytical methodologies. The results show a striking, and highly statistically significant, clustering between the ages of 13 and 18 years, though not at younger or older ages. The use of data over a long time period (here 1953-1987) is recommended. Application of this technique to HD would be a relatively simple method of confirming that exposures during late childhood and adolescence are critical.

Clustering and EBV

No published studies have reported investigations of the possible association between virological outcomes and case clustering. This is surprising since both may be interpreted as pointers to a role for infectious agents in the aetiology of HD.

Results of a small pilot study have recently become available (Alexander *et al.*, 1995). One of the key objectives of this study was to investigate associations between HD clustering, EBV and HHV-6 antibody titres and EBV-status of the tumour cells. The total study population was 494 cases of HD diagnosed in the Yorkshire Health Region (UK) between 1985 and 1989 but biological samples were collected from just two small subsets of the cases. These were:

(i)   first, a sample (n=39) diagnosed 1988-1989 which included all cases arising in young people (any subtype) or HDNS (any age) who resided in areas where clustering had been observed previously, and a random sample of other cases of the same age and Rye type; and

(ii) secondly, all other cases (n=14) from the total series which were in clusters which contained EBV-RS+ cases from the first sample.

No association was found between antibody titres to EBV-VCA, EBV-EA or HHV-6 and case clustering. However, the results (Table 2) indicate an association between case clustering and EBV status which is statistically significant.

Table 2:   EBV-RS status and case clustering.

	Clustering Classification			
Sample	Clustered[2] (EBV-RS+/Total)	Peripheral[2] (EBV-RS+/Total)	Random[2] (EBV-RS+/Total)	p[1]
First sample	2/6 (33%)	2/8 (25%)	1/25 (4%)	0.04
Second sample	3/10 30%	0/4 (0%)	-	-
Total	5/16 (31%)	2/12 (17%)	1/25 (4%)	0.017

1   exact 1-sided p-value for trend across the levels of clustering.
2   clustered and peripheral cases are all in young people (ages ≤34 years); the random group includes some older onset nodular sclerosis Hodgkin's disease cases.

The total numbers are small and the results require confirmation on other, larger, data sets, but provide preliminary support for a hypothesis that EBV-RS+ cases tend to occur within case clusters. On the other hand, no clusters consisted entirely of EBV-RS+ cases and the EBV-RS+ cases did not appear to be restricted to specific clusters; it follows that clustering is unlikely to be attributable to exposure to EBV.

In a further analysis, EBV-RS status was ascertained in HD cases in a suspected 'school cluster' in Yorkshire. Six cases have been identified in pupils, alumni and their siblings since 1980. Expected numbers cannot be derived exactly but use of a conservative estimate of 2.6 gives p<0.05. EBV-RS+ status was ascertained for 5 of the cases and, here also, the cases included a mixture of EBV-RS+ and EBV-RS- cases (Fig. 5).

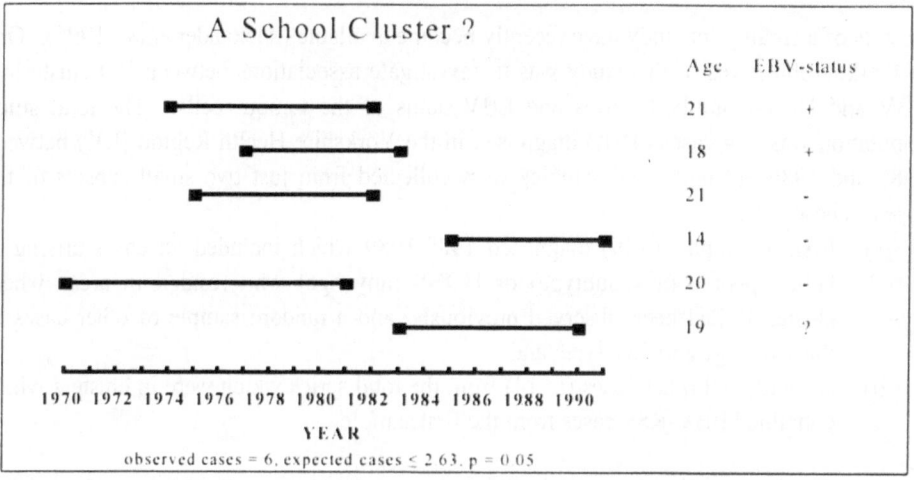

Fig. 5:   Period of attendance (index or sibling) at the school, age at diagnosis of Hodgkin's disease and EBV status for 6 subjects.

These data are consistent with the interpretation that clustering is attributable to shared exposure to another infectious agent which is a co-factor in EBV-RS+ (and some other) HD. This model is similar to the accepted one for endemic Burkitt's lymphoma where EBV is present in almost all tumour cells, is believed to be causally involved, but is not responsible for clustering, which is attributable to malaria and possibly other co-factors (Facer and Playfair, 1989).

CONCLUSIONS

HD in children and young adults presents the epidemiologist with several highly intriguing but potentially tractable problems. It appeared that epidemiology and molecular biology were harmonious (Mueller, 1991), but that was before it became apparent that the age- and subtype- pattern for EBV-association did not conform to the epidemiologists evidence for a viral aetiology. For young adults, the key question is still the validity of the late-host-response model. If this model holds, then the agent(s), the relevant ages and circumstances of exposure and predisposing host factors all require identification. Under the model it is interaction between inherited, and other host factors and infectious agents which can lead to

HD. New epidemiological methodology (Hopper and Carlin, 1992) is now available for investigating interactions of this nature and it would be appropriate to apply these to HD in young people.

The role of EBV requires clarification; it is possible that there are one or more other infectious aetiological agents (Glaser, 1990; Dalgleish and McElwain, 1986) and the new data included here support this. The polio model may hold for one of these. In any event, further study of HD in children is required and, for this, international collaboration may be necessary to ensure adequate numbers. Testing large population-based series for evidence of spatial clustering of childhood HD will assist in determining the role of infections for the disease in this age group and, hence, the validity of the polio model.

Further integrated study of virological and epidemiological outcomes should be encouraged and these should include investigation of clustering at the time of diagnosis and at earlier periods. The preliminary data reported here suggest that an infection other than EBV may be responsible for case clustering (determined either at time of diagnosis or during schooling) and that this may be a co-factor in EBV-associated as well as other cases of HD in young people. Testing of this hypothesis on large data sets is urgently required and is one of the ways in which inter-disciplinary research should lead to the kind of understanding of HD which can lead to preventive measures.

REFERENCES

Alexander FE, Williams J, McKinney PA, Ricketts TJ, Cartwright RA (1989) A specialist leukaemia/lymphoma registry in the UK. Part 2: clustering of Hodgkin's disease. *Br J Cancer* **60**:948

Alexander FE (1990) Clustering and Hodgkin's disease. *Br J Cancer* **62**:708

Alexander FE, McKinney PA, Williams J, Ricketts TJ, Cartwright RA (1991a) Epidemiological evidence for the "two-disease hypothesis" in Hodgkin's disease. *Int J Epidemiol* **20**:354

Alexander FE, Ricketts TJ, McKinney PA, Cartwright RA (1991b) Community lifestyle characteristics and incidence of Hodgkin's disease in young people. *Int J Cancer* **48**:10

Alexander FE, Cuzick J (1992) Methods for the assessment of disease clusters. Geographical and Environmental Epidemiology. *Methods for Small-Area Studies.* Elliot P, Cuzick J, English D, Stern R, eds (New York: Oxford University Press) p238

Alexander FE, Daniel CP, Armstrong AA, Clark DA, Onions DE, Cartwright RA, Jarrett RF (1995) Case clustering, EBV-RS status and Herpesvirus serology in Hodgkin's disease: results of a case-control study. *Eur J Cancer* (in press)

Chen R, Mantel N, Klingberg MA (1984) A study of three techniques for time-space clustering in Hodgkin's disease. *Stats in Med* **3**:173

Clark DA, Alexander FE, McKinney PA, Roberts BE, O'Brien C, Jarrett RF, Cartwright RA, Onions DE (1990) The seroepidemiology of human herpesvirus-6 (HHV-6) from a case-control study of leukaemia and lymphoma. *Int J Cancer* **45**:829

Correa P, O'Conor GT (1971) Epidemiologic patterns of Hodgkin's disease. *Int J Cancer* **8**:192

Dalgleish AG, McElwain T (1986) A viral etiology for Hodgkin's disease? *Aust NZ J Med* **16**:823

Davis S (1986) Case aggregation in young adults Hodgkin's disease, etiological evidence from a population experience. *Cancer* **57**:1602

Doll R, Payne P, Waterhouse JAH (1970) *Cancer Incidence in Five Continents*, Vol 11. (Berlin: Springer-Verlag)

Dorkin H (1987) Zur epidemiologie des morbus Hodgkin: Das familiare vorkommen: exogene faktoren? *Med Klinik* **82**:551

Evans AS, Gutensohn NM (1984) A population-based case-control study of EBV and other viral antibodies among persons with Hodgkin's disease and their siblings. *Int J Cancer* **34**:149

Facer AC, Playfair JHL (1989) Malaria, Epstein-Barr virus and the genesis of lymphomas. *Adv Cancer Res* **53**:33

Glaser S (1986) Recent incidence and secular trends in Hodgkin's disease and its histologic subtypes. *J Chronic Dis* **39**:789

Glaser SL (1990) Spatial clustering of Hodgkin's disease in the San Francisco Bay area. *Am J Epidemiol* **132**:S167

Glaser SL, Swartz WG (1990) Time trends in Hodgkin's disease incidence. *Cancer* **66**:2196

Glaser SL (1994) Reproductive factors in Hodgkin's disease in women: a review. *Am J Epidemiol* **139**:237

Grufferman S (1977) Clustering and aggregation of exposures in Hodgkin's disease. *Cancer* **39**:1829

Grufferman S, Cole P, Levitan TR (1979) Evidence against transmission of Hodgkin's disease in high schools. *N Engl J Med* **300**:1006

Grufferman S, Delzell E (1984) Epidemiology of Hodgkin's disease. *Epidemiol Revs* **6**:76

Gutensohn N, Cole P (1977) Epidemiology of Hodgkin's disease. *Int J Cancer* **19**:595

Gutensohn NM, Shapiro DS (1982) Social class risk factors amongst children with Hodgkin's disease. *Int J Cancer* **30**:433

Gutensohn NM (1982) Social class and age at diagnosis of Hodgkin's disease: new epidemiologic evidence for the "Two-Disease Hypothesis". *Cancer Treat Reports* **66**:689

Hartge P, Devesa SS, Fraumeni JF Jr (1994) Hodgkin's and non-Hodgkin's lymphoma. *Cancer Surveys* **19-20**:423

Henle W, Henle G (1973) Epstein-Barr virus related serology in Hodgkin's disease. *Natl Cancer Inst Monogr* **36**:79

Hopper JL, Carlin JB (1992) Familial aggregation of a disease consequent upon correlation between relatives is a risk factor measured on a continuous scale. *Am J Epidemiol* **136**:1138

James WH (1990) Further evidence for the hypothesis that one cause of childhood leukaemia is infection. *Paediatr Perinatal Epidemiol* **4**:113

Jehn U, Sauer H, Wolf H, Wilmanns W (1981) Hodgkin's disease following infectious mononucleosis: a case report. *Eur J Cancer* **17**:477

Kinlen LJ, Clarke K, Hudson C (1990) Evidence from population mixing in British New Towns 1946-1985 of an infective basis for childhood leukaemia. *Lancet* **336**:557

Knox EG (1964) The detection of space-time epidemics. *Applied Statistics*, 25

Kravdal Ø, Hansen S (1993) Hodgkin's disease: the protective effect of childbearing. *Int J Cancer* **55**:909

Levine PH, Ablashi DV, Saxinger WC, Connelly RR (1992) Antibodies to human herpes virus-6 in patients with acute lymphocytic leukemia. *Leukemia* **6**:1229

Mack TM, Cozen W, Shibata DK, Weiss LM, Nathwani BN, Hernandez AM, Taypor CR, Hamilton AS, Deapen DM, Rappaport EB (1995) Concordance for Hodgkin's disease in identical twins suggesting genetic susceptibility to the young-adult form of the disease. *N Engl J Med* **332**:413

MacFarlane GJ, Evstifeeva T, Boyle P, Grufferman S (1995) International patterns in the occurrence of Hodgkin's disease in children and young adult males. *Int J Cancer* **61**:165

McKinney PA, Alexander FE, Ricketts TJ, Williams J, Cartwright RA (1989) A specialist leukaemia/lymphoma registry in the UK. Part 1: incidence and geographical distribution of Hodgkin's disease. *Br J Cancer* **60**:942

McKinney PA, Alexander FE, Roberts BE, O'Brien C, Bird CC, Cartwright RA (1990) Yorkshire case control study of leukaemias and lymphomas, parallel multivariate analyses of seven disease categories. *Leuk Lymphoma* **2**:67

MacMahon B (1966) Epidemiology of Hodgkin's disease. *Cancer Res* **26**:1189

Merk, K, Bjorkholm M, Rengifo E, Gavilondo J, Holm G, Rivas H (1990) Epidemiological study of Hodgkin's disease in Cuba and Sweden. *Oncol* **47**:246

Mueller N, Evans A, Harris NL, Comstock GW, Jellum E, Magnus K, Orentreich N, Polk F, Vogelman J (1989) Hodgkin's disease and Epstein-Barr virus. *New Engl J Med* **320**:689

Mueller N (1991) An epidemiologist's view of the new molecular biology findings in Hodgkin's disease. *Ann Oncol* **2**:23

Munoz N, Davidson RJL, Witthoff B, Ericsson JE, de-The G (1978) Infectious mononucleosis and Hodgkin's disease. *Int J Cancer* **22**:10

Paffenberger RS, Wing AL, Hyde RT (1977) Characteristics in youth indicative of adult-onset Hodgkin's disease. *J Natl Cancer Inst* **58**:1489

Parkin DM, Muir CS, Whelan SL, Gao Y-T, Ferlay J, Powell J (1992) *Cancer Incidence in Five Continents*, Vol VI, IARC No.120. (Lyon)

Pike MC, Henderson BE, Casagrande J, Smith PG, Kinlen LJ (1974) Infectious aspects of Hodgkin's disease. *New Eng J Med* 734:341

Pike MC, Smith PG (1974) Clustering of cases of Hodgkin's disease and leukemia. *Cancer* **34**:1390

Riise T, Gronning M, Klauber MR, Barrett-Connor E, Nyland H, Albrektsen G (1991) Clustering of residence of multiple sclerosis patients at age 13-20 years in Hordaland, Norway. *Am J Epidemiol* **133**:932

Ross A, Davis S (1990) Point pattern analysis of Hodgkin's disease residences prior to diagnosis. *Am J Epidemiol* **132**:551

Rothman KJ (1987) Clustering of disease (editorial). *Am J Pub Hlth* **77**:13

Serraino D, Franceschi S, Talamini R, Barra S, Egi E, Carbone A, La Vecchia C (1991) Socio-economic indicators, infectious diseases and Hodgkin's disease. *Int J Cancer* **47**:352

Smith PG (1978) Current assessment of "Case Clustering" of lymphomas and leukaemias. *Cancer* **42**:1026.

Stalsberg H (1973) Hodgkin's disease in Western Europe: a review. *NCI Monogr* **36**:31

Smith PG, Pike MC, Kinlen LJ, Jones A (1977) Contacts between young patients with Hodgkin's disease. *Lancet* **ii**:59

Urquhart J, Black R, Buist E (1989) Exploring small area methods. Methodology of Enquiries into Disease Clustering. Proceedings of a meeting held on 22 April 1988 at the London School of Hygiene and Tropical Medicine. Elliot P, ed (Crown)

Vianna NJ, Greenwald P, Davies JNP (1971) Extended epidemic of Hodgkin's disease in high-school students. *Lancet* **i**:1209

Vianna NJ, Polan AK (1973) Epidemiologic evidence for transmission of Hodgkin's disease. *N Engl J Med* **289**:499

Zack MM, Heath CW, Andrews MDeW, Grivas AS, Christine BW (1977) High school contact among persons with leukemia and lymphoma. *J Natl Cancer Inst* **59**:1343

EVIDENCE OF AN INCREASED FREQUENCY OF HLA-DPB1*0301 IN HODGKIN'S DISEASE SUPPORTS AN INFECTIOUS AETIOLOGY

G Malcolm Taylor[1], David A Gokhale[1], Derek Crowther[2],
Penella Woll[2], Freda Alexander[3] & Ruth F Jarrett[4]

Immunogenetics Laboratory, St. Mary's Hospital[1] and Department of Medical
Oncology, Christie Hospital[2], Manchester, UK; Department of Public Health,
University of Edinburgh[3], and LRF Virus Centre, University of Glasgow[4], UK

SUMMARY

Hodgkin's disease (HD) may be the rare outcome of a common infection which is
influenced by host genetic susceptibility. We have analysed this hypothesis by determining
the frequency of HLA-DPB1 alleles in two series of HD patients using molecular typing
methods. One series consisted of a retrospective/prospective group of 118 patients over the
age of 15 years, and the other a prospective group of 45 patients between the ages of 16 and
24 years. In both series, the proportion of patients typing for HLA-DPB1*0301 was greater
than that for the controls, suggesting that this may be an HD-susceptibility allele. Analysis
of DPB1 alleles in relation to HD subtype also showed that the increase in *0301 was
present in nodular sclerosing HD (HDNS) patients, as well as in mixed cellularity HD
(HDMC) and lymphocyte predominant HD (HDLP) patients. Preliminary evidence was
obtained suggesting an increase in *0401, and possibly *0501, in HDMC and HDLP.
Analysis of *0301-like hypervariable region (HVR) associations with HD subtypes
indicated an increase in an *0301-like HVR-C motif in HDNS, but not in non-HDNS. The
frequency of *0301- and *0401-like HVR-C and HVR-F amino acid residues also differed
in HDNS and non-HDNS patients. The frequency of the HVR-C amino acid residues
Asp55, Glu56 (*0301) was increased in HDNS, but not in non-HDNS patients, whereas the
frequency of Ala55, Ala56 (*0401) was increased in non-HDNS, but not in HDNS patients.
In addition, the frequency of the HVR-F amino acid residues Asp84, Glu85 and Ala86
(*0301) was greater in the HDNS than non-HDNS patients, whereas there was no increase

in Gly84, Gly85, Pro86 (*0401). Although these are preliminary findings, they suggest that genetic susceptibility to an infectious aetiology in HD may reside at the level of DPB peptide-binding residues rather than with a specific allele.

INTRODUCTION

It has been suggested that some cases of Hodgkin's disease (HD), such as those occurring in children and young adults, are caused by an infectious agent (MacMahon, 1966), and that the agent concerned is a virus of low virulence and infectivity which enters through the upper-respiratory tract (Vianna and Greenwald, 1971). Evidence of an increased risk of HD with higher social class and smaller family size, and epidemiological similarities to paralytic poliomyelitis in the pre-vaccine era have suggested that HD may be the rare outcome of a common infection (Gutensohn and Cole, 1981). Although support for an infectious aetiology in HD comes from case-clustering (Alexander *et al.*, 1991), as well as geographical and ethnic incidence variations (Glaser, 1991), the identity of the HD agent remains unknown.

There has been much recent interest in the role of Epstein-Barr virus (EBV) in HD, following the detection of EBV genomes in Reed-Sternberg (RS) cells (Weiss *et al.*, 1989; Wu *et al.*, 1990; Khan *et al.*, 1992), since it seems to fulfil the criteria of an agent likely to cause the disease. However, the diagnostic heterogeneity of HD makes it difficult to envisage EBV as the only causative agent. This conclusion is supported by the finding that EBV genomes are present in only a minority of 15-34 year old HD cases (Jarrett *et al.*, 1991), and argues in favour of some other infection as the causative agent, at least in this age group.

In populations exposed to common infections, the relative risk of disease is probably influenced by host genetic susceptibility. Since resistance and susceptibility to infections appear to be under the partial control of HLA genes (Hill *et al.*, 1991), it could be predicted that any association between HD and HLA would support the theory of an infectious aetiology. Although initial studies pointed to an association with HLA-B locus alleles (Amiel *et al.*, 1967), further analyses showed that associations were weak and somewhat inconsistent (Hors and Dausset, 1983). On the basis that this might be due to recombination between HLA-B and an HD susceptibility gene, Bodmer *et al.* (1989) used RFLP-based molecular analysis to type a series of HD patients for HLA-DPw associated polymorphisms. Although they reported a decrease in HLA-DPw2 and an increase in a polymorphism associated with DPw3, 5 and 6 in HD, the resolution of DP alleles using RFLP analysis was technically difficult. A further study was therefore carried out as part of the 11th International HLA Workshop, using the polymerase chain reaction (PCR) to amplify the polymorphic β1 domain of DPB1, and sequence specific oligonucleotide (SSO) probes to type DPB1 alleles. Results of this study revealed a highly significant association between the DPB1*0301 and HD in a combined white Caucasian series of patients, with a relative risk (RR) of 1.95 (Tonks *et al.*, 1992a, 1992b).

This series included 39 HD patients and 40 normal controls from Manchester which alone showed an increase in *0301, with a RR value of 1.55 (Tonks *et al.*, 1992a). The Manchester patients had the advantage of being from a single centre, but the number of patients was small. We have continued to collect and type HD patients, and also to study a separate prospective series. We report here the interim results for the two series which further confirms the finding of an *0301 association with HD. However, in contrast with previous findings, we suggest that the crucial association is with functional motifs within DPB1*0301, that this may be influenced by HD subtype, and that these motifs may control the binding and presentation of peptides derived from an infectious agent to the patient's T-cells.

MATERIALS AND METHODS

Patients and controls

The two series of HD patients included in the study involve a retrospective/prospective series of patients aged 15 years and over from Manchester (the Manchester series), and a preliminary prospective series from the Young Hodgkin's Case Control Study (YHHCCS series), aged between 16 and 24 years, from the north-east and south-west of England. Further details of the Manchester series are shown in Table 1. For the comparison of DPB1 allele frequencies the two series were analysed separately in comparison with two control groups, one of which consisted of adult blood donors and the other cord blood samples from newborn infants.

Table 1: Details of the Manchester Hodgkin's disease patients.

Histology	Age range	Males	Females	Total
Nodular sclerosing	16-69	35	34	69
Lymphocyte depleted	63	0	1	1
Lymphocyte predominant	18-65	17	2	19
Mixed cellularity	15-45	18	3	21
Unclassified	17-75	7	1	8
Total	**15-75**	**77**	**41**	**118**

HLA-DPB1 typing

Genomic DNA was extracted from whole blood using established methods, and a 288 bp fragment of DPB1 exon 2 was amplified by PCR (Bugawan *et al.*, 1988) using the oligonucleotide primers DPB-AMP5' (5'-CCT CCC CGC AGA GAA TTA C-3') and DPB-AMP3' (5'-AGC CCT CAC TCA CCT CGG CG-3') kindly supplied by Julia Bodmer (ICRF, London). Primers amplifying a slightly larger region of exon 2 were supplied by the British Society for Histocompatibility and Immunogenetics (BSHI). Amplifications were performed in a total volume of 20 μL containing 50mM KCl, 10mM Tris-HCl (pH 8.3), 2mM $MgCl_2$, 0.1% Triton X-100, 0.2mM each dNTP, 0.25μM DPB1 primers, 0.5 Units *Taq* polymerase (Promega, Madison, USA), and 50 ng genomic DNA. PCR amplifications consisted of an initial denaturation step of $94^{\circ}C$ for 2 min 30 s, followed by 36 cycles (1 min at $94^{\circ}C$, 45 s at $62^{\circ}C$, 45 s at $74^{\circ}C$), with a final extension step of 3 min at

74°C. A 5 µL aliquot of each PCR product was monitored for amplification by agarose gel electrophoresis.

HLA-DPB1 typing (Bugawan *et al.*, 1990) was carried out using the 19 sequence specific oligonucleotide (SSO) probes supplied with the BSHI typing kit, and 5 from the 11th HLA Workshop panel. Following the addition of 200 µL TE and extraction with 200 µL chloroform, 15 µL aliquots of PCR-amplified DNA were slot-blotted onto nylon filters (Hybond N+, Amersham, UK) using a vacuum manifold (Hoefer, UK). The immobilised DNA was fixed and denatured by soaking the filters in 0.4M NaOH, neutralised in 1.5M NaCl, 0.5M Tris-HCl (pH7.2), 1mM EDTA, rinsed briefly in 2X SSPE (0.36M NaCl, 20 mM Na_2HPO_4, 2mM EDTA), and incubated, with gentle shaking, in pre-hybridisation buffer (6X SSPE, 5X Denhardt's solution, 5% SDS, 100 µg/mL denatured herring sperm DNA). The probes were end-labelled with γ^{32}P-ATP using T4 polynucleotide kinase (Promega) or with digoxygenin-ddUTP using terminal deoxynucleotidyl transferase (Boehringer, Lewes, Sussex, UK). Hybridisations were carried out with gentle agitation at 58°C, using 1 pmol labelled probe/mL pre-hybridisation solution. The filters were first washed once in 2X SSPE, 0.1% SDS at room temperature for 15 min, then in 3.0M tetramethyl ammonium chloride, 50mM Tris-HCl (pH 8.0), 2mM EDTA, 0.1% SDS for 10 min at $58\text{-}68^{\circ}$C and either rinsed in 2X SSPE (γ^{32}P-labelled probes) before autoradiography for 4-12 hours, or treated according to the manufacturers instructions (DIG-labelled probes), using the chemiluminescent detection substrate Lumigen-PPD (Boehringer), followed by autoradiography for 0.5-2 hours. DPB1 alleles and nomenclature were assigned according to the 11th HLA Workshop (Tait *et al.*, 1992). RRs were calculated by the cross-product odds ratio method (Levitan, 1988), and statistical evaluation was carried out using 2x2 χ^2 analysis using the SIMFIT programme, kindly provided by WG Bardsley. Since the analysis tested the prior hypothesis of an association between HD and DPB1*0301 (Tonks *et al.*, 1992a, b) no correction for the number of alleles tested was considered necessary.

RESULTS

In addition to the 39 Manchester HD patients and 40 adult controls typed for the 11th HLA Workshop (Tonks *et al.*, 1992a), we have typed 79 additional patients from Manchester (n=118), 52 additional adult controls (n=92) and 82 infant controls for the present study. Details of this series, shown in Table 1, indicate that 58% (n=69) of the patients had HDNS, with a male:female (M:F) ratio of about 1:1, whereas 42% were non-HDNS patients (n=49), with a M:F of 6:1. Only 1 patient had lymphocyte depleted HD (HDLD), and 8 could not be assigned to a specific subtype. The age ranges of the HDNS and non-HDNS patients are similar for each subtype. Since the number of non-HDNS patients in each category is small, the DPB1 analysis has largely been confined to the combined series of Manchester non-HDNS patients. In addition to the Manchester series, we also typed 45 prospective young adult HD patients, aged 16-24 years, from the YHHCCS series. These included 29 HDNS patients (64%), 6 HDMC patients, 1 patient with HDLP and 9 patients with an unassigned subtype.

There were no marked differences in the *number* of DPB1 alleles in the two series of HD patients (Manchester and YHHCCS) compared with the controls (HD: 18 alleles; adult controls: 15; infant controls: 16) or in the *number* of DPB1 genotypes (HD: 34 genotypes; adult controls: 31; infant controls: 28). However, in agreement with previous results we found an increase in DPB1*0301 allele frequency in the Manchester HD patients (13.5%) compared with controls (adult: 8.2%; infants: 8.5%), and this increase was also present in the YHHCCS series (12.2%). In contrast there were no marked differences in the frequency of other DPB1 alleles in patients and controls.

DPB1 phenotype analysis confirmed that more of the Manchester HD patients (24.5%) than adult (14.1%) and infant controls (17.1%) typed for *0301 (RR_{adult}: 1.98, p=0.08; RR_{infant}: 1.56, p=0.27). This increase was also seen in the YHHCCS patients (24.4%). Even though the number of patients in the two series is insufficient to achieve significance, they are in agreement with the results of the 11th HLA Workshop, and continue the trend seen in the first Manchester series (Tonks *et al.*, 1992a). In contrast to the increase in *0301, there was no marked difference in the frequency of other DPB1 phenotypes in HD patients compared with controls.

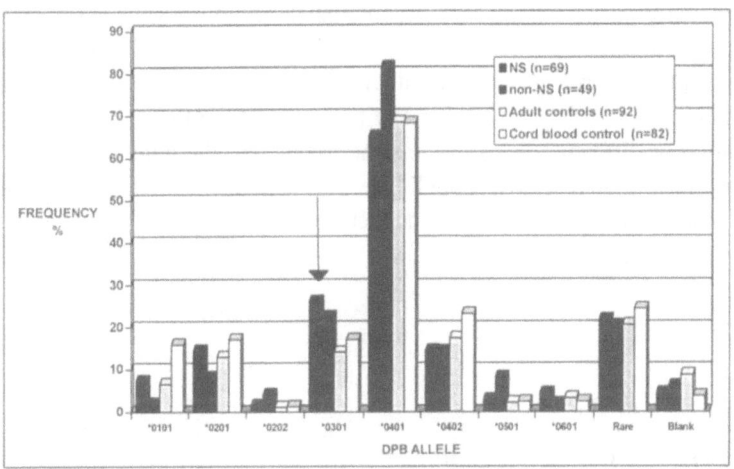

Fig. 1:   DPB1 phenotype analysis of Manchester Hodgkin's disease patient subtypes. NS, nodular sclerosis Hodgkin's disease; non-NS, histological subtypes other than nodular sclerosis Hodgkin's disease; arrow indicates *0301.

To determine whether the DPB1 allele frequency is affected by HD subtype, the Manchester patients were split into HDNS or non-HDNS groups. In HDNS, the percentage *0301 allele frequency increased from 13.5% to 15.2%, but it decreased to 11.2 % in the non-HDNS patients. Fig. 1 shows the subtype analysis in relation to the DPB1 phenotype frequency, and confirms that more HDNS patients (26.0%) than non-HDNS patients (22.4%) have *0301 ($RR_{adult\ control}$. HDNS=2.14, p=0.08). Although the results are only of borderline significance, they show a more marked association than any of the other DPB1 phenotypes. Further analysis of the non-HDNS patients revealed that none of the 8 unclassified and 1 HDLD patients typed for *0301. If these are removed and the HDMC and HDLP cases are analysed separately, the *0301 frequencies are 28.6% and 26.3% respectively, and thus similar to the HDNS group. Fig. 1 also shows that more of the combined non-HDNS patients type for *0401 (79.6% vs 68%), and *0501 (8.2% vs 2.3%)

than the two control series. When the HDMC and HDLP patients are analysed separately, the *0401 frequency in the HDMC patients is 85.7%, and 73.7% in the HDLP patients.

Analysis of DPB1 genotype frequencies in the 69 Manchester HDNS and 49 non-HDNS patients compared with the two series of controls revealed a striking increase in patients with the *0301/*0401 genotype (Fig. 2). In the total patient series, 13.6% had this genotype compared with 4.3% of adult and 7.3% of infant controls, whereas 11.5% of HDNS compared with 16.3% of non-HDNS patients were heterozygous for *0301/*0401. Although the increase was bordering on significance in the total patient group compared with adult controls (χ^2=3.70, p=0.054; RR=3.45), the increase was significant in the non-HDNS patients (χ^2=4.09; p=0.043; RR=4.29), but not in the HDNS patients (RR=1.81). No other DPB1 genotype, including *0301 homozygotes showed this degree of association with HD subtype. A similar increase in *0301/*0401 heterozygotes was seen in the YHHCCS patients, but the numbers are too small to analyse in detail.

HLA-DPB1 alleles are defined by 6 hypervariable (HVR) motifs in exon 2 which are not individually allele specific. Since HVRs only define an allele when arranged in specific combinations, it is possible to identify different alleles with the same HVR, and to look for associations with individual HVRs, rather than with alleles. Therefore, to determine whether susceptibility to HD is associated with a specific HVR, we analysed the frequency of *0301-like HVRs in the Manchester HD patients and controls. Where the same HVR motif occurred in different alleles in the same individual, the individual is assumed to be homozygous for that motif, even though heterozygous for the allele. For example, *0301 and *0601 share *VYQLRQ* in HVR-A, and therefore any individual who is an *0301/*0601 heterozygote is an HVR-A homozygote.

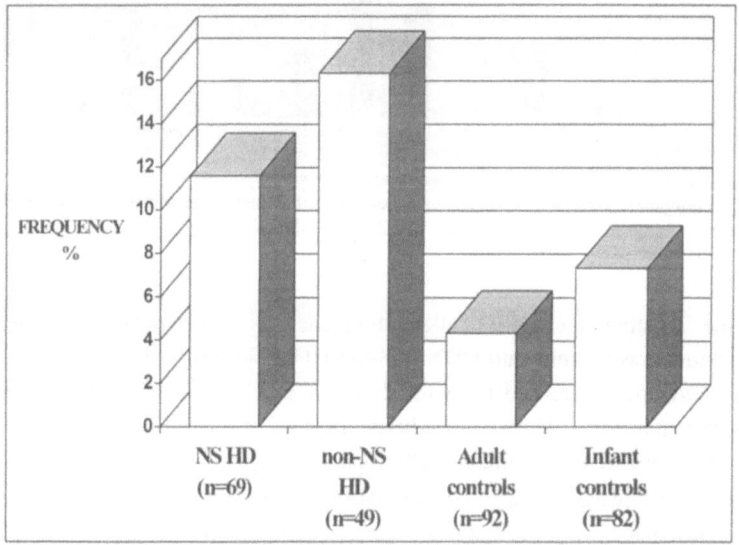

Fig. 2:   DPB1 *0301/*0401 genotype analysis of Manchester Hodgkin's disease patient subtypes. NS HD, nodular sclerosis Hodgkin's disease; non-NS HD, other subtypes of Hodgkin's disease.

Although the results of this analysis (Table 2) show an increase in all *0301-like motifs in the HDNS patients compared with controls, this is more striking for some HVRs than others. For instance, the amino acid motifs *REEFVR* (HVR-B) and *RPDEDYW* (HVR-C) are increased in HDNS patients, but not in non-HDNS patients. In contrast, the motif *DLLEEK* (HVR-D) is increased in both HDNS and non-HDNS compared with controls.

Table 2:   DPB1*0301-like amino acid motif frequency (%) in Hodgkin's disease patients and controls.

HVR*	Amino acid motif**	MANCHESTER			YHHCCS	CONTROLS	
		ALL CASES (n=118)	HDNS (n=69)	non-HDNS (n=49)	(n=45)	ADULT (n=92)	INFANT (n=82)
A	VYQLRQ	33.9	33.3	34.7	33.3	22.8	30.5
B	REEFVR	56.8	59.4	53.1	66.7	50.0	54.9
C	RPDEDYW	33.1	37.7	26.5	37.7	26.1	23.2
D	SQKDLLE	31.4	31.9	30.6	37.7	26.1	29.3
D	DLLEEK	28.0	27.5	28.6	31.0	18.5	18.3

Abbreviations: * HVR = DPB1 exon 2 hypervariable region
**Amino acid motif = *0301-like sequence motif coded by each HVR. Bold type indicates amino acid in each HVR common to all *0301-like alleles, but differing from non-*0301 like alleles.

Following the publication of the three-dimensional crystallographic structure of HLA-DRB1*0101 (Brown *et al.*, 1988, 1993), it is now possible to predict the amino acid residues in HLA-DPB1 which may be important in binding antigenic peptides. Comparison of the amino acid sequences of DRB1 and DPB1 shows that the polymorphic $\beta1$ domains of the two proteins are 71% identical. Most of the non-identical residues share common hydrophobic properties, indicating that the primary structures are very similar. If the amino acid sequences of the putative HD susceptibility allele, DPB1*0301, and the commonest DPB1 allele, *0401, are compared, they are found to be 86% identical, again indicating structural similarities. As with DRB1, most of the *0301 and *0401 residues which are non-identical also show structural similarity, but 2 HVR motifs (HVR-C, positions 55-56, and HVR-F, positions 84-86) show no such conservation.

To determine whether these motifs differ in frequency in HD patients and controls, a similar analysis to that shown in Table 2 was carried out. In this case however, the calculation is based on a comparison of the *0301 and *0401-like HVR-C amino acid residues Asp55, Glu56 (DE) and Ala55, Ala56 (AA) respectively, at positions 55 and 56. In addition, we compared the *0301-like HVR-F amino acid residues Asp84, Glu85, Ala86 (DEA) with the *0401-like residues Gly84, Gly85, Pro86 (GGP) at positions 84-86. Table 3 shows that the frequency of Asp55, Glu56 is increased in the Manchester HDNS and the YHHCCS patients compared with the controls, but that the increase in Ala55, Ala56 is confined to non-HDNS patients. For Asp84, Glu85, Ala86, there is an increased frequency in the Manchester HDNS and YHHCCS cases, only in comparison with the adult controls. It is noteworthy that there is a 10% difference in the frequency of these residues in adult and infant controls. Of additional interest is the observation that the frequency of the Asp84, Glu85, Ala86/ Gly84, Gly85, Pro86 genotypes is significantly higher in YHHCCS-HDNS patients (69%) compared with adult and infant controls (29% and 40% respectively).

Table 3: Phenotypic frequency (%) of amino acid residues 55-57 and 84-86 in Hodgkin's disease patients and controls.

Amino acid positions*	Residues**	Manchester HDNS (n=69)	Manchester non-HDNS (n=49)	YHHCCS (n=45)	CONTROLS	
					ADULT (n=92)	INFANT(n=82)
55-56	DE	60.9	51.0	62.2	50.0	54.9
55-56	AA	73.9	89.8	82.2	75.0	82.9
84-86	DEA	50.7	38.8	57.7	39.1	50.0
84-86	GGP	82.6	81.6	93.3	81.5	86.6

    *   Amino acid positions 55-56 in HVR C, and 84-86 in HVR F

  **   Residue DE (Asp55, Glu56) is found in alleles 0201, **0301**, 0402, 0601, 0801, 0901, 1001, 1401, 1701 and 2001, whereas residue AA (Ala55, Ala56) is found in 0101, **0401**, 1101, 1301 and 1501. Residue DEA (Asp84, Glu85, Ala86) is found in 0101, **0301**, 0501-1401, 1601, 1701, 1910 and 2001, whilst residue GGP (Gly84, Gly85, Pro86) is found in 0201, 0202, **0401**, and 0402.

DISCUSSION

The results of this study confirm those of the 11th HLA Workshop in which an increased frequency of DPB1*0301 was found when a combined series of 544 white Caucasian patients was compared with 464 controls. The Workshop study included a preliminary series of 39 HD patients and 40 adult controls from Manchester, which despite the small number, showed a relative risk for *0301 of 1.55. If *0301 is an HD-susceptibility allele we would expect to find this level of relative risk in a larger patient group. The present study includes an additional 79 Manchester patients and 52 controls, as well as 82 newborn infant controls. The results do indeed show that the trend in the *0301 association is maintained, with a relative risk for *0301 of 1.98 compared with adult controls, and 1.56 with infants. We also typed a preliminary prospective series of 45 patients between 16 and 24 years which are part of a young HD case-control study (YHHCCS). We are confident that the increased frequency of *0301 in HD is real, for the following reasons: 1. the increase was seen in both groups of HD patients, 2. the result agrees with previous studies of DPB1 in HD using PCR-SSO typing, 3. the association first seen in the early Manchester series was still present in the extended series of patients.

It could be argued that the increased frequency of *0301-positive Manchester HD patients is due to the selective survival of patients with this phenotype. However, the fact that *0301 frequency is also increased in the prospective YHHCCS series suggests that this is probably not the explanation, and that *0301 is indeed a susceptibility allele. Patients with HD generally have a high remission rate and long remission duration and it is likely that any selective effect in the Manchester series is minimal. In the 11th HLA Workshop study only patients with *0901 had a significantly shorter remission duration, whereas there was no difference in remission duration in patients with or without *0301 (Tonks *et al.*, 1992a).

Of the 118 Manchester patients in this retrospective/prospective series over the age of 15 years, about three-fifths are of the HDNS subtype, and two-fifths are HDMC, HDLP, HDLD and unassigned patients. In contrast to the results of the 11th HLA Workshop study, our analysis of the Manchester series shows an increase in *0301 in HDNS, and this is supported by recent results from Klitz et al. (1994). In the Manchester series, the lower frequency of *0301 in the non-HDNS cases disappeared when the unassigned cases were removed from the analysis. This suggests that *0301 is also increased in both the HDMC and HDLP subtypes. Although there was no increase in the number of patients who typed for *0401 in HDNS, this phenotype was present at an increased frequency in HDMC and HDLP patients. The increased frequency of *0401 in some but not all of the series in the 11th HLA Workshop (Tonks et al., 1992a, 1992b) may have been due to different proportions of HDMC and HDLP patients in the different series. Diagnostic subtyping of all of the Manchester patients was carried out in a single cancer histopathology department, and we are confident that the trend in subtype association is real, even though the numbers are too small to achieve significance.

Of considerable interest is our finding of a significant increase in the frequency of *0301/*0401 heterozygous non-HDNS Manchester patients. Although this could be due to the sum of the separate risks of *0301 and *0401 in non-HDNS, it may also be due to the product of the risks due to this genotype. This notion is supported by the lack of associations with either *0301 or *0401 homozygotes.

If DPB1*0301 is associated with susceptibility to HD, it could be acting as a surrogate marker of an infectious agent, since HLA class II alleles are known to be promiscuous antigen receptors (Chicz et al., 1993). Although it is tempting to speculate that this agent may be EBV, the present evidence, though indirect, argues against this idea. Since most of the YHHCCS cases are young adult HDNS patients, they would be expected on the basis of previous evidence (Jarrett et al., 1991) to exhibit a low frequency of EBV. If this is correct, DPB1 *0301 may be a marker of some other infectious agent. Whether this involves the reduced capacity of this allele to bind and present peptides derived from the unknown infectious agent to T-cells, an enhanced capacity to bind but not to signal T-cells, or finally to overstimulate T-cells has yet to be established by functional studies.

HLA-DPB1 is unusual among the HLA class II genes since it has no allele-specific sequences (Bugawan et al., 1988 and 1990). Instead, alleles consist of strings of 6 HVRs linked in the form of haplotypes by conserved inter-HVR sequences (Moonsamy et al., 1992). Each allele therefore consists of an array or patchwork of HVRs, some of which are shared by other alleles so that alleles may resemble each other for specific HVRs. In addition to allele frequency analysis it is thus possible to compare the frequency of individual HVRs in patients and controls. We used this approach in HD and found that the frequency of some *0301-like HVRs is indeed increased. This applies particularly to the DED motif in HVR-C which shows an increase in HDNS. We also noted differences in the frequency of HVR motifs in the two control groups, which may be due to the selection of the adults for a healthy phenotype.

We also compared the frequency of the two most important HVRs (C and F) in HD patients and controls by selecting the *0301-like amino acid motif DE and the *0401-like motif AA at positions 55 and 56 (HVR-C), and the *0301-like motif DEA and the *0401-like motif GGP at positions 84-86 of the β1 domain of DPB1 for analysis. Preliminary conclusions from these results suggest that that there is an increase in Asp55, Glu56, and Asp84, Glu85, Ala86 in HDNS. Although it is not yet possible to make conclusive predictions about how DPB1 motifs bind antigenic peptides, studies of the crystallographic structure of DRB1 suggest that positions 55-56 are involved in the formation of a salt bridge between the α and β subunits, and positions 84-86 are responsible for anchoring the immunogenic peptides required for the signalling of T-cells (Brown *et al.*, 1993).

It is clear from these and other studies (Tonks *et al.*, 1992a, 1992b; Klitz *et al.*, 1994) that though the susceptibility to HD conferred by the DPB1*0301 is relatively weak, elucidation of its functional role may have important implications for the identification of an aetiological agent in HD.

ACKNOWLEDGEMENTS

We thank the Leukaemia Research Fund, (DAG, FA, RJ), Kay Kendall Leukaemia Fund (GMT) and CRC (DC, PW) for research grant support, and WG Bardsley for the SIMFIT programme.

REFERENCES

Alexander FE, Ricketts TJ, McKinney PA, Cartwright RA (1991) Community lifestyle characteristics and incidence of Hodgkin's disease in young people. *Int J Cancer* **48**:10

Amiel JL (1967) Study of the leukocyte phenotypes in Hodgkin's disease. In *Histocompatibility Testing*. Curtoni ES, Mattiuz PL, Tosi RM, eds (Copenhagen: Munksgaard) p79

Bodmer JG, Tonks S, Oza AM, Lister TA, Bodmer WF (1989) HLA-DP based resistance to Hodgkin's disease. *Lancet* i:1455

Brown JH, Jardetsky T, Saper MA, Samraoui B, Bjorkman PJ, Wiley DC (1988) A hypothetical model of the foreign antigen binding site of class II histocompatibility molecules. *Nature* **332**:845

Brown JH, Jardestsky TS, Gorga JC, Stern LJ, Urban RG, Strominger JL, Wiley DC (1993) Three-dimensional structure of the human class II histocompatibility antigen HLA-DR1. *Nature* **364**:33

Bugawan TL, Horn GT, Long CM, Mickelson E, Hansen JA, Ferrara GB, Angelini G, Erlich HA (1988) Analysis of HLA-DP allelic sequence polymorphism using the *in vitro* enzymatic DNA amplification of DP-α and DP-β loci. *J Immunol* **141**:4024

Bugawan TL, Begovich AB, Erlich HA (1990) Rapid HLA-DPB typing using enzymatically amplified DNA and nonradioactive sequence-specific oligonucleotide probes. *Immunogenetics* **32**:231

Chicz RM, Urban RG, Gorga JC, Vignalli DAA, Lane WS, Strominger JL (1993) Specificity and promiscuity among naturally processed peptides bound to HLA-DR alleles. *J Exp Med* **178**:27

Glaser SL (1991) Black-white differences in Hodgkin's disease incidence in the United States by age, sex, histology subtype and time. *Int J Epidemiol* **20**:68

Gutensohn N, Cole P (1981) Childhood social environment and Hodgkin's disease. *New Eng J Med* **304**:135

Hill AVS, Allsopp CEM, Kwiatowski D, Anstey NM, Twumasi P, Rowe PA, Bennett S, Brewster D, McMichael AJ, Greenwood BM (1991) Common West African HLA antigens are associated with protection from severe malaria. *Nature* **352**:595

Hors J, Dausset J (1983) HLA and susceptibility to Hodgkin's disease. *Immunol Rev* **70**:167

Jarrett RF, Gallagher A, Jones DB, Alexander FE, Krajewski AS, Kelsey A, Adams J, Angus B, Gledhill S, Wright DH, Cartwright RA, Onions DE (1991) Detection of Epstein-Barr virus genomes in Hodgkin's disease: relation to age. *J Clin Path* **44**:844

Khan G, Coates PJ, Gupta RK, Kangro HO, Slavin G (1992) Presence of Epstein-Barr virus in Hodgkin's disease is not exclusive to Reed-Sternberg cells. *Am J Pathol* **140**:757

Klitz W, Aldrich CL, Fildes N, Horning SJ, Begovich AB (1994) Localisation of predisposition to Hodgkin disease in the HLA class II region. *Am J Hum Genet* **54**:497

Moosamy PV, Suraj VC, Bugawan TL, Saiki RK, Stoneking, M, Roudier J, Magzoub MMA, Hill AVS, Begovich AB (1992) Genetic diversity within the HLA class II region: ten new DPB1 alleles and their population distribution. *Tissue Antigens* **40**:153

Tait BD, Bodmer JG, Erlich HA, Ferrara GB, Albert E, Begovich A, Kimura A, Varney MD, Klitz W (1992) DNA typing: DPA and DPB analysis. In *HLA 1991*. Tsuji K, Aizawa M, Sasazuuki T, eds (Oxford University Press) **Volume 1** p485

Tonks S, *et al.* including Gokhale D, Taylor GM, Crowther D, Woll P (1992a). An international study of the association between HLA-DP and Hodgkin's disease. In *HLA 1991*. Tsuji K, Aizawa M, Sasazuuki T, eds (Oxford University Press) **Volume 2** p539

Tonks S, Oza AM, Lister TA, Bodmer JG (1992b) Association of HLA-DPB with Hodgkin's disease. *Lancet* **340**:968.

Vianna NJ, Greenwald P, Davies JNP (1971) Nature of Hodgkin's disease agent. *Lancet* **i**:733

Weiss LM, Movahed LA, Warnke RA, Sklar J (1989) Detection of Epstein-Barr viral genomes in Reed-Sternberg cells of Hodgkin's disease. *New Eng J Med* **320**:502

Wu T-C, Mann RB, Charache P, Hayward SD, Staal S, Lambe BC, Ambinder RF (1990) Detection of EBV gene expression in Reed-Sternberg cells of Hodgkin's disease. *Int J Cancer* **46**:801

WHAT CAN WE LEARN ABOUT THE AETIOLOGY OF HODGKIN'S DISEASE FROM FAMILY STUDIES?

Paul H Levine,[1] Albert Lin[2] and Margaret A Tucker[2]

[1]Viral Epidemiology Branch and [2]Genetic Epidemiology Branch, Epidemiology and Biostatistics Program, National Cancer Institute, Bethesda, Md. 20892, USA

INTRODUCTION

Family studies have given us an important tool in understanding the aetiology of cancer. There are many examples, such as malignant melanoma and nasopharyngeal carcinoma, where the interaction of genetic and environmental factors is important in producing the observed pattern of disease. In the case of malignant melanoma, for example, a subset of the population has been identified with hereditary predisposition to melanoma associated with dysplastic nevi (Clark *et al.*, 1978; Reimer *et al.*, 1978). The pattern of inheritance appears to be dominant with approximately 90% penetrance. Environment plays a role in this genetically determined malignancy since the risk of melanoma in people with dysplastic nevi is related to the number of sunburns. Education and counselling of susceptible family members regarding sun exposure apparently changes their exposure patterns. Within several years of less sun exposure, the number of new melanomas and dysplastic nevi decrease in the individuals following skin care guidelines (Tucker, 1986). In nasopharyngeal carcinoma, the genetic aspects and the environmental aspects remain speculative. HLA loci have been linked to a putative disease-susceptibility gene in some specific multiple case families. The putative environmental factors include Epstein-Barr virus (EBV), exposure to nitrosamines, cigarette smoke, and other agents (Hildesheim and Levine, 1993). Nasopharyngeal carcinoma appears to be in the middle of a spectrum ranging from ataxia telangiectasia and xeroderma pigmentosum, where genetics appear to be the most important contributing factor, to adult T-cell leukaemia/lymphoma where the exposure to the oncogenic virus, HTLV-I, at an early age is clearly far more important than genetic susceptibility (Murphy *et al.*, 1989). In virtually every malignancy in which there is an interaction between genetics and the environment, which would include breast cancer

Etiology of Hodgkin's Disease
Edited by R. Jarrett, Plenum Press, New York, 1995

and colon cancer, general epidemiological studies often provide a clue as to where scientists should focus their attention. Geographical patterns and migrant studies have led us to focus on sunlight as a risk factor for melanoma and diet as a risk factor for breast cancer and colon cancer. For Hodgkin's disease (HD), the emphasis has been on geography and socio-economic factors (Correa and O'Conor, 1971; Cole *et al.*, 1968; Gutensohn 1982; see other chapters).

FAMILIAL HODGKIN'S DISEASE

Regardless of how one examines the data, whether it be by age at onset, geographical variation, or histopathological subtype, the evidence continues to accumulate supporting MacMahon's hypothesis (MacMahon, 1966) that HD consists of more than one aetiological entity (see Alexander, this volume). In this context, it is useful to evaluate family studies in view of the current aetiological hypotheses as an approach to further distinguishing subgroups where genetic predisposition, rather than environment, plays an increasingly important role.

The risk of HD developing in a first degree relative of a patient with HD is increased, but the precise risk is uncertain and the data have been interpreted as supporting both a genetic and environmental (shared childhood experiences) hypothesis. In 1959, Razis *et al.* reported a relative risk of three in first degree relatives of patients with HD (Razis *et al.*, 1959), but Grufferman reanalysed the data and noted a relative risk of 17.2 in those with confirmed histology (Grufferman, 1982). Grufferman *et al.* (1977) also noted a relative risk of 7.1 in siblings of young adult cases (less than 45 years of age) but none if diagnosed after 45. Siblings of the same sex apparently had twice the risk of developing HD than siblings of the opposite sex, which could be interpreted as being due either to environment, because of sharing the same childhood rooms, toys, etc., or genetics, being due to a chromosome abnormality somewhere near but not too closely linked to the sex chromosomes. Grufferman has also observed (Grufferman, 1977) that parent-child pairs occurred more commonly than spouse pairs, but this does not necessarily differentiate between environmental and genetic impact.

The family studies group of the Epidemiology and Biostatistics Program, National Cancer Institute, which has been investigating multiple-case families referred by a number of physicians and family members, has currently registered 45 families with 108 cases of HD in its database. Early on, the NCI group reported an association between HD and specific HLA alleles within the families (Greene *et al.*, 1979). This preliminary finding was followed by an analysis of 41 multiplex families (Chakravarti *et al.*, 1986), some of which had been previously published. The results of the genetic analyses provided strong evidence of linkage between a susceptibility gene tightly linked to the HLA locus and HD. The genetic model which provided the best fit for the data was a recessive model. This putative locus could potentially account for 60% of familial cases. The remaining 40% of cases are more likely to be due to other familial/polygenic or environmental factors.

A recent report (Lin *et al.*, 1993), based on the 45 families included in this current analysis, noted that 11 of the 21 sibling pairs were discordant for gender, a higher percentage than that reported by Grufferman *et al.* (Grufferman *et al.*, 1977). The median age of familial cases was 26, most cases being diagnosed between 20 and 40 years of age.

In view of the recent emphasis on EBV detection in Reed-Sternberg (RS) cells of HD biopsies (Weiss *et al.*, 1989; Bignon *et al.*, 1990; Jarrett *et al.*, 1991; Ambinder *et al.*, 1993; Pallesen *et al.*, 1991; see other chapters), we have been investigating the impact of EBV on our HD families (Lin *et al.*, 1994). In this study, which will be reported in detail subsequently, we found only 6 of 18 evaluable families studied (biopsies available for EBV genome detection in at least two sibling pairs) with EBV-associated disease in at least one member, and of these 6 families two were concordant for EBV and four were discordant. Therefore, an EBV-associated patient was more likely to have a sibling with non-EBV-associated disease, emphasising the relative unimportance of this virus in familial HD disease.

As noted by Jarrett *et al.*, (1991) the highest proportion of EBV-positive cases was in our youngest and oldest age groups (Fig. 1). Finally, we noted a close correlation of histological subtypes in our families, all except one sibling pair having identical histologies.

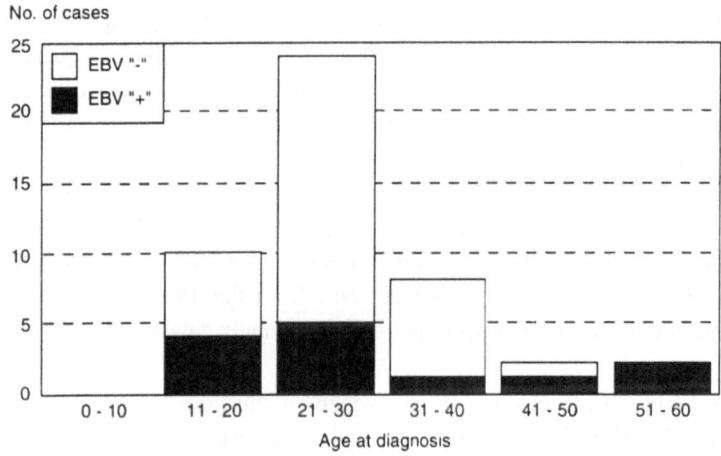

Fig. 1: Familial Hodgkin's disease: EBV genome positivity by age at diagnosis.

DISCUSSION

Several useful conclusions can be made from our data accumulated thus far. First, it is clear from these data that there is no concordant EBV positivity in these families with multiple cases of HD, and that familial HD is generally not EBV-associated. Since EBV-associated cases are less common in young adults than children or older adults (Jarrett *et al.*, 1991), this is not surprising since the predominant age group in familial HD is the young adult group (ages 20-40).

Second, we have noted that approximately half of our sibling case-pairs are discordant for gender. This finding is in contrast to the association reported by Grufferman *et al.* (1977). A more detailed investigation of shared childhood environment in each family as well as evaluation of genetic markers, may help disentangle the relative importance of these potential risk factors.

Third, the close concordance of histological subtype within family members is of interest. Whether the risk for a particular histological subtype is determined by genetic or environmental factors is a potential hypothesis to be investigated. Evaluation of specific parameters such as age of onset may be useful in providing additional hypotheses for investigation in non-familial cases.

Family studies of HD deserve emphasis because of the opportunity to compare the effects of genetics and environment in well-defined groups with generally similar socio-economic characteristics. There are techniques evaluating the patterns of cancer in multiple case families, primarily segregation analysis, that fit different models of inheritance patterns and environmental exposure. While it is apparent that environment is more important than genetics in nonfamilial HD, HLA studies suggest that 60% of familial cases may result from a genetic predisposition (Chakravarti *et al.,* 1986).

Familial HD is relatively rare, approximately 10% in published series. This may be due to the small family size associated with upper socio-economic groups at higher risk of developing HD, and in part to the putative recessive pattern. The infrequent detection of HD families, therefore, indicates the importance of multiple groups working together on similar protocols so that similar data are collected. This is particularly true in identifying histological subtypes associated with familial HD. It may well be important to set aside a special collection of blocks, or representative slides of such blocks, for future virological and genetic studies as new tests become available. Developing a list of hypotheses, a basic questionnaire, and a system for collecting and storing biological specimens for common data analysis could be a very important outcome from this meeting. Common data sets could eventually allow a metanalysis to put individual study data in perspective.

There are relevant data from other studies which should now be considered in regard to analysis of familial HD. For example, what is the frequency of lymphocyte predominance familial HD vs. sporadic? This is particularly important because the data from immunological and clinical studies suggest that the nodular lymphocyte predominance form may be different in aetiology and pathogenesis from the other forms of HD (Wright, 1989).

In addition to HD, are there any other malignancies which are particularly likely to occur in relatives of HD patients? This report only includes the families with more than one case of HD but it is clear from our data, as well as data from Israel (Shpilberg *et al.,* 1991) and the United Kingdom (Cartwright, personal communication), that haematological malignancies tend to aggregate in families without any particular histopathological entity being predominant. It would be important to look at these multiple case haematological malignancy families for particular markers of immune response that might be genetically determined. There are some interesting reports indicating that genetics are responsible for depressed natural killer cell activity, for example (Dubey *et al.*, 1994). These types of studies need to be pursued in a familial setting.

Young adult HD needs to be the target of research in developed countries because this is the type most associated with clustering and does not appear to be EBV related. In regard to familial cancer, it is this young adult group and particularly the nodular sclerosis subtype that appears to emerge as the prototype on which to focus the search for genetic markers of susceptibility. We would propose that collected series of HLA patterns in HD be reanalysed concentrating on patients between ages 20-40 with particular emphasis on the nodular sclerosis group.

In summary, familial aggregation of HD represents a relatively unusual occurrence that deserves intensive investigation. Development of a registry of familial cases associated with a biospecimen repository may facilitate our understanding of the role of genetics in at least one group of HD patients.

Note added in proof
Since the submission of this manuscript, Mack *et al.* (1995) reported a study of HD in twins. Their finding of concordancy in 10 of 179 pairs of monozygotic twins under the age of 50 years where one was affected by HD, in contrast to none of 187 pairs of dizygotic twins with one affected by HD, supports genetic susceptibility as an important factor in young adult HD. The finding of only one of the nine twin pairs being concordant for EBV-genome detection is in agreement with our findings, noted above, that EBV is not an important factor in familial HD.

REFERENCES

Ambinder RF, Browning PJ, Lorenzana I, Leventhal BG, Cosenza H, Mann RB, MacMahon EME, Medina R, Cardona V, Grufferman S, Olshan A, Levin A, Petersen EA, Blattner W, Levine PH (1993) Epstein-Barr virus and childhood Hodgkin's disease in Honduras and the United States. *Blood* **81**:462

Bignon Y-J, Bernard D, Curé H, Fonck Y, Pauchard J, Travade P, Legros M, Dastugue B, Plagne R (1990) Detection of Epstein-Barr viral genomes in lymph nodes of Hodgkin's disease patients. *Mol Carcinogenesis* **3**:9

Chakravarti A, Halloran SL, Bale SJ, Tucker MA (1986) Etiological heterogeneity in Hodgkin's disease: HLA linked and unlinked determinants of susceptibility independent of histological concordance. *Genetic Epidemiol* **3**:407

Clark WH Jr, Reimer R, Greene MH, Ainsworth A, Mastrangelo M (1978) Origin of familial malignant melanoma from heritable melanocytic lesions. *Arch Derm* **114**:732

Cole P, MacMahon B, Aisenberg A (1968) Mortality from Hodgkin's disease in the U.S. *Lancet* **ii**:1371

Correa P, O'Conor GT (1971) Epidemiologic patterns of Hodgkin's disease. *Int J Cancer* **8**:192

Dubey DP, Alper CA, Mirza NM, Awdeh Z, Yunis EJ (1994) Polymorphic Hh genes in the HLA-B(C) region control natural killer cell frequency and activity. *J Exp Med* **179**:1193

Greene MH, McKeen EA, Li FP, Blattner WA, Fraumeni JF Jr (1979) HLA antigens in familial Hodgkin's disease. *Int J Cancer* **23**:777

Grufferman S, Cole P, Smith PG, Lukes RJ (1977) Hodgkin's disease in siblings. *N Engl J Med* **296**:248

Grufferman S (1977) Clustering and aggregation of exposures in Hodgkin's disease. *Cancer* **39**:1829

Grufferman S. (1982) Hodgkin's disease. In *Cancer Epidemiology and Prevention*. Schottenfeld D, Fraumeni Jr, JF, eds (Philadelphia: WB Saunders, Co.) p739

Gutensohn NM (1982) Social class and age at diagnosis of Hodgkin's disease: new epidemiologic evidence for the "two-disease hypothesis". *Cancer Treat Rep* **66**:689

Hildesheim A, Levine PH (1993) Etiology of nasopharyngeal carcinoma: a review. *Epidemiol Reviews* **15**:466

Jarrett RF, Gallagher A, Jones DB, Alexander FE, Krajewski AS, Kelsey A, Adams J, Angus B, Gledhill S, Wright DH, Cartwright RA, Onions DE (1991) Detection of Epstein-Barr virus genomes in Hodgkin's disease: relation to age. *J Clin Pathol* **44**:844

Lin A, Whitehouse J, Shaw G, Tucker A (1993) Familial aggregation of Hodgkin's disease in a cohort of 48 families. *Proc Annu Meet Am Soc Clin Oncol* **12**:184

Lin A, Kingma D, Lennette E, Jaffe E, Ambinder R, Levine P, Tucker M (1994) Epstein-Barr virus is not associated with familial Hodgkin's disease. *Proc Annu Meet Am Soc Clin Oncol* **13**:5

Mack TM, Cozen W, Shibata DK, Weiss LM, Nathwani BN, Hernandez AM, Taypor CR, Hamilton AS, Deapen DM, Rappaport EB (1995) Concordance for Hodgkin's disease in identical twins suggesting genetic susceptibility to the young-adult form of the disease. *N Engl J Med* **332**:413

MacMahon B (1966) Epidemiological evidence on the nature of Hodgkin's disease. *Cancer Res* **26**:1189

Murphy EL, Hanchard B, Figueroa JP, Gibbs WN, Lofters WS, Campbell M, Goedert JJ, Blattner WA (1989) Modelling the risk of adult T-cell leukemia/lymphoma in persons infected with human T-lymphotropic virus type I. *Int J Cancer* **43**:250

Pallesen G, Hamilton-Dutoit SJ, Rowe M, Young LS (1991) Expression of Epstein-Barr virus latent gene products in tumour cells of Hodgkin's disease. *Lancet* **337**:320

Razis DV, Diamond HD, Carver LF (1959) Familial Hodgkin's disease: its significance and implications. *Ann Intern Med* **51**:933

Reimer RR, Clark WH Jr, Greene MH, Ainsworth AM, Fraumeni JF Jr (1978) Precursor lesions in familial melanoma: a new genetic preneoplastic syndrome. *JAMA* **239**:744

Shpilberg O, Modan M, Chetrit A, Modan B, Ramot B (1991) Familial aggregation of hematological malignancies among patients with Hodgkin's disease. Second International Symposium on Hodgkin's disease, October 3-5, 1991, Cologne, Germany, p29 (abstr)

Tucker MA (1988) Individuals at high risk of melanoma. *Pigment Cell* **9**:95

Weiss LM, Movahed LA, Estmir RA (1989) Detection of Epstein-Barr viral genomes in Reed-Sternberg cells of Hodgkin's disease. *N Engl J Med* **320**:502

Wright DH (1989) Pathology of Hodgkin's disease: Anything New? In *New Aspects in the Diagnosis and Treatment of Hodgkin's disease*. *Recent Results in Cancer Research*. Diehl V, Pfreundschuh M, Loeffler M, eds (Berlin, New York: Springer-Verlag) p2

EPSTEIN-BARR VIRUS AND LYMPHOMAS: AN OVERVIEW

Lawrence S Young[1] and Gerald Niedobitek[2]

Institute for Cancer Studies[1] and Department of Pathology[2],
University of Birmingham, Birmingham B15 2TT, UK

INTRODUCTION

Epstein-Barr virus (EBV) is a ubiquitous human herpes virus which is found as a predominantly asymptomatic infection in all human communities. Primary EBV infection usually occurs in childhood and, once infected, individuals become life-long virus carriers. The virus is the causative agent of infectious mononucleosis (IM), a self-limiting lymphoproliferative disease resulting from delayed primary EBV infection, and is also associated with a number of malignant tumours, including Burkitt's lymphoma (BL) and lymphomas arising in immunocompromised patients (Miller, 1990). The B lymphotropic nature of EBV is evidenced by its association with these lymphoproliferations and by the ability of the virus to immortalise normal resting B lymphocytes *in vitro*, converting them into permanently growing lymphoblastoid cell lines (LCLs) (Nilsson *et al.*, 1971). When peripheral blood lymphocytes from EBV seropositive individuals are placed in culture, the few virus-infected B-cells that are present regularly give rise to spontaneous outgrowth of EBV-transformed LCLs provided that immune T-cells are either removed or inhibited by addition of cyclosporin A to the culture (Rickinson *et al.*, 1984). This phenomenon highlights the importance of EBV-specific cytotoxic T lymphocytes (CTLs) in controlling EBV-induced B-cell transformation (Fig. 1).

EBV is orally transmitted and infectious virus, measured by its ability to immortalise B-cells *in vitro*, can be detected in oropharyngeal secretions from IM patients, from patients who are immunosuppressed and, at lower levels, from healthy EBV seropositive individuals (Gerber *et al.*, 1972; Strauch *et al.*, 1974; Yao *et al.*, 1985).

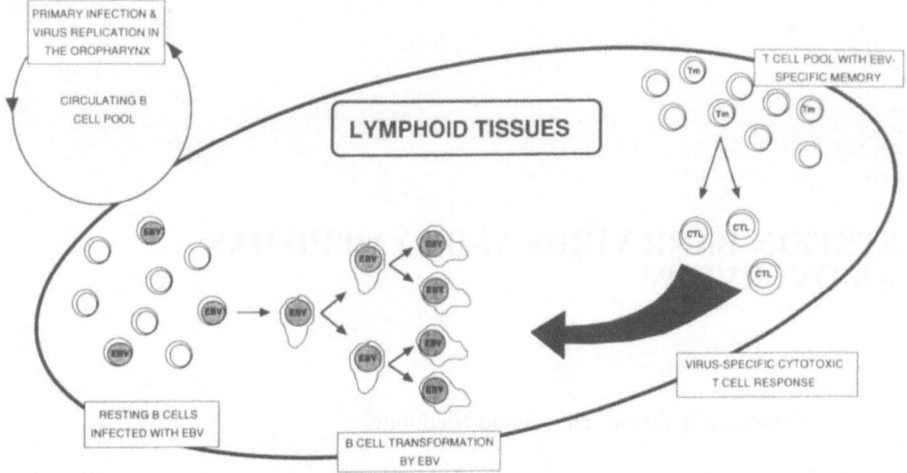

Fig. 1:   EBV infection in normal healthy virus carriers. Primary EBV infection occurs in the oropharynx and is mediated either through B-cells or epithelial cells. Following primary infection of B lymphocytes, a chronic virus carrier state is established in which the outgrowth of EBV-infected B-cells is controlled by an EBV-specific cytotoxic T-cell response. At certain sites, presumably in the oropharynx, latently infected B-cells may become permissive for lytic EBV infection. Infectious virus released from these cells may be shed directly into the saliva or may infect epithelial cells and other B-cells.   In this way a virus-carrier state is established which is characterised by persistent, latent infection in circulating B-cells and occasional EBV replication in B-cells and epithelial cells. Whilst the exact mode of primary and persistent EBV infection and the relative contributions of B-cells and epithelial cells are uncertain, recent data point to the B-cell compartment as the main mediator of primary as well as persistent infection.

Under certain circumstances EBV is also able to infect epithelial cells as demonstrated by the regular detection of the virus in the tumour cells of undifferentiated nasopharyngeal carcinomas (NPCs) (Miller, 1990).   Replicating EBV has been demonstrated in desquamated oropharyngeal epithelial cells from IM patients (Sixbey *et al.*, 1984) and, more recently, in the AIDS-associated epithelial lesion, oral hairy leukoplakia (HL) (Greenspan *et al.*, 1985).   These findings have been used to argue that oropharyngeal epithelial cells are the target of primary EBV infection and that epithelial cells are the site of viral persistence and replication in normal virus carriers (Allday *et al.*, 1988). However, this concept has been challenged recently by several lines of evidence all pointing to B lymphocytes as mediators of primary and persistent EBV infection (see below) (Niedobitek and Young, 1994). The potential interactions between lymphocytes and epithelial cells are illustrated in Fig. 1.

EBV BIOLOGY

Models of EBV latency
The ability of EBV to immortalise efficiently primary, resting B-cells *in vitro* provides a useful model for examining the EBV gene expression and its effects on cellular phenotype.

In these LCLs EBV replication is an infrequent event and the virus is in a predominantly latent state (Kieff *et al.,* 1990). Thus, every cell in an LCL carries multiple copies of the viral episome and constitutively expresses a limited set of viral gene products, the so-called latent proteins, consisting of six nuclear antigens (EBNAs -1, -2, -3A, -3B, -3C and -LP) and three latent membrane proteins (LMPs -1, -2A and -2B). The relative positions and orientations of these latent genes on the large (172 kilobases), covalently-closed EBV episome are shown schematically in Fig. 2. The different EBNAs are encoded by individual mRNAs generated by differential splicing of the same long 'rightward' primary transcript expressed from one of two promoters (Cp and Wp) located close together in the BamHI C and W region of the genome (Fig. 2) (Speck and Strominger, 1989). The LMP transcripts are expressed from separate promoters in the BamHI N region of the EBV genome, with the leftward LMP-1 and rightward LMP-2B mRNAs apparently controlled by the same bi-directional promoter sequence (Fig. 2) (Speck *et al.,* 1989; Laux *et al.,* 1989). The interaction of EBNA-1 with the origin of replication (oriP) of the EBV genome is an essential function as it mediates the replication and extrachromosomal maintenance of the EBV episome. LMP-1 is oncogenic in rodent fibroblasts (Wang *et al.,* 1985) and a mutant EBV strain lacking the EBNA-2 gene is unable to immortalise B-cells (Bornkamm 1982; Rabson *et al.,* 1982), suggesting that these two latent proteins are particularly important in the immortalisation process. However, the general consensus is that EBV-induced immortalisation of B-cells involves the co-ordinated action of several latent gene functions and recent work with recombinant EBVs suggests that EBNA-1, EBNA-2, EBNA-3A, EBNA-3C and LMP-1 are all necessary for LCL generation. In addition to the latent proteins, LCLs also show abundant expression of the small non-polyadenylated (and therefore non-coding) RNAs, EBERs 1 and 2 (Fig. 2), whose function is not clear but whose expression is a constant feature of all types of latent EBV infection (Kieff *et al.,* 1990).

The consistent pattern of EBV latent protein expression in LCLs is matched by an equally consistent and characteristic cellular phenotype with high level expression of the B-cell activation markers CD23, CD30, CD39 and CD70 and of the cellular adhesion molecules LFA-1 (CD11a/18), LFA-3 (CD58) and ICAM-1 (CD54) (Rowe *et al.,* 1985; Gregory *et al.,* 1988). That these markers are either absent or expressed at low levels on resting B-cells, but are transiently induced to high levels when these cells are activated into short-term growth by antigenic or mitogenic stimulation, suggests that EBV-induced immortalisation may be elicited through the constitutive activation of the same cellular pathways that drive physiological B-cell proliferation. The ability of EBNA-2, EBNA-3C and LMP-1 to induce LCL-like phenotypic changes when expressed individually in human B-cell lines implicates these viral proteins as key effectors of the immortalisation process (Wang *et al.,* 1987; Wang *et al.,* 1990).

The pattern of latent EBV gene expression in LCLs is referred to as the "latency III" (Lat III) form of EBV infection. A second form of EBV infection in B-cells, referred to as "latency I" (Fig. 3), has been identified in BL tumour biopsy cells and in early passage BL cell lines where abundant EBER transcription is found and EBNA-1 is selectively expressed in the absence of the other EBNA and LMP proteins (Rowe *et al.,* 1987; Gregory *et al.,* 1990). The selective expression of EBNA-1 involves a different mRNA transcript expressed from a novel EBNA-1 promoter (F/Qp) in the BamHI F/Q region of the viral genome which is independent of Cp or Wp promoters (Fig. 2) (Sample *et al.*, 1991; see note added in proof). In culture, BL cells grow as a carpet of dispersed cells as opposed to the multicellular aggregates that are observed in LCL cultures. Furthermore, BL cells display a distinct cell surface marker phenotype characterised by expression of CD10 (CALLA) and CD77 (BLA) but no or little expression of the cellular activation antigens and adhesion molecules that are regularly expressed at high levels in LCLs (Rowe *et al.*, 1987; Gregory *et al.*, 1990). The Lat I form of latency observed in BL cell lines is not always stably maintained *in vitro*, and on serial passage a drift to a Lat III pattern of gene expression can be observed concomitant with a change in the cellular phenotype towards that seen in LCLs (Fig. 3) (Rowe *et al.*, 1985; Gregory, *et al.*, 1990).

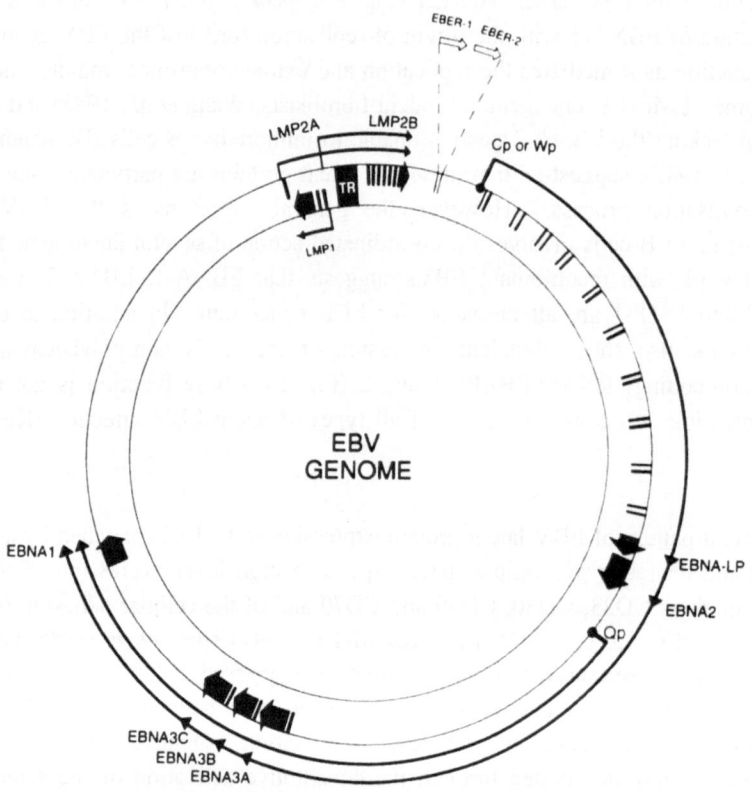

Fig. 2:   Location and transcription of the EBV latent genes on the double-stranded viral DNA episome. The large solid arrows represent coding exons for each of the latent proteins and the direction in which they are transcribed. EBNA-LP is transcribed from variable numbers of repetitive exons in the BamHI W fragments. LMP-2 is composed of multiple exons located either side of the terminal repeat regions which are fused during the circularisation of the linear DNA to produce the viral episome. The open arrows represent the highly transcribed non-polyadenylated RNAs, EBER-1 and EBER-2, which are a constant feature of latent EBV infection. The outer long arrowed line represents EBV transcription in Lat III where all the EBNAs are transcribed from either the Cp or Wp promoter; the different EBNAs are encoded by individual mRNAs generated by differential splicing of the same long primary transcript. The inner shorter arrowed line represents the EBNA-1 transcript originating from the Qp promoter located in the Ban HI Q region; this is transcribed in latency types I and II.

Another form of EBV latency, Lat II (Fig. 3), is characterised by selective expression of the Qp-driven EBNA-1 mRNA, of LMP-1, -2A and -2B transcripts, and of the EBERs; this form of infection was first identified in biopsies of nasopharyngeal carcinoma (Brooks *et al.*, 1992) but is clearly not restricted to epithelial cells since it is also observed in EBV-positive cases of Hodgkin's disease (HD) (Herbst *et al.*, 1993) and can under some circumstances be experimentally induced in BL cells *in vitro* (Rowe *et al.*, 1992). All three forms of EBV latency can be interconverted in somatic cell hybrids between LCLs and either BL cells or certain non-lymphoid lines (Kerr *et al.*, 1992). These transitions are influenced by the cell phenotype of the resultant hybrids, thus emphasising the complex interplay between cellular factors and the resident pattern of EBV latent gene expression.

EBV types 1 and 2

Two major types of EBV isolate, originally referred to as A and B and now called types 1 and 2, can be defined which appear to be identical over the bulk of the EBV genome but show polymorphism (with 50-80% sequence homology depending on the locus) in a subset of latent genes, namely those encoding EBNA-LP, EBNA-2, EBNA-3A, EBNA-3B and EBNA-3C (Dambaugh *et al.*, 1984; Rowe *et al.*, 1989; Sample *et al.*, 1990). A combination of virus isolation and seroepidemiological studies suggest that type 1 virus isolates are predominant (but not exclusively so) in many Western countries, whereas both types are widespread in equatorial Africa, New Guinea and perhaps certain other regions (Zimber *et al.*, 1986; Young *et al.*, 1987; Sixbey *et al.*, 1989; Yao *et al.*, 1991). The balance of evidence to date suggests that healthy individuals are only infected with one virus type (Yao *et al.*, 1991), although this may well change in immunologically compromised patients (Sixbey *et al.*, 1989).

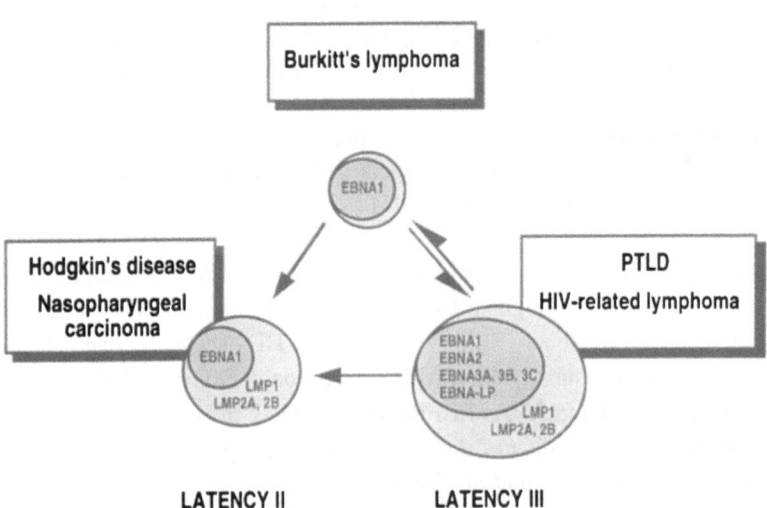

Fig. 3: Three different forms of EBV latency have been described as exemplified by certain human tumours. All three forms can be interconverted in somatic cell hybrids between lymphoblastoid cell lines and Burkitt's lymphoma cell lines or certain non-lymphoid cell lines. PTLD, post-transplant lymphoproliferative disease.

Work in the cell culture model systems described above indicates that both types of EBV can establish the same spectrum of latent infections. Interestingly, however, type 1 isolates are more potent than type 2 in achieving B-cell immortalisation *in vitro*; the type 2 virus-transformed LCLs characteristically showing much slower growth especially in early passage (Rickinson *et al.*, 1987). In view of this biological difference in experimental systems, it is important to determine the relative contributions of the two virus types to lymphoid malignancies. This question is by no means fully resolved at present. The available evidence suggests that both virus types are equally represented amongst BL tumours in Africa and New Guinea, consistent with the relative incidence of the two virus types in the healthy populations in these areas (Zimber *et al.*, 1986; Young *et al.*, 1987). More interesting is the relative association of the two types of EBV with lymphomas in the immunosuppressed; surprisingly, recent reports suggest that type 2 virus isolates can be detected with unusually high frequency amongst immunologically compromised patients and also in their accompanying lymphomas (Sixbey *et al.*, 1989; Boyle *et al.*, 1991). In this context it is interesting that HD cases occurring in HIV-positive individuals appear to be invariably EBV-positive and that an increased proportion of these tumours carry type 2 EBV (Herndier *et al.*, 1993; Boyle *et al.*, 1993).

EBV AND MALIGNANT LYMPHOMAS

EBV and Hodgkin's disease

Over the last few years evidence has been accumulating which implicates EBV in the aetiology of HD (see Jarrett, this volume). The malignant population of Hodgkin and Reed-Sternberg cells (H-RS cells) found in HD lymph nodes are unusual in that they form only a minority of the tumour mass and are of uncertain cell lineage (for review see Herbst *et al.*, 1993). Seroepidemiological studies have demonstrated elevated antibody titres to EBV antigens in HD patients at the time of presentation as well as in serum samples taken some years prior to the onset of the disease (Mueller *et al.*, 1989). There is also a three-fold increased risk of developing HD following IM (Munoz *et al.*, 1978). Recently, more direct evidence linking EBV with HD has been published. Thus, EBV genomes have been detected in tumour material in 19-50% of HD cases and *in situ* hybridisation has localised these viral genomes to the malignant H-RS cells (for review see Herbst *et al.*, 1993). The demonstration of monoclonal EBV episomes in HD suggests that clonal cellular proliferation is initiated subsequent to EBV infection. Expression of the small EBV-encoded non-polyadenylated RNAs EBER-1 and EBER-2 in H-RS cells has been reported and is consistent with the finding of these abundant transcripts in all situations where latent EBV infection is sustained (Wu *et al.*, 1990; Herbst *et al.*, 1992). An aetiological role for EBV in the pathogenesis of HD is further supported by the finding that the oncogenic LMP-1 viral protein is highly expressed in the malignant H-RS cells of EBV-positive HD cases in the absence of EBNA-2 expression (Pallesen *et al.*, 1991; Herbst *et al.*, 1991a). This restricted pattern of EBV latent protein expression is similar to that found in NPC. In particular, analysis of EBV transcription in fresh HD biopsies identified EBNA-1 mRNA initiating from the BamHI F/Q region of the viral genome (so-called Q promoter) as in NPC and BL, rather than the Cp and Wp promoters which are used to direct transcription of all the EBNAs in LCLs (Deacon *et al.*, 1993).

EBV and non-Hodgkin's lymphomas

Since the discovery of EBV in the tumour cells of endemic BL, a vast body of work has implicated the virus in the pathogenesis of various lymphoproliferations.

In view of the important role of T-cells in the control of EBV infection in chronic virus carriers, it is not surprising that EBV-associated lymphoproliferative disorders frequently occur in immunocompromised individuals. The most relevant clinical settings in which such tumours arise are immunosuppressed transplant patients and HIV-infected persons. However, there are differences between these two groups both with respect to lymphoma morphology and virus association.

Post-transplant lymphoproliferative disorders (PTLD) represent a spectrum of diseases ranging from benign polymorphic and polyclonal lymphoproliferations to monomorphic and monoclonal high grade B-cell non-Hodgkin's lymphomas (NHLs) (Cleary *et al.*, 1988; Locker and Nalesnik, 1989; Randhawa *et al.*, 1989; Swerdlow, 1992). The vast majority of these lesions appears to be EBV-associated and to express the full range of latent viral gene products (latency III) (Young *et al.*, 1989; Thomas *et al.*, 1990).

AIDS-related lymphomas (ARLs) are mainly high grade B-cell NHLs with a tendency to present at extranodal sites. These lymphomas fall broadly into two groups, Burkitt-type lymphomas and large cell lymphomas with many immunoblasts (Hamilton-Dutoit *et al.*, 1993a). EBV has been detected in most immunoblast-rich lymphomas (77%) but only in a minority of Burkitt-type lymphomas (34%) and this difference appears to be mirrored in the different degree of immune system impairment, patients with immunoblast-rich lymphomas having the more severely impaired immunity (Pedersen *et al.*, 1991; Hamilton-Dutoit *et al.*, 1993a). Whilst in some ARLs a type III pattern of virus latency is observed, others show a more restricted pattern of EBV gene expression corresponding to types I or II of EBV latency (Hamilton-Dutoit *et al.*, 1993b; Shibata *et al.*, 1993).

EBV is found in virtually all cases of endemic BL. Whilst this would suggest an important role for the virus in the development of this neoplasm, the exact contribution of EBV is uncertain particularly because those viral gene products with known transforming potential are not expressed in BL cells. In this context it is of interest that more recent studies have suggested a possible transforming function for EBNA-1, the only latent viral protein expressed in BL cells (Wilson and Levine, 1992; Magrath *et al.*, 1993).

In contrast to its regular presence in endemic cases, EBV is detectable in only about 20% of sporadic BL cases (Hummel *et al.*, 1993). However, detection rates in other B-cell NHLs are even lower. In a recent study, Hummel *et al.* (1993) found EBV in 3 of 65 (4.5%) low grade B-cell NHLs and in 9 of 77 (13%) high grade NHLs. In some of these cases LMP-1 expression was detected in the absence of EBNA-2, suggestive of a type II latency (Hummel *et al.*, 1993). It appears therefore that in immunocompetent individuals, EBV is only rarely associated with B-cell NHLs other than BL. Moreover, whilst EBV is usually present in virtually all tumour cells of BL (endemic or sporadic) it appears that other B-cell NHLs often harbour the virus in only a sub-population of tumour cells (see below).

EBV infection of T lymphocytes has been demonstrated in some atypical lymphoproliferative disorders often associated with primary EBV infection (Kikuta *et al.*, 1988; Yoneda *et al.*, 1990; Mori *et al.*, 1992; Kawaguchi *et al.*, 1993). Thus, EBV-positive T-cells have been observed in a case of Kawasaki-like disease (Kikuta *et al.*, 1988) and more recently in cases of EBV-associated haemophagocytic syndrome (Kawaguchi *et al.*, 1993). Since the late 1980s, evidence has emerged suggesting an association of EBV with a proportion of peripheral T-cell NHLs (PTLs) (Jones *et al.*, 1988). Recent surveys of larger series have demonstrated the frequent presence of the virus in such cases and have suggested that PTLs are more frequently EBV-associated than B-cell NHLs (Korbjuhn *et al.*, 1993; de Bruin *et al.*, 1993; Pallesen *et al.*, 1994). Interpretation of these data is difficult, however, for several reasons. Morphological classification of T-cell lymphoproliferations is difficult and particularly the differential diagnosis with HD may prove problematic. Some T-lymphoproliferative disorders such as angioimmunoblastic lymphadenopathy (AILD) do not represent a homogeneous entity but rather a spectrum of lymphoproliferations. Furthermore, whilst the overall detection rate of EBV in PTLs may be high (up to 47%) closer examination of the data reveals that significant numbers of EBV-positive tumour cells are found in only a minority of cases (Korbjuhn *et al.*, 1993; Pallesen *et al.*, 1994). In some PTLs, only a small fraction of tumour cells appears infected whilst in others the virus is present only in reactive "bystander" cells. This observation, however, raises the additional problem of identifying on morphological grounds whether an EBV-positive cell is a tumour cell or a non-neoplastic lymphocyte, particularly in lesions composed of different cell types. It appears that EBV is regularly detected in nasal angiocentric T-cell lymphomas of the lethal midline granuloma type and in these tumours virtually all tumour cells harbour the virus (Harabuchi *et al.*, 1990; Borisch *et al.*, 1993). AILD cases have been reported to be frequently EBV-positive (Anagnostopoulos *et al.*, 1992; Weiss *et al.*, 1992). However, in these lesions the virus is often present in B-cells rather than in T-cells raising the possibility of a secondary proliferation of EBV-positive B-cells due to an underlying immune defect (Weiss *et al.*, 1992). There appears to be agreement in the literature that, with the exception of upper respiratory tract lesions, extranodal PTLs, particularly those of the skin, are only infrequently EBV-associated (Pallesen *et al.*, 1994).

The frequent detection of EBV in HD cases has prompted studies into a possible EBV-association of CD30[+] anaplastic large cell (ALC) lymphomas. It appears that the overall detection rate of EBV in ALC lymphomas is lower than in HD (Herbst *et al.*, 1991b). The virus is more frequently detectable in those ALC lymphomas showing a B-cell phenotype than in T-cell cases (Herbst *et al.*, 1991b; Pallesen *et al.*, 1994). Moreover, the pattern of viral gene expression observed in ALC lymphomas is more heterogeneous than in HD with some cases of B-cell ALC lymphomas demonstrating a type III latency (Herbst *et al.*, 1991b).

In the paradigmatic EBV-associated tumours BL and undifferentiated NPC as well as in HD, the virus is present in all tumour cells and the viral episomes are monoclonal (Raab-

Traub and Flynn, 1986). These findings suggest that in these cases EBV infection has taken place before expansion of the malignant cell clone and suggest an important role for the virus in the development of these tumours. By contrast, there is increasing evidence to suggest that in other virus-associated lymphomas of B- and T-cell phenotype EBV is frequently detectable only in a fraction of tumour cells (Lewin *et al.*, 1988; Hamilton-Dutoit *et al.*, 1993b; Korbjuhn *et al.*, 1993; Hummel *et al.*, 1993; Pallesen *et al.*, 1994). The most likely explanation for this phenomenon appears to be an infection of established tumour cells with the virus. EBV infection at this late stage may not alter the course of the tumour or may confer an additional growth advantage to the virus-carrying cells leading to the outgrowth of one or more EBV-positive clone. This latter scenario could then lead to the detection of monoclonal viral genomes in Southern blots, wrongly suggesting EBV infection early in the neoplastic process. An alternative explanation for the detection of EBV in only a proportion of tumour cells would be the loss of viral genomes from initially EBV-positive neoplasms. Whilst this has been described for some NPC cell lines *in vitro* (Glaser *et al.*, 1989) there is currently no evidence to suggest that this may occur *in vivo*. Before drawing firm conclusions, however, technical reasons for this observation have to be excluded. Expression of the EBERs can show some heterogeneity (Herbst *et al.*, 1992). Furthermore, degradation of tissue RNA may lead to a reduction of the available EBER copies. In most studies today, non-radioactive *in situ* hybridisation techniques are employed. Whilst such methods are usually sufficient to detect the abundant EBER transcripts there remains some doubt as to their level of sensitivity in situations where the amount of target RNA is reduced. Thus, further studies are required to clarify this issue.

PERSISTENT EBV INFECTION

EBV appears to have at least two natural target cell types, B-lymphocytes and epithelial cells (Niedobitek and Young, 1994). The ability of EBV to infect lymphoid cells is illustrated by the regular detection of the virus in endemic BL and by its ability to transform peripheral blood B-cells into permanently growing LCLs. That EBV can also infect epithelial cells is demonstrated by the presence of the virus in the tumour cells of virtually all undifferentiated NPCs and by the detection of replicating EBV in HL an AIDS-associated benign epithelial lesion of the tongue (Greenspan *et al.*, 1985; Miller, 1990).

Primary infection with EBV occurs through the oropharynx by contact with saliva from an EBV-infected individual. Following primary infection, the virus establishes a life-long persistence in the infected host. The questions as to how primary infection occurs and how persistent infection is established are of central importance for our understanding of the role of the virus in virus-associated neoplasms.

The presence of EBV in the tumour cells of undifferentiated NPCs and the detection of replicating virus in exfoliated oropharyngeal epithelial cells from IM patients and in HL have been taken as evidence for a central role of oropharyngeal epithelial cells in primary and persistent EBV infection (Allday *et al.*, 1988). It has been hypothesised that oropharyngeal epithelial cells are the primary target of EBV infection. According to this model, the virus would replicate in the more differentiated epithelial cell layers and persist in the basal epithelial cells with infection of B-lymphocytes being a secondary event.

However, this concept has been challenged recently by a growing body of evidence. In tissues from IM patients, EBV is readily detectable in B-lymphoid cells but not in epithelial cells suggesting that tonsillar B-cells may be the primary target of EBV infection (Niedobitek et al., 1989; Niedobitek et al., 1992a).

In chronic virus carriers, variable numbers of EBV-harbouring lymphoid cells are detectable in normal and hyperplastic lymphoid tissues (Niedobitek et al., 1992a). Expansion of EBV-positive cells has been observed in isolated germinal centres, suggesting that EBV-carrying B-cells may participate in physiological germinal centre reactions (Niedobitek et al., 1992a). By contrast, there is little evidence to suggest that latent EBV infection of epithelial cells is a frequent event. In HL, the best studied model of EBV infection in non-neoplastic epithelial cells, replicating virus is readily detected in the upper epithelial cell layers but viral gene products associated with latent infection are consistently absent from the basal epithelial cells in such lesions (Young et al., 1991; Niedobitek et al., 1991a; Thomas et al., 1991). It appears therefore, that HL represents a focus of lytic virus infection without a detectable latent phase.

The detection of EBV DNA in epithelial cells of dysplastic lesions in the nasopharynx has been reported recently (Yeung et al., 1993). However, more studies are required to assess the significance of this report and the frequency with which such an event may occur. We have recently demonstrated the absence of epithelial EBV infection in nasopharyngeal biopsies from patients at high risk of NPC (Sam et al., 1993). Furthermore, EBV has not been detected in normal mucosa next to EBV-positive NPCs or gastric carcinomas (Niedobitek et al., 1991b; Shibata and Weiss, 1992; Rowlands et al., 1993). Thus, these findings would seem to suggest again that latent EBV infection of normal epithelial cells is rare.

In bone marrow transplant recipients whose resident haemopoietic tissue had been destroyed, eradication of EBV has been demonstrated in some cases (Gratama et al., 1988). As one would not expect a complete loss of epithelial cells in this context, this observation provides further circumstantial evidence against the model of EBV persistence in the epithelial cell compartment. Furthermore, it has been demonstrated that treatment of IM patients and of chronic virus carriers with acyclovir leads to reduced oropharyngeal virus shedding but leaves the number of circulating virus-harbouring lymphoid cells unaffected (Yao et al., 1989a; Yao et al., 1989b).

Based on these observations, we have proposed that primary and persistent EBV infection are mainly mediated through B-lymphocytes (Niedobitek and Young, 1994). Occasional amplification in germinal centre reactions could provide a mechanism facilitating long-term survival of virus-carrying B-cells in the infected host. It is envisaged that EBV is delivered by B-lymphocytes to distant mucosal sites where reactivation of the virus and subsequent infection of epithelial cells occur. This model would explain the detection of EBV in various carcinomas including some arising at sites where direct epithelial EBV infection is not possible, e.g., the thymus. A similar mechanism could also be responsible for the delivery of the virus to the precursor cells of other neoplasms such as HD. According to this hypothesis, virus infection of epithelial cells would be accidental and not relevant to virus persistence. The significance of HL as a model for virus replication in the immunocompetent host has probably been exaggerated. EBV infection of exfoliated oropharyngeal epithelial cells has been reported in IM patients (Sixbey et al., 1984), but the significance of this observation for EBV shedding in chronic virus carriers is uncertain.

Recently, EBV has been detected in a surprisingly large number of peripheral T-cell lymphomas, and EBV infection of non-neoplastic T-lymphocytes has been demonstrated in certain conditions. Whether T-cells are susceptible to EBV infection in the normal host or participate in the maintenance of virus persistence is uncertain and requires further investigation.

CONCLUSIONS

Progress in molecular biological techniques has led to the identification of a rapidly growing number of EBV-associated human tumours. By contrast, our understanding of the part played by the virus in the development of these neoplasms is still fragmentary. Several points have emerged recently which are likely to be of significance in this respect.

The spectrum of virus-associated tumours is much broader than anticipated and now includes not only BL, NPC, HD and lymphomas in immunocompromised patients but also a proportion of PTLs and some gastric adenocarcinomas. It is likely that there will be further additions to this list.

In some virus-associated tumours, such as BLs and NPCs, monoclonal viral episomes are present in all tumour cells indicating that EBV infection is an early event in the oncogenic process. However, there is a growing body of evidence suggesting that in some tumours, e.g., PTLs and some B-cell NHLs, the virus may be present only in a proportion of tumour cells. This would seem to suggest that at least in these cases EBV infection occurred late in the oncogenic process.

It has become clear that the operational definitions of different forms of EBV latency are not easily applied to human tumours. Thus, it appears that different tumour cells of one EBV-associated neoplasm may display different patterns of viral gene expression. Furthermore, the effects EBV infection has on the phenotype of different cells *in vitro* are not completely reproduced *in vivo*. The only known molecules whose expression *in vivo* may be related to the expression of certain viral proteins are the CD30 and CD70 lymphocyte activation antigens (Niedobitek et al., 1992b; Hamilton-Dutoit et al., 1993b; Kanavaros et al., 1993).

Recent research has re-emphasised the role of B-lymphocytes as mediators of primary and persistent EBV infection. Epithelial EBV infection appears to be an accidental phenomenon with little if any significance for virus persistence in the normal host. The potential role of T lymphocytes remains to be elucidated.

Prospects for the immunotherapy of Hodgkin's disease

Disturbances of the immune system may result in the unchecked growth of LCL-like (Lat III) tumours (PTLDs, ARLs) and it is these tumours which should be the most amenable to immunotherapeutic intervention. This notion has been confirmed by a recent study demonstrating the successful treatment of EBV-positive PTLDs developing in allogeneic bone marrow transplant recipients with lymphocytes from the bone marrow donor (Papadopoulos *et al.*, 1994).

The development of other EBV-positive tumours in the face of immune responses that are apparently sufficient to prevent the outgrowth of normal EBV-transformed B-cells (i.e., PTLD) suggests that these neoplasms have evolved mechanisms for evading EBV-specific immunosurveillance. In BL, one crucial factor appears to be the restricted expression of EBV genes (Lat I) since to date every one of the EBV latent proteins, with the notable exception of EBNA-1, has been shown to be a potential target for EBV-specific CTLs (Murphy *et al.*, 1990; Murphy *et al.*, 1992; Khanna *et al.*, 1992). In addition, a significant proportion of BLs shows selective down-regulation of certain HLA class I alleles (Torsteinsdottir *et al.*, 1988; Andersson *et al.*, 1991). The generally low levels of adhesion molecules on the BL cell surface may also help these tumours to remain immunologically silent (Gregory *et al.*, 1988). Together, these various facets of the BL phenotype may enable the tumour cells to survive and grow in the face of the normal EBV-specific CTL response observed in many BL patients. Furthermore, they may well explain why the phenotypic drift that BL cell lines display *in vitro* does not occur *in vivo*; drifting cells adopting an LCL-like phenotype *in vivo* would presumably be selectively removed by virus-specific CTLs.

The relative roles of EBV-specific cellular immunity in host defences against PTLD/ARL and against BL become easier to discern in view of what is now known about the target antigens for EBV-specific CTLs. However, since both LMP-1 and LMP-2 can provide target peptides for EBV-specific CTL responses in the context of at least some common HLA types (Lee *et al.*, 1993), instances of Lat II forms of infection in tumours that arise in apparently immunocompetent individuals are more difficult to understand. Thus, whilst HD patients (and NPC patients) show some evidence of impaired T-cell immunity, it is unlikely that this is sufficient in itself to allow outgrowth of the EBV-positive tumours. In this context the recent observation of a reversible case of EBV-positive HD arising in a patient on mild immunosuppressive therapy is of potential significance (Kamel *et al.*, 1993). In the case of HD it has not yet been possible to test whether or not the malignant cells are actually susceptible to recognition and lysis by LMP-1- and LMP-2-specific CTLs. *In vivo* it is conceivable that the release of cytokines by H-RS cells themselves, or by the non-malignant cells that constitute the bulk of the tumour mass, might provide a microenvironment that inhibits the recruitment and/or function of appropriate CTLs. The analysis of H-RS cells with regard to HLA expression and antigen processing/presenting capacity should be a priority. The recent demonstration of B7/BB1 antigen expression in H-RS cells suggests that these cells may have accessory cell function (Delabie *et al.*, 1993).

Likewise, the possibility that sequence variation over CTL epitopes present within LMP-1 or LMP-2 may contribute to the development of EBV-positive HD needs to be investigated.

Recent studies on PTLDs herald the advent of EBV-specific immunotherapy. The application of this approach to the treatment of HD will require a more in depth analysis of EBV-specific immune function in HD patients as well as studies aimed at elucidating the origin and immunological characteristics of H-RS cells.

Note added in proof

Recent work has identified a transcription initiation site within the Q region of the viral genome which drives EBNA-1 expression in latency forms I and II. This promoter is now referred to as Qp. The previously defined Fp appears to be a lytic promoter (Schaefer BC, Strominger JL, Speck SH. The Epstein-Barr virus BamHI F promoter is an early lytic promoter: lack of correlation with EBNA-1 gene transcription in group I Burkitt's lymphoma cell lines. *J Virol* (in press); Schaefer BC, Strominger JL, Speck SH. Redefining the EBNA-1 gene promoter and transcription initiation site in group I Burkitt's lymphoma cell lines. *Proc Natl Acad Sci USA* (in press)).

REFERENCES

Allday MJ, Crawford DH (1988) Role of epithelium in EBV persistence and pathogenesis of B-cell tumours. *Lancet* i:855

Anagnostopoulos I, Hummel M, Finn T, Tiemann M, Korbjuhn P, Dimmler C, Gatter K, Dallenbach F, Parwaresch M, Stein H (1992) Heterogenous Epstein-Barr virus infection patterns in peripheral T cell lymphoma of angioimmunoblastic lymphadenopathy type. *Blood* **80**:1804

Andersson ML, Stam NJ, Klein G, Pleg HL, Masucci MG (1991) Aberrant expression of HLA Class-1 antigens in Burkitt lymphoma cells. *Int J Cancer* **47**:544

Borisch B, Hennig I, Laeng RH, Waelti ER, Kraft R, Laissue J (1993) Association of the subtype 2 of the Epstein-Barr virus with T-cell non-Hodgkin lymphoma of the midline granuloma type. *Blood* **82**:858

Bornkamm GW, Hudewentz J, Freese UK, Zimber U (1982) Deletion of the non-transforming Epstein-Barr virus strain P3HR-1 causes fusion of the large internal repeat to the DSL region. *J Virol* **43**:952

Boyle MJ, Sewell WA, Sculley TB, Apolloni A, Turner JJ, Swanson CE, Penny R, Cooper DA (1991) Subtypes of Epstein-Barr virus in human immunodeficiency virus-associated non-Hodgkin lymphomas. *Blood* **78**:3004

Boyle MJ, Vasak ET, Schuchnigg M, Turner JJ, Sculley T, Penny R, Cooper DA, Tindall B, Sewell WA (1993) Subtypes of Epstein-Barr virus (EBV) in Hodgkin's disease: association between B-type EBV and immunocompromise. *Blood* **81**:468

Brooks L, Yao QY, Rickinson AB, Young LS (1992) Epstein-Barr virus latent gene transcription in nasopharyngeal carcinoma cells: coexpression of EBNA1, LMP1, and LMP2 transcripts. *J Virol* **66**:2689

Cleary ML, Nalesnik MA, Shearer WT, Sklar J (1988) Clonal analysis of transplant-associated lymphoproliferations based on the structure of genomic termini of the Epstein-Barr virus. *Blood* **72**:349

Dambaugh T, Hennessy K, Chamnankit L, Kieff E (1984) U2 region of Epstein-Barr virus DNA may encode Epstein-Barr nuclear antigen 2. *Proc Natl Acad Sci USA* **81**:7632 de Bruin PC, Jiwa NM, van der Valk P, van Heerde P, Gordijn R, Ossenkoppele GJ, Walboomers JMM, Meijer CJLM (1993) Detection of Epstein-Barr virus nucleic acid sequences and protein in nodal T-cell lymphomas: relation between latent membrane protein-1 positivity and clinical course. *Histopathol* **23**:509

Deacon EM, Pallesen G, Niedobitek G, Crocker J, Brooks L, Rickinson AB, Young LS (1993) Epstein-Barr virus and Hodgkin's disease: transcriptional analysis of virus latency in the malignant cells. *J Exp Med* **177**:339

Delabie J, Ceuppens JL, Vandenberghe P, de Boer M, Coorevits L, de Wolf-Peters C (1993) The B7/BB1 antigen is expressed by Reed-Sternberg cells of Hodgkin's disease and contributes to the stimulating capacity of Hodgkin's disease-derived cell lines. *Blood* **82**:2845

Gerber P, Nonoyama M, Lucas S, Perlin E, Goldstein LI (1972) Oral excretion of Epstein-Barr virus by healthy subjects and patients with infectious mononucleosis. *Lancet* **ii**:988

Glaser R, Zhang HY, Yao K, Zhu HC, Wang FX, Li GY, Wen DS, Li YP (1989) Two epithelial tumour cell lines (HNE-1 and HONE-1) latently infected with Epstein-Barr virus that were derived from nasopharyngeal carcinomas. *Proc Natl Acad Sci USA* **86**:9524

Gratama JW, Oosterveer MAP, Zwaan FE, Lepoutre J, Klein G, Ernberg I (1988) Eradication of Epstein-Barr virus by allogeneic bone marrow transplantation: implications for the site of viral latency. *Proc Natl Acad Sci USA* **85**:8693

Greenspan JS, Greenspan D, Lennette ET, Abrams DI, Conant MA, Petersen V, Freese UK (1985) Replication of Epstein-Barr virus within the epithelial cells of oral "hairy" leukoplakia an AIDS-associated lesion. *N Engl J Med* **313**:1564

Gregory CD, Murray RJ, Edwards CF, Rickinson AB (1988) Downregulation of cell adhesion molecules LFA-3 and ICAM-1 in Epstein-Barr virus-positive Burkitt's lymphoma underlies tumour cell escape from virus-specific T cell surveillance. *J Exp Med* **167**:1811

Gregory CD, Rowe M, Rickinson AB (1990) Different Epstein-Barr virus (EBV)-B cell interactions in phenotypically distinct clones of a Burkitt lymphoma cell line. *J gen Virol* **71**:1481

Hamilton-Dutoit SJ, Raphael M, Audouin J, Diebold J, Lisse I, Pedersen C, Oksenhendler E, Marelle L Pallesen G (1993a) *In situ* demonstration of Epstein-Barr virus small RNAs (EBER1) in acquired immunodeficiency syndrome-related lymphomas: correlation with tumour morphology and primary site. *Blood* **82**:619

Hamilton-Dutoit SJ, Rea D, Raphael M, Sandvej K, Delecluse HJ, Gisselbrecht C, Marelle L, van Krieken HJ, Pallesen G (1993b) Epstein-Barr virus-latent gene expression and tumour cell phenotype in acquired immunodeficiency syndrome-related non-Hodgkin's lymphoma. Correlation of lymphoma phenotype with three distinct patterns of viral latency. *Am J Pathol* **143**:1072

Harabuchi Y, Yamanaka N, Kataura A, Imai S, Kinoshita T, Mizuno F, Osato T (1990) Epstein-Barr virus in nasal T-cell lymphomas in patients with midline granuloma. *Lancet* **335**:128

Herbst H, Dallenbach F, Hummel M, Niedobitek G, Pileri S, Müller-Lantzsch N, Stein H (1991a) Epstein-Barr virus latent membrane protein expression in Hodgkin- and Reed-Sternberg cells. *Proc Natl Acad Sci USA* **88**:4766

Herbst H, Dallenbach F, Hummel M, Niedobitek G, Finn T, Young LS, Rowe M, Müller-Lantzsch N, Stein H (1991b) Epstein-Barr virus DNA and latent gene products in Ki-1 (CD30)-positive anaplastic large cell lymphomas. *Blood* **78**:2666

Herbst H, Steinbrecher E, Niedobitek G, Young LS, Brooks L, Müller-Lantzsch N, Stein H (1992) Distribution and phenotype of Epstein-Barr virus-harboring cells in Hodgkin's disease. *Blood* **80**:484

Herbst H, Stein H, Niedobitek G (1993) Epstein-Barr virus and CD30+ malignant lymphomas. *Crit Rev Oncog* **4**:191

Herndier BG, Sanchez HC, Chang KL, Chen YY, Weiss LM (1993) High prevalence of Epstein-Barr virus in the Reed-Sternberg cells of HIV-associated Hodgkin's disease. *Am J Pathol* **142**:1073

Hummel H, Anagnostopoulos I, Korbjuhn P, Dallenbach F, Stein H (1993) Epstein-Barr virus infection patterns in malignant lymphomas. In *The Epstein-Barr virus and associated diseases*. Tursz T, Pagano JS, Ablashi DV, de The G, Lenoir G, Pearson GR, eds (Colloque INSERM/John Libbey Eurotext) **225**:433

Jones JF, Shurin S, Abramowsky C, Tubbs RR, Sciotto CG, Wahl R, Sands J, Gottman D, Katz BZ, Sklar J (1988) T-cell lymphomas containing Epstein-Barr viral DNA in patients with chronic Epstein-Barr virus infections. *N Engl J Med* **318**:733

Kamel OW, van de Rijn M, Weiss LM, Del Zoppo GJ, Hench PK, Robbins BA, Montgomery PG, Warnke RA, Dorfman RF (1993) Reversible lymphomas associated with Epstein-Barr virus occurring during methotrexate therapy for rheumatoid arthritis and dermatomyositis. *N Engl J Med* **328**:1317

Kanavaros P, Jiwa M, van der Valk P, Walboomers J, Horstman A, Meijer CJLM (1993) Expression of Epstein-Barr virus latent gene products and related cellular activation and adhesion molecules in Hodgkin's disease and non-Hodgkin's lymphomas arising in patients without overt pre-existing immunodeficiency. *Hum Pathol* **24**:725

Kawaguchi H, Miyashita T, Herbst H, Niedobitek G, Asada M, Tsuchida M, Hanada R, Kinoshita A, Sakurai M, Kobayashi N, Mizutani S (1993) Subclinical proliferation of Epstein-Barr virus infected T-lymphocytes in Epstein-Barr virus associated hemophagocytic syndrome (EBV-AHS). *J Clin Invest* **92**:1444

Kerr BM, Lear AL, Rowe M, Croom-Carter D, Young LS, Rookes SM, Gallimore PH, Rickinson AB (1992) Three transcriptionally distinct forms of Epstein-Barr virus latency in somatic cell hybrids: cell phenotype dependence of virus promoter usage. *Virology* **187**:189

Khanna R, Burrows SR, Kurilla MG, Jacob CA, Misko IS, Sculley TB, Kieff E, Moss DJ (1992) Localisation of Epstein-Barr virus cytotoxic T-cell epitopes using recombinant vaccinia - implications for vaccine development. *J Exp Med* **176**:169

Kieff E, Liebowitz D (1990) Epstein-Barr virus and its replication. In *Virology*. Fields BN, Knipe DM, eds (New York: Raven Press) p1889

Kikuta H, Taguchi Y, Tomizawa K, Kojima K, Kawamura N, Ishizaka A, Sakiyama Y, Matsumoto S, Imai S, Kinoshita T, Koizumi S, Osato T, Kobayashi I, Hamada I, Hirai K (1988) Epstein-Barr virus genome-positive T lymphocytes in a boy with chronic active EBV infection associated with Kawasaki-like disease. *Nature* **333**:455

Korbjuhn P, Anagnostopoulos I, Hummel M, Tiemann M, Dallenbach F, Parwaresch MR, Stein H (1993) Frequent latent Epstein-Barr virus infection of neoplastic T cells and bystander B cells in human immunodeficiency virus-negative European peripheral pleomorphic T-cell lymphomas. *Blood* **82**:217

Laux G, Economou A, Farrell P (1989) The terminal protein gene 2 of Epstein-Barr virus is transcribed from a bi-directional latent promoter region. *J gen Virol* **70**:3079

Lee SP, Thomas WA, Murray RJ, Khanim F, Kaur S, Young LS, Rowe M, Kurilla M, Rickinson AB (1993) HLA A2.1-restricted cytotoxic T cells recognizing a range of Epstein-Barr virus isolates through a defined epitope in latent membrane protein LMP2. *J Virol* **67**:7428

Lewin N, Aman P, Mellstedt H, Zech L, Klein G (1988) Direct outgrowth of *in vivo* Epstein-Barr virus (EBV)-infected chronic lymphocytic leukemia (CLL) cells into permanent lines. *Int J Cancer* **41**:892

Locker J, Nalesnik M (1989) Molecular genetic analysis of lymphoid tumours arising after organ transplantation. *Am J Pathol* **135**:977

Magrath I, Jain V, Bhatia K (1993) Molecular epidemiology of Burkitt's lymphoma. In *The Epstein-Barr virus and associated diseases*. Tursz T, Pagano JS, Ablashi DV, de The G, Lenoir G, Pearson GR, eds (Colloque INSERM/John Libbey Eurotext) **225**:377

Miller G (1990) Epstein-Barr virus - Biology pathogenesis and medical aspects. In *Virology*. Fields BN, Knipe DM, eds (New York: Raven Press) p1921

Mori M, Kurozumi H, Akagi K, Tanaka Y, Imai S, Osato T (1992) Monoclonal proliferation of T cells containing Epstein-Barr virus in fatal infectious mononucleosis. *N Engl J Med* **327**:58

Mueller N, Evans A, Harris NL, Comstock GW, Jellum E, Magnus K, Orentreich N, Polk F, Vogelman J, (1989) Hodgkin's disease and Epstein-Barr virus - Altered antibody pattern before diagnosis. *N Engl J Med* **320**:689

Munoz N, Davidson RJL, Witthoff B, Ericsson JE, De-The G (1978) Infectious mononucleosis and Hodgkin's disease. *Int J Cancer* **22**:10

Murray RJ, Kurilla MG, Griffin HM, Brooks JM, Mackett M, Arrand JR, Rowe M, Burrows SR, Moss DJ, Kieff E, Rickinson AB (1990) Human cytotoxic T cell responses against Epstein-Barr virus nuclear antigens demonstrated using recombinant vaccinia viruses. *Proc Natl Acad Sci USA* **87**:2906

Murray RJ, Kurilla MG, Brooks JM, Thomas WA, Rowe M, Kieff E, Rickinson AB (1992) Identification of target antigens for the human cytotoxic T-cell response to Epstein-Barr virus (EBV) - implications for the immune control of EBV-positive malignancies. *J Exp Med* **176**:157

Niedobitek G, Hamilton-Dutoit S, Herbst H, Finn T, Vetner M, Pallesen G, Stein H (1989) Identification of Epstein-Barr virus infected cells in tonsils of acute infectious mononucleosis by *in situ* hybridisation. *Hum Pathol* **20**:796

Niedobitek G, Young LS, Lau R, Brooks L, Greenspan D, Greenspan J, Rickinson AB (1991a) Epstein-Barr virus infection in oral hairy leukoplakia: virus replication in the absence of a detectable latent phase. *J gen Virol* **72**:3035

Niedobitek G, Hansmann ML, Herbst H, Young LS, Dienemann D, Hartmann CA, Finn T, Pitteroff S, Welt A, Anagnostopoulos I, Friedrich R, Lobeck H, Sam CK, Araujo I, Rickinson AB, Stein H (1991b) Epstein-Barr virus and carcinomas: undifferentiated carcinomas but not squamous cell carcinomas of the nasopharynx are regularly associated with the virus. *J Pathol* **165**:17

Niedobitek G, Herbst H, Young LS, Brooks L, Masucci MG, Crocker J, Rickinson AB, Stein H (1992a) Patterns of Epstein-Barr virus infection in non-neoplastic lymphoid tissue. *Blood* **79**:2520

Niedobitek G, Fahraeus R, Herbst H, Latza U, Ferszt A, Klein G, Stein H (1992b) The Epstein-Barr virus encoded membrane protein (LMP) induces phenotypic changes in epithelial cells. *Virchows Arch B* **62**:55

Niedobitek G, Young LS (1994) Persistence of Epstein-Barr virus and the pathogenesis of virus-associated tumours. *Lancet* **343**:333

Nilsson K, Klein G, Henle W, Henle G (1971) The establishment of lymphoblastoid cell lines from adult and from foetal human lymphoid tissue and its dependence on EBV. *Int J Cancer* **8**:443

Pallesen G, Hamilton-Dutoit SJ, Rowe M, Young LS (1991) Expression of Epstein-Barr virus latent gene products in tumour cells of Hodgkin's disease. *Lancet* **337**:320

Pallesen G, Hamilton-Dutoit SJ, Zhou X (1994) The association of Epstein-Barr virus (EBV) with T cell lymphoproliferations and Hodgkin's disease: two new developments in the EBV field. *Adv Cancer Res* **62**:179

Papadopoulos EB, Ladanyi M, Emanuel D, MacKinnon S, Boulad F, Carabasi MH, Castro-Malaspina H, Childs BH, Gillio AP, Small TN, Young JW, Kernan NA, O'Reilly RJ (1994) Infusions of donor leukocytes to treat Epstein-Barr virus-associated lymphoproliferative disorders after allogeneic bone marrow transplantation. *N Engl J Med* **330**:1185

Pedersen C, Gerstoft J, Lundgren JD, Skinhoj P, Bottzauw J, Geisler C, Hamilton-Dutoit SJ, Thorsen S, Lisse I, Ralfkiaer E, Pallesen G (1991) HIV-associated lymphoma: histopathology and association with Epstein-Barr virus genome related to clinical, immunological and prognostic factors. *Eur J Cancer* **27**:1416

Raab-Traub N, Flynn K (1986) The structure of the termini of the Epstein-Barr virus as a marker of clonal cellular proliferation. *Cell* **47**:883

Rabson M, Gradoville L, Heston L, Miller G (1982) Non-immortalizing P3J-HR-1 Epstein-Barr virus: a deletion mutant of its transforming parent. *J Virol* **44**:834

Randhawa PS, Yousem SA, Paradis IL, Dauber JA, Griffith BP, Locker J (1989) The clinical spectrum, pathology, and clonal analysis of Epstein-Barr virus-associated lymphoproliferative disorders in heart-lung transplant recipients. *Am J Clin Pathol* **92**:177

Rickinson AB, Rowe M, Hart IJ, Yao QY, Henderson LE, Rabin H, Epstein MA (1984) T-cell-mediated regression of "spontaneous" and of Epstein-Barr virus-induced B cell transformation *in vitro*: studies with cyclosporin A. *Cell Immunol* **87**:646

Rickinson AB, Young LS, Rowe M (1987) Influence of the Epstein-Barr virus nuclear antigen EBNA 2 on the growth phenotype of virus-transformed B cells. *J Virol* **61**:1310

Rowe M, Rooney CM, Rickinson AB, Lenoir GM, Rupani H, Moss DJ, Stein H, Epstein MA (1985) Distinctions between endemic and sporadic forms of Epstein-Barr virus-positive Burkitt's lymphoma. *Int J Cancer* **35**:435

Rowe M, Rowe DT, Gregory CD, Young LS, Farrell PJ, Rupani H, Rickinson AB (1987) Differences in B cell growth phenotype reflect novel patterns of Epstein-Barr virus latent gene expression in Burkitt's lymphoma cells. *EMBO J* **6**:2743

Rowe M, Young LS, Cadwallader K, Petti L, Kieff E, Rickinson AB (1989) Distinction between Epstein-Barr virus type A (EBNA 2A) and type B (EBNA 2B) isolates extends to the EBNA 3 family of nuclear proteins. *J Virol* **63**:1031

Rowe M, Lear AL, Croom-Carter D, Davies AH, Rickinson AB (1992) Three pathways of Epstein-Barr virus gene activation from EBNA1-positive latency in B lymphocytes. *J Virol* **66**:122

Rowlands DC, Ito M, Mangham DC, Reynolds G, Herbst H, Hallissey MT, Fielding JWL, Newbold KM, Jones EL, Young LS, Niedobitek G (1993) Epstein-Barr virus and carcinomas: rare association of the virus with gastric adenocarcinomas. *Br J Cancer* **68**:1014

Sam CK, Brooks LA, Niedobitek G, Young LS, Prasad U, Rickinson AB (1993) Analysis of Epstein-Barr virus infection in nasopharyngeal biopsies from a high nasopharyngeal carcinoma risk group. *Int J Cancer* **53**:957

Sample J, Young L, Martin B, Chatman T, Kieff E, Rickinson A, Kieff E (1990) Epstein-Barr virus types 1 and 2 differ in their EBNA-3A, EBNA-3B and EBNA-3C genes. *J Virol* **64**:4084

Sample J, Brooks L, Sample C, Young LS, Rowe M, Rickinson A, Kieff E (1991) Restricted Epstein-Barr virus protein expression in Burkitt lymphoma is reflected in a novel EBNA-1 mRNA and transcriptional initiation site. *Proc Natl Acad Sci USA* **88**:6343

Shibata D, Weiss LM (1992) Epstein-Barr virus-associated gastric adenocarcinoma. *Am J Pathol* **140**:769

Shibata D, Weiss LM, Hernandez AM, Nathwani BN, Bernstein L, Levine AM (1993) Epstein-Barr virus-associated non-Hodgkin's lymphoma in patients infected with the human immunodeficiency virus. *Blood* **81**:2102

Sixbey JW, Nedrud JG, Raab-Traub N, Hanes RA, Pagano JS (1984) Epstein-Barr virus replication in oropharyngeal cells. *N Engl J Med* **310**:1225

Sixbey JW, Shirley P, Chesney PJ, Buntin DM, Resnick L (1989) Detection of a second widespread strain of Epstein-Barr virus. *Lancet* **ii**:761

Speck SH, Strominger JL (1989) Transcription of Epstein-Barr virus in latently infected, growth-transformed lymphocytes. *Adv Viral Oncol* **8**:133

Strauch B, Andrews LL, Siegel N, Miller G (1974) Oropharyngeal excretion of Epstein-Barr virus by renal transplant recipients and other patients with immunosuppressive drugs. *Lancet* i:234

Swerdlow SH (1992) Post-transplant lymphoproliferative disorders: a morphologic, phenotypic and genotypic spectrum of disease. *Histopathol* **20**:373

Thomas JA, Hotchin NA, Allday MJ, Amlot P, Rose M, Yacoub M, Crawford DH (1990) Immunohistology of Epstein-Barr virus-associated antigens in B cell disorders from immunocompromised individuals. *Transplantation* **49**:944

Thomas JA, Felix DH, Wray D, Southam JC, Cubie H, Crawford DH (1991) Epstein Barr virus gene expression and epithelial cell differentiation in oral hairy leukoplakia. *Am J Pathol* **139**:1369

Torsteinsdottir S, Brautbar C, Ben Bassat H, Klein E, Klein G (1988) Differential expression of HLA antigens on human B-cell lines of normal and malignant origin: a consequence of immune surveillance or a phenotype vestige of the progenitor cells? *Int J Cancer* **41**:913

Wang D, Liebowitz D, Kieff E (1985) An EBV membrane protein expressed in immortalized lymphocytes transforms established rodent cells. *Cell* **43**:831

Wang F, Gregory CD, Rowe M, Rickinson AB, Wang D, Birkenbach M, Kikutani H, Kishimoti T, Kieff E (1987) Epstein-Barr virus nuclear antigen 2 specifically induces expression of the B-cell activation antigen CD23. *Proc Natl Acad Sci USA* **84**:3452

Wang F, Gregory C, Sample C, Rowe M, Liebowitz D, Murray R, Rickinson A, Kieff E (1990) Epstein-Barr virus latent membrane protein (LMP-1) and nuclear proteins 2 and 3c are effectors of phenotypic changes in B lymphocytes: EBNA-2 and LMP-1 cooperatively induce CD23. *J Virol* **64**:2309

Weiss LM, Jaffe ES, Liu XF, Chen YY, Shibata D, Medeiros LJ (1992) Detection and localisation of Epstein-Barr viral genome in angioimmunoblastic lymphadenopathy and angioimmunoblastic lymphadenopathy-like lymphoma. *Blood* **79**:1789

Wilson JB, Levine AJ (1992) The oncogenic potential of Epstein-Barr virus nuclear antigen 1 in transgenic mice. *Curr Top Microbiol Immunol* **182**:375

Wu TC, Mann RB, Charache P, Hayward SD, Staal S, Lambe BC, Ambinder RF (1990) Detection of EBV gene expression in Reed-Sternberg cells of Hodgkin's disease. *Int J Cancer* **46**:801

Yao QY, Rickinson AB, Epstein MA (1985) A re-examination of the Epstein-Barr virus carrier state in healthy seropositive individuals. *Int J Cancer* **35**:35

Yao QY, Ogan P, Rowe M, Wood M, Rickinson AB (1989a) The Epstein-Barr virus:host balance in acute infectious mononucleosis patients receiving Acyclovir anti-viral therapy. *Int J Cancer* **43**:61

Yao QY, Ogan P, Rowe M, Wood M, Rickinson AB (1989b) Epstein-Barr virus-infected B cells persist in the circulation of Acyclovir-treated virus carriers. *Int J Cancer* **43**:67

Yao QY, Rowe M, Martin B, Young LS, Rickinson AB (1991) The Epstein-Barr virus carrier state: Dominance of a single growth-transforming isolate in the blood and in the oropharynx of healthy virus carriers. *J gen Virol* **72**:1579

Yeung WM, Zong YS, Chiu CT, Sham JST, Choy DTK, Ng MH (1993) Epstein-Barr virus carriage by nasopharyngeal carcinoma *in situ*. *Int J Cancer* **53**:746

Yoneda N, Tatsumi E, Kawanishi M, Teshigawara K, Masuda S, Yamamura Y, Inui A, Yoshino G, Oimomi M, Baba S, Yamaguchi N (1990) Detection of Epstein-Barr virus genome in benign polyclonal proliferative T cells of a young male patient. *Blood* **76**:172

Young LS, Yao QY, Rooney CM, Sculley TB, Moss DJ, Rupani H, Laux G, Bornkamm GW, Rickinson AB (1987) New type B isolates of Epstein-Barr virus from Burkitt's lymphoma and from normal individuals in endemic areas. *J gen Virol* **68**:2853

Young L, Alfieri C, Hennessy K, Evans H, O'Hara C, Anderson KC, Ritz J, Shapiro RS, Rickinson A, Kieff E, Cohen JI (1989) Expression of Epstein-Barr virus transformation-associated genes in tissues of patients with EBV lymphoproliferative disease. *N Engl J Med* **321**:1080

Young L, Lau R, Rowe M, Niedobitek G, Packham G, Shanahan F, Rowe DT, Greenspan D, Greenspan JS, Rickinson AB, Farrell PJ (1991) Differentiation-associated expression of the Epstein-Barr virus BZLF1 transactivator protein in oral hairy leukoplakia. *J Virol* **65**:2868

Zimber U, Adldinger HK, Lenoir GM, Vuillaume M, Knebel-Doeberitz M, Laux G, Desgranges C, Wittmann P, Freese UK, Schneider U, Bornkamm GW (1986) Geographical prevalence of two types of Epstein-Barr virus. *Virology* **154**:56

HODGKIN'S DISEASE AND EPSTEIN-BARR VIRUS

Ruth F Jarrett

LRF Virus Centre, Department of Veterinary Pathology,
University of Glasgow, Glasgow G61 1QH, UK

INTRODUCTION

An association between Epstein-Barr virus (EBV) and Hodgkin's disease (HD) was first described almost 25 years ago, however it is only during the last 8 years that a firm molecular basis for this association has been established. The literature relating to EBV and HD has increased exponentially over the latter years and laboratories throughout the world have contributed to a body of data consistently linking this virus with HD. The majority of participants at this Workshop have studied the role of EBV in HD which is mentioned in many chapters of this book. This chapter summarises the key findings and attempts to place them chronologically, in order to highlight the development and gradual acceptance of the relationship.

EARLY STUDIES LINKING EBV AND HODGKIN'S DISEASE

Seroepidemiological studies performed in the early 1970s first linked EBV with HD. Antibody titres to EBV early antigens, viral capsid antigens and nuclear antigens were found to be elevated in HD cases compared to controls (Levine *et al.*, 1971). In contrast to the results of serological studies on other herpesviruses (Hesse *et al.*, 1977; Henderson *et al.*, 1973; Langenhuysen *et al.*, 1974), such case-control differences were consistently observed by a number of different laboratories (Henderson *et al.*, 1973; Henle and Henle, 1973; Langenhuysen *et al.*, 1974; Hesse *et al.*, 1977; Evans *et al.*, 1978). Further evidence for the association came from epidemiological studies showing that persons with a past history of infectious mononucleosis, which is associated with late primary infection by EBV, had an increased risk of developing HD (Munoz *et al.*, 1978). Gutensohn and Cole

(1980) pooled the results of several large studies and found that overall the risk of developing HD was increased 3-fold in such individuals.

Early molecular studies failed to detect viral genomes in tissues affected by HD (Pagano *et al.*, 1973; Lindahl *et al.*, 1974). In retrospect it is easy to understand the reasons for this. The assays available in the mid-1970s were insensitive, relying on the use of radiolabelled viral genomes or cRNA as probes and on liquid hybridisation or early filter hybridisation techniques. Although such methodologies were sufficiently sensitive to detect viral genomes in tumours such as Burkitt's lymphoma, where the tumour cells constitute a large proportion of the tumour mass, it is likely that they did not have sufficient sensitivity to detect genomes present in the Reed-Sternberg (RS) cells of HD. In addition, since HD is probably a heterogeneous condition comprising more than one entity, it is necessary to examine large case series before excluding involvement of any particular agent (MacMahon, 1966; see Alexander, this volume).

In 1985, Poppema *et al.* published a case report in which they described EBV nuclear antigen (EBNA)-positivity in RS cells in a patient with 'lymphoma with all histologic characteristics of Hodgkin's disease' which had developed following chronic or reactivated EBV infection (Poppema *et al.*, 1985). This is probably the first report of the expression of EBV gene products in RS cells in HD. However, reviewers of this manuscript were reluctant to accept that EBV could be detected in HD and therefore suggested that the lesion be described as 'morphologically consistent with Hodgkin's disease' rather than simply Hodgkin's disease.

MOLECULAR STUDIES

In 1987 Weiss and colleagues reported the detection of EBV genomes in the affected tissues of 4 out of 21 cases of HD (Weiss *et al.*, 1987). This was facilitated by the use of Southern blot analysis and a cloned probe representing the major internal repeat sequence of the EBV genome. It was in some ways a fortuitous finding, as the study was initiated following the detection of immunoglobulin gene rearrangements in HD biopsies (Weiss *et al.*, 1986) and the authors wished to exclude the possibility that there were clones of EBV-driven B-cells in the reactive component of the lesions. There was, however, no correlation between the detection of EBV and the presence of clonal immunoglobulin gene rearrangements prompting speculation about a more direct relationship between virus and disease (Weiss *et al.*, 1987). Examination of the EBV terminal repeats demonstrated that the EBV was clonal in 3 of the 4 positive cases in this study, indicating that the infected cell population had arisen from a single EBV-infected cell and that polyclonal infection of cells within the reactive component was not responsible for the findings (Weiss *et al.*, 1987).

The report by Weiss *et al.* (1987), inspired many similar studies which gave rise to remarkably consistent findings given the relatively small sizes of the case series examined. EBV genomes were detected in 17-41% cases and there was a suggestion that cases of mixed cellularity disease were more often EBV-positive than nodular sclerosis cases (Anagnostopoulos *et al.*, 1989; Boiocchi *et al.*, 1989; Staal *et al.*, 1989; Weiss *et al.*, 1989;

Uccini *et al.*, 1990; Gledhill *et al.*, 1991). In keeping with the original study most workers showed that the EBV infection was clonal; we investigated the clonality of EBV in 28 cases which were EBV-positive on Southern blot analysis and in all cases clonality could be demonstrated (Jarrett *et al.*, 1991; Armstrong *et al.*, 1992b).

IN SITU STUDIES

The possibility that EBV was directly involved in the pathogenesis of HD was greeted with either great scepticism or extreme caution. EBV is a ubiquitous virus and establishes a latent infection in B lymphocytes (see Young and Niedobitek, this volume), therefore it was feasible that the above assays were simply detecting latent EBV present in B-cells in the lesions. Since HD patients have immune dysfunction it seemed possible that the viral load might be increased (Slivnick *et al.*, 1990). In order to show that EBV played any direct role in disease pathogenesis it was imperative to determine the cellular localisation of the viral sequences. The first studies to address this issue employed DNA *in situ* hybridisation (Weiss *et al.*, 1989; Anagnostopoulos *et al.*, 1989; Uccini *et al.*, 1990). Although they provided evidence for infection of RS cells, the signal to noise ratios were not good and the number of cases which could be examined was small. Definitive evidence for the localisation of the EBV came from later studies using antibodies directed against EBV latent gene products and probes derived from the EBV EBER RNA sequences (Pallesen *et al.*, 1991a; Wu *et al.*, 1990).

In 1991, Pallesen and co-workers published the results of a series of HD cases examined immunohistochemically using monoclonal antibodies directed against epitopes on the LMP-1 and EBNA-2 proteins (for description of EBV latent genes see preceding chapter) (Pallesen *et al.*, 1991a). Clear reactivity with the LMP-1 antibodies, which was restricted to RS cells, was observed in 40/84 (46%) cases. This provided clear evidence that EBV was present within the RS cells, and also that an EBV latent gene product with known oncogenic potential was expressed by these cells. Expression of EBNA-2 was not detected. Similar findings were soon reported by other groups (Herbst *et al.*, 1991; Armstrong *et al.*, 1992a; Murray *et al.*, 1992; Delsol *et al.*, 1992). It was found that the LMP-1 antibody cocktail (CS1-4) used in many of these studies could be applied to routinely processed paraffin-embedded material, thus opening up the possibility of performing larger epidemiological studies. Differences in rates of EBV-association by histological subtype now reached statistical significance; such differences are described in greater detail in later chapters.

The application of the EBV EBER *in situ* hybridisation assay to the study of EBV in HD was first reported by Wu *et al.* in 1990, but its widespread use lagged slightly behind the use of immunohistochemical techniques. This assay relies on the detection of the EBER (for EBV encoded RNA) RNAs which are two small, RNA polymerase III transcripts, expressed at high levels in all cells latently infected by EBV (Howe and Steitz, 1986; Howe and Mei-Di Shu, 1989). It is estimated that 10^6 copies of the EBER RNAs are expressed for every viral genome. The EBERs therefore provide a very good target for probes used in *in situ* hybridisation analyses and the assay is exquisitely sensitive. Various assays incorporating antisense oligonucleotides or RNA probes are in widespread use and a kit is commercially available (Dako). The assay is readily applicable to the study of routinely fixed paraffin-embedded material (Fig. 1). Studies performed using EBER *in situ* hybridisation confirmed the presence of EBV within RS cells in 32-50% of HD cases (Weiss *et al.*, 1991; Armstrong *et al.*, 1992b; Hummel *et al.*, 1992; Khan *et al.*, 1992; Herbst *et al.*, 1992; Khan *et al.*, 1993; Armstrong *et al.*, 1994). Using this methodology EBV was also detectable in occasional small lymphoid cells in the reactive component of lesions (Weiss *et al.*, 1991; Armstrong *et al.*, 1992b; Hummel *et al.*, 1992; Khan *et al.*, 1992; Herbst *et al.*, 1992). The sensitivity of this assay, coupled with the failure to detect EBER transcripts in the RS cells of all cases, rules out the possibility that HD is always associated with EBV infection of RS cells.

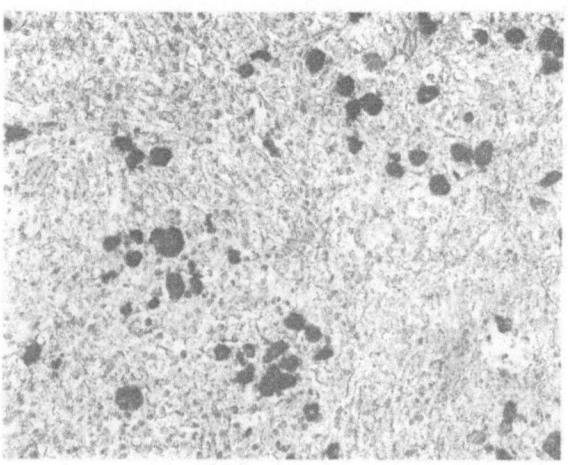

Fig. 1:   EBV EBER *in situ* hybridisation applied to a case of childhood Hodgkin's disease. Note staining of nuclei of cells with the morphology of Reed-Sternberg cells.

In a case of EBV-associated HD it is possible to demonstrate EBV infection of all, or virtually all, the abnormal cells (Weiss *et al.*, 1991; Armstrong *et al.*, 1992b). In addition, there is complete concordance between the results of EBV analyses when multiple samples, taken from different sites at different time points, from an individual case are examined (Coates *et al.*, 1991; Boiocchi *et al.*, 1993; Brousset *et al.*, 1994; Weiss, this volume). In all cases in which we have demonstrated clonal EBV genomes by Southern blot analysis we have localised the EBV genomes to RS cells, thereby providing indirect evidence that RS cells represent a clonal population of cells (Armstrong *et al.*, 1992b).

COMPARISON OF ASSAYS USED TO DETECT EBV INFECTION

EBV can be detected using a variety of methods including Southern blot analysis, the polymerase chain reaction (PCR), DNA and RNA *in situ* hybridisation and immunohistochemistry using monoclonal antibodies or polyclonal sera. The ideal assay for the analysis of HD samples should be reproducible, sensitive, permit cellular localisation and be applicable to the study of archival material. The EBER *in situ* hybridisation technique appears to meet all of these requirements. Comparative studies by ourselves and others have confirmed that this assay is the most useful way of determining the EBV status of HD lesions (Armstrong *et al.*, 1992b; Herbst *et al.*, 1992). There is, however, a very good correlation between the results of LMP-1 immunohistochemistry, Southern blot analysis and this assay (Armstrong *et al.*, 1992b; Brousset *et al.*, 1993).

PCR has also been used to detect EBV in HD lesions (Bignon *et al.*, 1990; Herbst *et al.*, 1990; Brocksmith *et al.*, 1991; Knecht *et al.*, 1991; Shibata *et al.*, 1991; Armstrong *et al.*, 1992b; Brousset *et al.*, 1993). Using a sensitive PCR assay it is possible to detect EBV in DNA samples from almost all cases, and in many of these it is clear that the EBV is present in the reactive component and not in the RS cells themselves. There is a relationship between the amount of EBV DNA in a lesion and the cellular localisation of the EBV and it is therefore possible to design a PCR assay or scoring system which gives similar results to those obtained in other assays (Herbst *et al.*, 1990; Armstrong *et al.*, 1992b); however, the relative ease of performing EBER *in situ* hybridisation makes this supererogatory.

The first indication that EBV might be linked to HD came from serological studies. However, the relationship between raised titres to EBV antigens and the presence of EBV in RS cells is not entirely clear at the present time. Our data show that elevated titres are not restricted to EBV-associated cases (unpublished data). Conversely, some cases of EBV-associated HD do not have detectable antibodies against EBV early antigens. There is a suggestion that titres against viral capsid antigens are higher in EBV-associated cases compared to non-EBV-associated cases although, in the relatively small series examined, differences were not statistically significant (our unpublished data). Brousset *et al.* (1992) were also unable to find an association between raised EBV antibody titres and EBV-positivity in RS cells. It remains to be seen whether a serological assay will identify EBV-associated cases of HD.

THE PATTERN OF EBV GENE EXPRESSION IN HD

As described in the preceding chapter, EBV infection may be lytic or latent. EBV infection in RS cells is almost exclusively latent. Two studies have examined HD biopsies using antibodies to EBV early antigens, viral capsid antigens and membrane antigens. Pallesen *et al.* (1991b) detected expression of the EBV BZLF1 protein, which controls the switch between latent and lytic infection, in 3/96 cases all of which were known to be EBV-associated. None of these cases was positive for other early, viral capsid or membrane antigens. Similarly, Brousset and co-workers (1993) detected BZLF1 expression in only 3/40 cases; in one of the latter cases, however, there was expression of early antigens, evidence of EBV DNA replication and expression of the BLLF1 RNA, which encodes the membrane antigen gp350/220. Joske *et al.* (1992) also found evidence of transcription of BLLF1 in 1/20 cases. In cases in which reactivity with BZLF1 antibodies was demonstrated, staining was confined to the nuclei of only a very small proportion of RS cells (Pallesen *et al.*, 1991b; Broussett *et al.*, 1993). In keeping with these findings, Southern blot analyses using probes for the EBV terminal repeat sequence have detected circular, and not linear, viral genomes consistent with a predominantly latent infection (Armstrong *et al.*, 1992b).

In EBV latent infection, 3 different patterns of EBV gene expression have been described (see Young and Niedobitek, this volume). Latency III (Lat III) corresponds to that seen in lymphoblastoid cell lines in which all the EBNAs are expressed in addition to LMP-1, -2A and -2B and the EBER RNAs. In the latency I pattern, which is observed in Burkitt's lymphoma, expression is restricted to the EBNA-1 protein and the EBER RNAs. An intermediate pattern (Lat II), in which EBNA-1, LMP-1, -2A and -2B and the EBER RNAs are expressed, has been described in B-cells *in vitro* and in nasopharyngeal carcinoma.

Immunohistochemical studies clearly show that the RS cells in HD express the LMP-1 protein in the absence of EBNA-2 (Pallesen *et al.*, 1991a; Herbst *et al.*, 1991). Immunoblotting studies have confirmed that the authentic LMP-1 protein is expressed; furthermore, given the paucity of tumour cells within affected tissues the ability to detect LMP-1 by immunoblotting substantiates previous impressions that RS cells express large amounts of this protein (Deacon *et al.*, 1993). Until recently good monoclonal antibodies reactive with EBNA-1 were not available, and there was only indirect evidence for expression of this latent gene product. However a recent study by Grasser *et al.* (1994) provides immunohistochemical evidence for expression of EBNA-1 within RS cells.

Analysis of EBV RNA transcription in HD biopsies using RT-PCR has shown expression of RNA encoding EBNA-1, LMP-1, -2A and -2B (Deacon *et al.*, 1993). It would seem likely that LMP-2 is expressed at the protein level, however this has not been confirmed due to the lack of suitable reagents. The EBNA-1 mRNA transcript in HD initiates from the BamHI Q region of the viral genome (so-called Q promoter, see Young and Niedobitek, this volume) as in nasopharyngeal carcinoma and Burkitt's lymphoma, rather than the Cp and Wp promoters which are used to direct transcription of all the EBNAs in lymphoblastoid cell lines (Deacon *et al.*, 1993). This promoter usage suggests that the other EBNAs are not transcribed in RS cells. Transcripts from the EBV BamHI A fragment are also detectable in HD biopsies; the significance of these latent gene transcripts is at present unclear but they are also detectable in nasopharyngeal carcinoma and B-cell lines (Deacon *et al.*, 1993).

The pattern of EBV latent gene expression in HD therefore corresponds to the Lat II pattern which is also seen in nasopharyngeal carcinoma. In contrast to other lymphomas associated with EBV, such as anaplastic large cell lymphoma and AIDS-related lymphomas, which can show some heterogeneity in the pattern of EBV latent gene expression both within and between cases, this latency pattern appears to be a constant feature of HD. This holds true even in the context of HIV infection (see Hamilton-Dutoit, this volume).

EBV SUBTYPES IN HD

As discussed in the preceding chapter, there are 2 subtypes of EBV designated type 1 and 2. Analysis of the subtype of EBV present in DNA samples from 35 HD biopsies from EBV-associated cases from the UK indicated that all were of type 1 (Gledhill *et al.*, 1991; Jarrett *et al.*, 1991; Armstrong *et al.*, 1992b). Similar results have been reported for Italian and US cases (De Re *et al.*, 1993; Ambinder *et al.*, 1993). In contrast, type 2 genomes and dual infection have been detected in HIV+ve cases and in HIV-ve cases from Algeria and Turkey (De Re *et al.*, 1993; Boyle *et al.*, 1993; Bouzid *et al.*, 1993; Bettina Borisch, personal communication). The significance of these results is not entirely clear since the epidemiology of type 2 infection is still somewhat controversial, and in cases of dual infection it is not clear which EBV subtype is present in RS cells. The results do suggest, however, that EBV-associated HD is not restricted to type 1 EBV.

THE ROLE OF EBV IN THE PATHOGENESIS OF HODGKIN'S DISEASE

The available data show conclusively that EBV is present and expressed in the RS cells in a proportion, around 40%, of cases of HD. The findings suggest that the virus is likely to be playing a role in disease pathogenesis. Arguments against the idea that EBV is simply a passenger virus are as follows. First, the EBV infection in HD is clonal and it is possible to demonstrate infection of all, or virtually all, abnormal cells. Therefore the tumour has arisen from a single EBV-infected cell and viral genomes have been retained by these cells. At the very least this suggests that the virus is conferring some growth advantage. Secondly, the EBV LMP-1 protein, which is known to have oncogenic potential (Wang *et al.*, 1985), is highly expressed by RS cells in all EBV-associated cases. This protein does not appear to be expressed by latently-infected lymphoid cells in normal healthy individuals (Tierney *et al.*, 1994). This suggests that LMP-1 may well play a functional role in the disease process. Furthermore since LMP-1 and -2 both elicit cytotoxic T-cell responses some mechanism of escape from immune destruction must be operative (see Poppema and Visser, this volume).

It has been suggested that RS cell precursors may simply be a site of viral latency and that the virus does not contribute to the subsequent clonal expansion of these cells. If this were the case, then one would have to postulate that around 40% of the RS cell precursors were EBV-infected. This would seem unlikely given the low proportions of lymphoid cells which are latently infected (Tierney *et al.*, 1994).

Despite these arguments, there is as yet no proof that EBV is playing a causal role and the function of the expressed latent genes in HD is not known. LMP-1 is known to up-regulate a variety of activation antigens, adhesion molecules and also Bcl-2 (Wang *et al.*, 1990; Kieff and Liebowitz, 1990; Henderson *et al.*, 1991). The induction of the activation antigen CD23 by LMP-1 and EBNA-2 is thought to be a critical event in the immortalisation of B-cells by EBV, however this antigen is only infrequently expressed by RS cells and expression does not correlate with LMP-1 expression (Armstrong *et al.*, 1992a; Sandvej *et al.*, 1992). Likewise although Bcl-2 protein can be detected at varying levels in some RS cells there is no evidence for the induction of this protein by LMP-1 (Armstrong *et al.*, 1992a; see Jiwa *et al.* and Weiss, this volume).

SUMMARY

There is clearly an association between EBV and a proportion, around 40%, of cases of HD. In these cases EBV gene products are detected in RS cells, the tumour cells of HD, and there is a uniform pattern of EBV latent gene expression. The EBV LMP-1 protein, which is known to have oncogenic potential, is highly expressed by RS cells. The function of this and other viral gene products in RS cells is not known at the present time.

The relationship between EBV and HD is now widely accepted and most current studies of HD include an assessment of the EBV status of the lesions. It is however a relatively short time since the idea that EBV might be involved in this malignancy was treated with great scepticism. The evolution of the proof of this association emphasises the need for the continuous re-appraisal of relationships between viruses and cancer as new and more sensitive techniques of detection emerge. If Poppema *et al.* (1985) had been permitted to describe their EBNA-positive lymphoma as typical HD would this whole field have been opened up 2 years earlier? If Weiss *et al.* (1987) had not investigated the occurrence of immunoglobulin gene rearrangements in HD would we still be in the dark?

Research aimed at showing an association between a virus and a cancer should attempt to establish closer and closer links between the virus and the malignancy. This is the way in which the investigation of the relationship between EBV and HD is evolving; it is hoped that future experiments will shed light on the way in which the virus is contributing to the malignant process and the reasons for the failure to mount an adequate immune response.

REFERENCES

Ambinder RF, Browning PJ, Lorenzana I, Leventhal BG, Cosenza H, Mann RB, MacMahon EM, Medina R, Cardona V, Grufferman S, Olshan A, Levin A, Petersen EA, Blattner W, Levine PH (1993) Epstein-Barr virus and childhood Hodgkin's disease in Honduras and the United States. *Blood* **81**:462

Anagnostopoulos I, Herbst H, Niedobitek G, Stein H (1989) Demonstration of monoclonal EBV genomes in Hodgkin's disease and KI-1-positive anaplastic large cell lymphoma by combined Southern blot and in situ hybridization. *Blood* **74**:810

Armstrong AA, Gallagher A, Krajewski AS, Jones DB, Wilkins BS, Onions DE, Jarrett RF (1992a) The expression of the EBV latent membrane protein (LMP-1) is independent of CD23 and bcl-2 in Reed-Sternberg cells in Hodgkin's disease. *Histopathology* **21**:72

Armstrong AA, Weiss LM, Gallagher A, Jones DB, Krajewski AS, Angus B, Brown G, Jack AS, Wilkins BS, Onions DE, Jarrett RF (1992b) Criteria for the definition of Epstein-Barr virus association in Hodgkin's disease. *Leukemia* **6**:869

Armstrong AA, Lennard A, Alexander FA, Angus BA, Proctor SJ, Onions DE, Jarrett RF (1994) Prognostic significance of Epstein-Barr virus association in Hodgkin's disease. *Eur J Cancer* **30**:1045

Bignon Y, Bernard D, Cure H, Fonck Y, Pauchard J, Travade P, Legros M, Dastugue B, Plagne R (1990) Detection of Epstein-Barr viral genomes in lymph nodes of Hodgkin's disease patients. *Molecular Carcinogenesis* **3**:9

Boiocchi M, Carbone A, De Re V, Dolcetti R (1989) Is the Epstein-Barr virus involved in Hodgkin's disease? *Tumori* **75**:345

Boiocchi M, Dolcetti R, De-Re V, Gloghini A, Carbone A (1993) Demonstration of a unique Epstein-Barr virus-positive cellular clone in metachronous multiple localizations of Hodgkin's disease. *Am J Pathol* **142:**33

Bouzid M, Belkaid MI, Colonna P, Bouguermouh AM, Ooka T (1993) Co-existence of the A and B types of Epstein-Barr virus DNA in lymph node biopsies from Algerian patients with Hodgkin's disease and non-Hodgkin's lymphoma. *Leukemia* **7:**1451

Boyle MJ, Vasak E, Tschuchnigg M, Turner JJ, Sculley T, Penny R, Cooper DA, Tindall B, Sewell WA (1993) Subtypes of Epstein-Barr virus (EBV) in Hodgkin's disease: association between B-type EBV and immunocompromise. *Blood* **81:**468

Brocksmith D, Angel CA, Pringle JH, Lauder I (1991) Epstein-Barr viral DNA in Hodgkin's disease: amplification and detection using the polymerase chain reaction. *J Pathol* **165:**11

Brousset P, Meggetto F, Chittal S, Bibeau F, Arnaud J, Rubin B, Delsol G (1993) Assessment of the methods for the detection of Epstein-Barr virus nucleic acids and related gene products in Hodgkin's disease. *Lab Invest* **69:**483

Brousset P, Schlaifer D, Meggetto F, Bachmann E, Rothenberger S, Pris J, Delsol G, Knecht H (1994) Persistence of the same viral strain in early and late relapses of Epstein-Barr virus-associated Hodgkin's disease. *Blood* **84:**2447

Brousset P, Knecht H, Rubin B, Drouet E, Chittal S, Meggetto F, Saati TA, Bachmann E, Denoyel G, Sergeant A (1993) Demonstration of Epstein-Barr virus replication in Reed-Sternberg cells in Hodgkin's disease. *Blood* **82:**872

Coates PJ, Slavin G, D'Ardenne AJ (1991) Persistence of Epstein-Barr virus in Reed-Sternberg cells throughout the course of Hodgkin's disease. *J Pathol* **164:**291

Deacon EM, Pallesen G, Niedobitek G, Crocker J, Brooks L, Rickinson AB, Young LS (1993) Epstein-Barr virus and Hodgkin's disease: transcriptional analysis of virus latency in the malignant cells. *J Exp Med* **177:**339

Delsol G, Brousset P, Chittal S, Rigal-Huguet F (1992) Correlation of the expression of Epstein-Barr virus latent membrane protein and in situ hybridization with biotinylated BamHI-W probes in Hodgkin's disease. *Am J Pathol* **140:**247

De Re V, Boiocchi M, Vita SD, Dolcetti R, Gloghini A, Uccini S, Baroni C, Scarpa A, Cattoretti G, Carbone A (1993) Subtypes of Epstein-Barr virus in HIV-1-associated and HIV-1-unrelated Hodgkin's disease cases. *Int J Cancer* **54:**895

Evans AS, Carvalho RPS, Frost P, Jamra M, Pozzi DHB (1978) Epstein-Barr virus infections in Brazil. 2. Hodgkin's disease. *J Natl Cancer Inst* **61:**19

Gledhill S, Gallagher A, Jones DB, Krajewski AS, Alexander FE, Klee E, Wright DH, O'Brien C, Onions DE, Jarrett RF (1991) Viral involvement in Hodgkin's disease: detection of clonal type A Epstein-Barr viral genomes in tumour samples. *Br J Cancer* **64:**227

Grasser FA, Murray PG, Kremmer E, Klein K, Remberger K, Feiden W, Reynolds G, Niedobitek G, Young LS, Mueller-Lantzsch N (1994) Monoclonal antibodies directed against the Epstein-Barr virus-encoded nuclear antigen 1 (EBNA1): immunohistologic detection of EBNA1 in the malignant cells of Hodgkin's disease. *Blood* **84:**3792

Gutensohn N, Cole P (1980) Epidemiology of Hodgkin's disease. *Semin Oncol* **7:**92

Henderson BE, Dworsky R, Menck H, Alena B, Henle W, Henle G, Terasaki P (1973) Case-control study of Hodgkin's disease. II. Herpesvirus group antibody titers and HL-A type. *J Natl Cancer Inst* **51:**1443

Henderson S, Rowe M, Gregory C, Croom-Carter D, Wang F, Longnecker R, Kieff E, Rickinson A (1991) Induction of bcl-2 expression by Epstein-Barr virus latent membrane protein 1 protects infected B cells from programmed cell death. *Cell* **65:**1107

Henle W, Henle G (1973) Epstein-Barr virus-related serology in Hodgkin's disease. *Natl Cancer Inst Monogr* **36**:79

Herbst H, Niedobitek G, Kneba M, Hummel M, Finn T, Anagnostopoulos I, Bergholz M, Krieger G, Stein H (1990) High incidence of Epstein-Barr virus genomes in Hodgkin's disease. *Am J Pathol* **137**:13

Herbst H, Dallenbach F, Hummel M, Niedobitek G, Pileri S, Mueller-Lantzsch N, Stein H (1991) Epstein-Barr virus latent membrane protein expression in Hodgkin and Reed-Sternberg cells. *Proc Natl Acad Sci USA* **88**:4766

Herbst H, Steinbrecher E, Niedobitek G, Young LS, Brooks L, Muller LN, Stein H (1992) Distribution and phenotype of Epstein-Barr virus-harboring cells in Hodgkin's disease. *Blood* **80**:484

Hesse J, Levine PH, Ebbesen P, Connelly RR, Mordhorst CH (1977) A case control study on immunity to two Epstein-Barr virus-associated antigens, and to Herpes simplex virus and adenovirus in a population-based group of patients with Hodgkin's disease in Denmark, 1971-73. *Int J Cancer* **19**:49

Howe JG, Mei-Di Shu (1989) Epstein-Barr virus small RNA (EBER) genes: unique transcription units that combine RNA polymerase II and III promoter elements. *Cell* **57**:825

Howe JG, Steitz JA (1986) Localization of Epstein-Barr virus-encoded small RNAs by in situ hybridization. *Proc Natl Acad Sci USA* **83**:9006

Hummel M, Anagnostopoulos I, Dallenbach F, Korbjuhn P, Dimmler C, Stein H (1992) EBV infection patters in Hodgkin's disease and normal lymphoid tissue: expression and cellular localization of EBV gene products. *Br J Haematol* **82**:689

Jarrett RF, Gallagher A, Jones DB, Alexander FE, Krajewski AS, Kelsey A, Adams J, Angus B, Gledhill S, Wright DH, Cartwright RA, Onions DE (1991) Detection of Epstein-Barr virus genomes in Hodgkin's disease: relation to age. *J Clin Pathol* **44**:844

Joske DJ, Emery GA, Bachmann E, Bachmann F, Odermatt B, Knecht H (1992) Epstein-Barr virus burden in Hodgkin's disease is related to latent membrane protein gene expression but not to active viral replication. *Blood* **80**:2610

Khan G, Coates PJ, Gupta RK, Kangro HO, Slavin G (1992) Presence of Epstein-Barr virus in Hodgkin's disease is not exclusive to Reed-Sternberg cells. *Am J Pathol* **140**:757

Khan G, Norton AJ, Slavin G (1993) Epstein-Barr virus in Hodgkin disease. Relation to age and subtype. *Cancer* **71**:3124

Kieff E, Liebowitz D (1990) Epstein-Barr virus and its replication. In *Virology*. Fields BN, Knipe DM, eds. (New York: Raven) p1889

Knecht H, Odermatt BF, Bachmann E, Teixeira S, Sahli R, Hayoz D, Heitz P, Bachmann F (1991) Frequent detection of Epstein-Barr virus DNA by the polymerase chain reaction in lymph node biopsies from patients with Hodgkin's disease without genomic evidence of B- or T-cell clonality. *Blood* **78**:760

Langenhuysen MMAC, Cazemier T, Houwen B, Brouwers TM, Halie MR, The TH, Nieweg HO (1974) Antibodies to Epstein-Barr virus, cytomegalovirus, and Australia antigen in Hodgkin's disease. *Cancer* **34**:262

Levine PH, Ablashi DV, Berard CW, Carbone PP, Waggoner DE, Malan L (1971) Elevated antibody titers to Epstein-Barr virus in Hodgkin's disease. *Cancer* **27**:416

Lindahl T, Klein G, Reedman BM, Johansson B, Singh S (1974) Relationship between Epstein-Barr virus (EBV) DNA and the EBV-determined nuclear antigen (EBNA) in Burkitt lymphoma biopsies and other lymphoproliferative malignancies. *Int J Cancer* **13**:764

MacMahon B (1966) Epidemiology of Hodgkin's disease. *Cancer Res* **26**:1189

Munoz N, Davidson RJL, Witthoff B, Ericsson JE, De-The G (1978) Infectious mononucleosis and Hodgkin's disease. *Int J Cancer* **22**:10

Murray PG, Young LS, Rowe M, Crocker J (1992) Immunohistochemical demonstration of the Epstein-Barr virus-encoded latent membrane protein in paraffin sections of Hodgkin's disease. *J Pathol* **166**:1

Pagano JS, Huang CH, Levine P (1973) Absence of Epstein-Barr viral DNA in American Burkitt's lymphoma. *N Engl J Med* **289**:1395

Pallesen G, Hamilton-Dutoit SJ, Rowe M, Young LS (1991a) Expression of Epstein-Barr virus latent gene products in tumour cells of Hodgkin's disease. *Lancet* **337**:320

Pallesen G, Sandvej K, Hamilton-Dutoit SJ, Rowe M, Young LS (1991b) Activation of Epstein-Barr virus replication in Hodgkin and Reed-Sternberg cells. *Blood* **78**:1162

Poppema S, van-Imhoff G, Torensma R, Smit J (1985) Lymphadenopathy morphologically consistent with Hodgkin's disease associated with Epstein-Barr virus infection. *Am J Clin Pathol* **84**:385

Sandvej KB, Hamilton-Dutoit SJ, Pallesen G (1992) Influence of Epstein-Barr virus encoded latent membrane protein 1 on the expression of CD23 antigen, ICAM-1 and LFA-3 in Hodgkin and Reed-Sternberg cells. A morphometric analysis. *Leuk Lymphoma* **9**:95

Shibata D, Hansmann M, Weiss LM, Nathwani BN (1991) Epstein-Barr virus infections and Hodgkin's disease: a study of fixed tissues using the polymerase chain reaction. *Hum Pathol* **22**:1262

Slivnick DJ, Ellis TM, Nawrocki JF, Fisher RI (1990) The impact of Hodgkin's disease on the immune system. *Semin Oncol* **17**:673

Staal SP, Ambinder R, Beschorner WE, Hayward GS, Mann R (1989) A survey of Epstein-Barr virus DNA in lymphoid tissue. Frequent detection in Hodgkin's disease. *Am J Clin Pathol* **91**:1

Tierney RJ, Steven N, Young LS, Rickinson AB (1994) Epstein-Barr virus latency in blood mononuclear cells: analysis of viral gene transcription during primary infection and in the carrier state. *J Virol* **68**:7374

Uccini S, Monardo F, Stoppacciaro A, Gradilone A, Agliano AM, Faggioni A, Manzari V, Vago L, Costanzi G, Ruco LP, Baroni CD (1990) High frequency of Epstein-Barr virus genome detection in Hodgkin's disease of HIV-positive patients. *Int J Cancer* **46**:581

Wang D, Liebowitz D, Kieff E (1985) An EBV membrane protein expressed in immortalized lymphocytes transforms established rodent cells. *Cell* **43**:831

Wang F, Gregory C, Sample C, Rowe M, Liebowitz D, Murray R, Rickinson A, Kieff E (1990) Epstein-Barr virus latent membrane protein (LMP-1) and nuclear proteins 2 and 3C are effectors of phenotypic changes in B lymphocytes: EBNA-2 and LMP1 cooperatively induce CD23. *J Virol* **64**:2309

Weiss LM, Strickler JG, Hu E, Warnke RA, Sklar J (1986) Immunoglobulin gene rearrangements in Hodgkin's disease. *Hum Pathol* **17**:1009

Weiss LM, Strickler JG, Warnke RA, Purtilo DT, Sklar J (1987) Epstein-Barr viral DNA in tissues of Hodgkin's disease. *Am J Pathol* **129**:86

Weiss LM, Movahed LA, Warnke RA, Sklar J (1989) Detection of Epstein-Barr viral genomes in Reed-Sternberg cells of Hodgkin's disease. *N Engl J Med* **320**:502

Weiss LM, Chen Y, Liu X, Shibata D (1991) Epstein-Barr virus and Hodgkin's disease. A correlative in situ hybridization and polymerase chain reaction study. *Am J Pathol* **139**:1259

Wu T, Mann RB, Charache P, Hayward D, Staal S, Lambe BC, Ambinder RF (1990) Detection of EBV gene expression in Reed-Sternberg cells of Hodgkin's disease. *Int J Cancer* **46**:801

RELATIONSHIP BETWEEN EXPRESSION OF CELLULAR GENES AND EPSTEIN-BARR VIRUS IN HODGKIN'S DISEASE

Mehdi N Jiwa, Joost J Oudejans, Adriaan JC van den Brule, Paul van der Valk, Jau MM Walboomers and Chris JLM Meijer

Department of Pathology, Free University Hospital, Amsterdam, the Netherlands

INTRODUCTION

The diagnosis of Hodgkin's disease (HD) is based on the presence of typical Reed-Sternberg (RS) cells in an appropriate cellular environment of "reactive" cells. The origin of the neoplastic cells in HD is still controversial (Hsu et al., 1985; Cabanillas et al., 1988; Hsu et al., 1992; Drexler, 1992). Recent data favour the hypothesis that the RS cells and their mononuclear variants (collectively called H-RS cells) originate from lymphoid cells, although conflicting data have been published about the exact lineage (Cibull et al., 1989; Schmid et al., 1991; Drexler, 1992; Herbst et al., 1993; Koduru et al., 1993). Expression of B or T-cell specific markers on H-RS cells has been described (Oka et al., 1988; Bjarni et al., 1989; Cibull et al., 1989; Schmid et al., 1991). The number of H-RS cells expressing lymphoid differentiation markers may vary considerably among the different histological types. In HD, nodular lymphocyte predominant type the neoplastic cells express B-cell markers (Coles et al., 1988; Kadin et al., 1988), whereas in other subtypes B and/or T-cell markers can be found on H-RS cells in a variable proportion of cases (Oka et al., 1988; Kadin et al., 1988; Bjarni et al., 1989; Cibull et al., 1989).

The fact that a considerable number of HD cases harbour the Epstein-Barr virus (EBV) also supports the view that H-RS cells are indeed originating from lymphoid cells, since EBV is a mainly lymphotropic virus. The presence of EBV varies among the different types of HD, i.e. up to 70% in the mixed cellularity type, 30-50% in the nodular sclerosing type and sporadically in other types of HD (Weiss et al., 1991; Pallesen et al., 1991; Hummel et al., 1992; see Jarrett and Armstrong, this volume). It is known that in vitro EBV has transforming properties and is also able to modulate the expression of certain cellular genes in lymphoblastoid cell lines in vitro. Here we will discuss certain aspects of the modulation of cellular genes by EBV and subsequently we discuss certain possible pathogenetic mechanisms of EBV in the genesis of HD.

EPSTEIN-BARR VIRUS

EBV is a human gamma herpesvirus; its genome is approximately 180 kb in length and encodes around 100 different genes. The genes can be divided into latent genes, the only genes transcribed during the latent, non-productive infection, and immediate early, early and late genes, which are used during the lytic phase. In EBV-associated lymphoproliferative malignancies in patients without an overt immunodeficiency only a limited number of the EBV latent genes are transcribed. The latent genes are divided into six EBNAs (EBV nuclear antigens), three latent membrane proteins (LMP-1, LMP-2A and -2B) and two non-coding small RNAs, EBER-1 and -2.

In vitro three different latent expression patterns are recognised: type I, the expression seen in Burkitt lymphoma cell lines in which only EBER-1 and -2 and EBNA-1 are expressed; type II, the expression pattern observed in nasopharyngeal carcinomas in which besides both EBERs and EBNA-1 also LMP-1 is expressed and type III, the lymphoblastoid cell line expression pattern in which all EBNAs and latent membrane proteins, as well as both EBERs are expressed (Rowe *et al.,* 1987; Speck and Strominger, 1989; Brooks *et al.,* 1992). More recently, some other genes have been found to be transcribed during certain latent infections, both *in vitro* and *in vivo*. One of these is derived from the BamHI A fragment, BARF0, the function of which is as yet unsolved. Moreover, some alternative transcripts of early genes can be found during latency. For example, the BHRF1 gene, also known as the viral bcl-2 homologue, is usually expressed during lytic infections, but transcripts driven by a separate promoter (located in the BamHI H fragment) are expressed during latency in certain cells (Lear *et al.,* 1992; Henderson *et al.,* 1993). Although EBV is supposed to infect B lymphocytes expressing the EBV receptor, CR2 or CD21, EBV is also found in certain T-cell non-Hodgkin's lymphomas (NHLs), especially extranodal nasal T-cell NHLs (De Bruin *et al.,* 1994). The role of EBV in the pathogenesis of HD is the subject of many investigations, but at present is not clear.

IN VITRO STUDIES

Up-regulation/induction of cellular (proto-)oncogenes
Many EBV gene products are able to modulate cellular gene expression *in vitro*. A brief overview is given in Table 1. From this Table it is clear that EBV might be directly involved in the transformation of epithelial and lymphoid cells. Furthermore, it can up-regulate the expression of cellular (proto-)oncogenes, thus providing certain growth advantages to EBV infected cells and modulate certain cellular genes necessary to circumvent immunosurveillance.

Certain genes of EBV are involved in the up-regulation of the cellular proto-oncogenes. LMP-1 is able to up-regulate the human proto-oncogene bcl-2 in certain lymphoblastoid cell lines *in vitro* (Henderson *et al.*, 1991), whereas in other lymphoblastoid cell lines no such up-regulation has been observed (Martin *et al.*, 1993). Also some other oncogenes are up-regulated by EBV. For example EBNA-2 is able to up-regulate c-fgr, whereas early-D stabilises the c-myc oncogene product.

Table 1: Relationship between EBV and the expression of cellular genes *in vitro*.

Effects on cellular gene (products)	EBV gene involved	References
binding to Rb protein	EBNA-LP	Szekely *et al.*, 1993
Induction/up-regulation of cellular (proto-)oncogenes		
bcl-2	LMP-1	Henderson *et al.*, 1993
c-myc	infection, early-D	Alfieri *et al.*, 1991
c-fgr	EBNA-2	Calender *et al.*, 1990
Induction/up-regulation of other cellular genes		
adhesion molecules, such as LFA-1, ICAM-1 and LFA-3	LMP-1	Calender *et al.*, 1990
up-regulation of the B-cell marker CD23	LMP-1/EBNA-2	Wang *et al.*, 1990 Cordierbussat *et al.*, 1993
up-regulation of CD21 (the 'EBV receptor')	EBNA-2	Cordierbussat *et al.*, 1993
CD30	infection	Rowe *et al.*, 1987
CD40	LMP-1	Wang *et al.*, 1990
CDw70	LMP-1	Niedobitek *et al.*, 1992
blocking of calcium modulation	LMP-2A	Miller *et al.*, 1993
Down-regulation of cellular genes		
B and T-cell markers, such as CD3 and CD20	infection	Bai *et al.*, 1994 Garnier *et al.*, 1993
Epithelial membrane antigen	LMP-1	Niedobitek *et al.*, 1992
Genomic/functional homology		
c-jun	BZLF1	Flemington & Speck, 1990
bcl-2	BHRF1	Henderson *et al.*, 1993
IL-10	BCRF1	Hsu *et al.*, 1990

Table 2: Expression of cellular genes in relationship to EBV in non-lymphocyte predominant types of Hodgkin's disease.

Cellular antigen	Expression	References
LFA-1	=	Kanavaros et al., 1993
ICAM-1	=	Kanavaros et al., 1993
CD3	↓	Bai et al., 1994
CD20	↓	Bai et al., 1994
CD21	↑	Jiwa et al., 1992
CD23	=	Kanavaros et al., 1993
CD30	=	Kanavaros et al., 1993
		Bai et al., 1994
CD40	=	Wang et al., 1990
EMA/115D8	↓	Bai et al., 1994
Bcl-2	=/↓	Jiwa et al., 1993
		Jiwa et al., 1995

↑, up-regulated; ↓, downregulated; =, no differences between EBV positive and EBV-negative cases of HD.

The c-myc oncogene protein is a nuclear protein that is involved in cell proliferation and differentiation and seems to play an important role in the cell cycle (Kelly et al., 1983). Activation of this oncogene may occur in different ways. One mechanism is translocation of the c-myc locus on chromosome 8 to the immunoglobulin loci on chromosomes 14, 2 or 22, as described in Burkitt's lymphoma and other B-cell NHLs, thereby placing the c-myc gene under the influence of the immunoglobulin promoter. Another mechanism is by EBV infection (Alfieri et al., 1991). Co-expression of c-myc and bcl-2 promotes the development of haematological malignancies in transgenic mice and humans (Vaux et al., 1988).

Expression of EBV genes sharing genomic and/or functional homology with cellular genes
Several of the latent and lytic genes of EBV share (functional) homology with human genes. The best known are BZLF1 (ZEBRA), the latent/lytic 'switch protein', which shares functional homology with the c-jun oncogene and the BHRF1 gene, which shares sequence and functional homology with the human bcl-2 proto-oncogene (Henderson et al., 1993).

A third EBV gene, BCRF1, is a viral homologue of interleukin (IL)-10. Although this gene was first indentified as a late gene, expressed only during the lytic phase of viral replication, recent evidence indicates that BCRF1 is also expressed during EBV latency. Expression of BCRF1 seems essential for the transforming potential of EBV. Moreover, expression of BCRF1 might have an influence on immunosurveillance, since IL-10 has known immunosuppressive properties (Stewart and Rooney, 1992; Miyazaki et al., 1993).

EBV proteins binding to human tumour suppressor genes
One of the EBV latent proteins, EBNA-LP or EBNA-5, not only has transforming properties (Mannick et al. 1991), but is able to form a complex with the Rb protein and also with p53, thereby impairing the biological function of these proteins (Szekely et al., 1993; see Piris et al., this volume).

Modulation of cellular activation markers and adhesion molecules by EBV

In vitro EBV is able to transform B lymphocytes and epithelial cells, and to modulate the expression of several cellular genes. For example, cellular markers, like CD23, CD30, CD39, CDw70 and certain adhesion molecules such as CD11a (LFA-1), CD58 (LFA-3) and CD54 (ICAM-1) can be up-regulated by EBV (Wang *et al.*, 1990; Calender *et al.*, 1990). Moreover, EBV is able to down-regulate *in vitro* the epithelial membrane antigen (EMA) (Niedobitek *et al.*, 1992).

IN VIVO STUDIES

Expression of EBV latent genes in Hodgkin's disease

As stated earlier, EBV is found in a substantial number of HD cases. The prevalence of EBV is dependent on the type of HD (see above). In HD only a limited number of the latent genes is expressed: both EBERs, EBNA-1, LMP-1 and sometimes also LMP-2A and -2B expression can be detected (Deacon *et al.*, 1993). Sporadically latent transcripts of BHRF1 were found (Jiwa *et al.*, 1995).

As mentioned earlier, the role of the virus in the pathogenesis is as yet unsolved. It is known that certain EBV genes, especially EBNA-2 and LMP-1 have transforming properties *in vitro*, of which only LMP-1 is expressed in HD. LMP-1 is expressed on the majority of the EBV carrying H-RS cells (Pallesen *et al.*, 1991; Jiwa *et al.*, 1993).

Expression of EBV genes with functional homology
to known oncogenes or growth factors

The EBV 'switch protein' ZEBRA is only sporadically expressed in the neoplastic cells in HD (Pallesen *et al.*, 1991), indicating that the virus is mainly present in a latent state. Expression of BHRF1, the viral bcl-2 homologue, was detected in certain EBV-positive B-cell NHLs in both immunocompromised and immunocompetent patients (submitted for publication). In HD in only 1 out of 10 cases BHRF1 latent transcripts could be detected by reverse-transcriptase PCR (Jiwa *et al.*, 1995). Since this technique is a non-morphological method, no information was obtained about the origin of the cells (neoplastic cells or reactive, 'innocent bystanders') expressing these transcripts. Other techniques, like *in situ* hybridisation are required to solve these questions.

Expression of cellular genes in Hodgkin's disease in relation to EBV

In HD the expression of certain cellular adhesion molecules, such as CD11a (LFA-1), CD58 (LFA-3) and CD54 (ICAM-1) and certain cellular activation markers, like CD23, CD30 and CD40 did not correlate with expression of EBV latent products (Kanavaros *et al.*, 1993; O'Grady *et al.*, 1994).

EBV is not only associated with up-regulation of cellular genes, but also with selective down-regulation. For example, down-regulation of CD20 and CD23 by EBV has been reported in the neoplastic cells of B-cell NHL arising in SCID mice following injection of EBV-positive peripheral blood B-cells (Garnier *et al.*, 1993). In HD expression of both B and T-cell lineage specific and associated markers has been observed frequently. In the EBV-positive cases decreased expression of CD20 and CD3 has been described (Bai *et al.*, 1994), whereas certain activation markers like the episialine epitopes MAM3 and MAM6, as detected by the anti-EMA and 115D8 antibodies respectively, are also down-regulated in the EBV-positive cases (Bai *et al.*, 1994).

Two important cellular (proto-)oncogenes, c-myc and bcl-2, are frequently expressed in HD (Jiwa *et al.*, 1993). Initially no differences between EBV-positive and negative HD cases were found (Jiwa *et al.*, 1993), since in nearly all cases the neoplastic cells expressed c-myc protein. Bcl-2 is also frequently expressed in a variable number of the neoplastic cells (see Weiss, this volume). However, more recent studies indicate that the number of neoplastic cells expressing bcl-2 protein is significantly lower in EBV-positive cases compared with EBV-negative cases of both the nodular sclerosing and mixed cellularity subtype (Jiwa *et al.*, 1995).

This prompted us to investigate whether the viral bcl-2 homologue, the BHRF1 gene, was activated in EBV-positive cases of HD and had taken over the function of the human bcl-2 protein. BHRF1 transcripts could be detected in only 1 out of 10 cases, indicating that BHRF1 is only sporadically expressed in HD. This suggests that mechanisms other than bcl-2 provide growth advantages to the neoplastic cell population in EBV-positive HD, although the mechanisms are as yet unknown.

CONCLUSIONS

From different *in vitro* and *in vivo* studies it has become clear that EBV is able to immortalise and transform different epithelial and lymphoid cells in different ways. Consequently, the selective growth advantage of EBV-transformed cells is also mediated in different ways.

Bcl-2 and c-myc are frequently expressed in both EBV-positive and negative cases of HD, although the number of bcl-2 positive H-RS cells in EBV-negative cases was significantly higher than in EBV-positive cases. In these EBV-positive cases it was supposed that the viral 'bcl-2 homologue', BHRF1, was expressed instead of the human bcl-2 protein. However initial results revealed that BHRF1 is expressed only sporadically in HD.

The decreased expression of CD3, CD20 and EMA observed in H-RS cells of EBV-positive cases might reflect a possible mechanism of circumventing immunosurveillance, thus providing H-RS cells with a selective growth advantage. Investigations concerning the immunophenotype of H-RS cells should take into account EBV status, since EBV is able to modulate the expression of cellular genes, including lymphoid lineage specific markers.

REFERENCES

Alfieri C, Birkenbach M, Kieff E (1991) Early events in Epstein-Barr virus infection of human B lymphocytes. *Virology* **181**:595

Bai MC, Jiwa NM, Horstman A, Vos W, Kluin PH, van der Valk P, Mullink H, Walboomers JMM, Meijer CJLM (1994) Decreased expression of cellular markers in Epstein-Barr virus positive Hodgkin's disease. *J Pathol* 174:49

Bjarni A, Agnarsson MD, Marshall E, Kadin ME (1989) The immunophenotype of Reed-Sternberg cells. A study of 50 cases of Hodgkin's disease using fixed frozen tissues. *Cancer* **63**:2083

Brooks L, Yao QY, Rickinson AB, Young LS (1992) Epstein-Barr virus latent gene transcription in nasopharyngeal carcinoma cells: coexpression of EBNA1, LMP1, and LMP2 transcripts. *J Virol* **66**:2689

Cabanillas F, Pathak S, Trujillo J, Grant G, Cork A, Hagemeister FB, Velasquez WS, McLaughlin P, Redman J, Katz R, Butler JJ, Freireich E J (1988) Cytogenetic features of Hodgkin's disease suggest possible origin from a lymphocyte. *Blood* **71**:1615

Calender A, Cordier M, Billaud M, Lenoir GM (1990) Modulation of cellular gene expression in B lymphoma cells following *in vitro* infection by Epstein-Barr virus (EBV). *Int J Cancer* **46**:658

Cibull ML, Stein H, Gatter KC, Mason DY (1989) The expression of the CD3 antigen in Hodgkin's disease. *Histopathol* **15**:597

Coles FB, Cartum RW, Pastuszak WT (1988) Hodgkin's disease lymphocyte-predominant type: immunoreactivity with B-cell antibodies. *Mod Pathol* **1**:274

Cordierbussat M, Billaud M, Calender A, Lenoir GM (1993) Epstein-Barr virus (EBV) nuclear-antigen-2-induced up-regulation of CD21 and CD23 molecules is dependent on a permissive cellular context. *Int J Cancer* **53**:153

Deacon EM, Pallesen G, Nieboditek G, Crocker J, Brooks L, Rickinson AB, Young LS (1993) Epstein-Barr virus and Hodgkin's disease: transcriptional analysis of virus latency in the malignant cells. *J Exp Med* **177**:339

De Bruin PC, Jiwa M, Oudejans JJ, van der Valk P, van Heerde P, Sabourin JC, Csanaky G, Gaulard P, Noorduyn AL, Willemze R, Meijer CJLM (1994) Presence of Epstein-Barr virus in extranodal T-cell lymphomas: differences in relation to site. *Blood* **83**:1612

Delsol G, Stein H, Pulford KAF, Gatter KC, Erber WN, Zinne K, Mason DY (1984) Human lymphoid cells express epithelial membrane antigen. *Lancet* **17**:1124

Drexler HG (1992) Recent results on the biology of Hodgkin and Reed-Sternberg cells. I. Biopsy material. *Leuk Lymphoma* **8**:283

Flemington E, Speck SH (1990) Identification of phorbol ester response elements in the promoter of Esptein-Barr virus putative lytic switch gene BZLF1. *J Virol* **64**:1217

Garnier JL, Cooper NR, Cannon MJ (1993) Low expression of CD20 and CD23 in Epstein-Barr virus induced B cell tumours in SCID/hu mice. *Am J Pathol* **142**:353

Henderson S, Rowe M, Gregory C, Croom-Carter D, Wang F, Longnecker R, Kieff E, Rickinson AB (1991) Induction of bcl-2 expression by Epstein-Barr virus latent membrane protein 1 protects infected B cells from programmed cell death. *Cell* **65**:1107

Henderson S, Huen D, Rowe M, Dawson C, Johnson G, Rickinson A (1993) Epstein-Barr virus-coded BHRF1 protein, a viral homologue of bcl-2, protects human B-cells from programmed cell death. *Proc Natl Acad Sci USA* **90**:8479

Herbst H, Stein H, Nieboditek G (1993) Epstein-Barr virus and CD30+ malignant lymphomas. *Critical Rev Oncol Hematol* **4**:191

Hsu, S, Xie S, El-Okda MO, Hsu P (1992) Correlation of c-fos/c-jun expression with histiocytic differentiation in Hodgkin's Reed-Sternberg cells. *Am J Pathol* **140**:155

Hsu SM, Yang K, Jaffe ES (1985) Phenotypic expression of Hodgkin's and Reed-Sternberg cells in Hodgkin's disease. *Am J Pathol* **118**:209

Hsu S, Jaffe ES (1984) Leu M1 and peanut agglutinin stain the neoplastic cells of Hodgkin's disease. *Am J Clin Pathol* **82**:29

Hsu DW, de Waal Malefyt R, Fiorentino DF, Dang MN, Vieira P, de Vries J, Spits H, Mosmann TR, Moore KW (1990) Expression of interleukin-10 activity by Epstein-Barr virus protein BCRF1. *Science* **250**:830

Hummel M, Anagnostopoulos I, Dallenbach F, Korbjuhn P, Dimmler C, Stein H (1992) EBV infection patterns in Hodgkin's disease and normal lymphoid tissue: expression and cellular localisation of EBV gene products. *Br J Haematol* **82**:689

Jiwa NM, van der Valk P, Mullink H, Vos W, Horstman A, Maurice MM, Olde-Weghuis DEM, Walboomers JMM, Meijer CJLM (1992) Epstein-Barr virus DNA in Reed-Sternberg cell of Hodgkin's disease is frequently associated with CR2 (EBV receptor) expression. *Histopathol* **21**:51

Jiwa NM, Kanavaros P, de Bruin PC, van der Valk P, Horstman A, Vos W, Mullink H, Walboomers JMM, Meijer CJLM (1993) Presence of Epstein-Barr virus harbouring small and intermediate-sized cells in Hodgkin's disease. Is there a relationship with Reed-Sternberg cells? *J Pathol* **170**:129

Jiwa NM, Kanavaros P, van der Valk P, Walboomers JMM, Horstman A, Vos W, Mullink H, Meijer CJLM (1993) Expression of c-myc and bcl-2 oncogene products in Reed-Sternberg cells independent of presence of Epstein-Barr virus. *J Clin Pathol* **46**:211

Jiwa NM, Oudejans JJ, Bai MC, van den Brule AJC, Horstman A, Vos W, van der Walk P, Kluin PhM, Walboomers JMM, Meijer CJLM (1995) Expression of bcl-2 protein and transcription of the Epstein-Barr virus bcl-2 homologue BHRF-1 in Hodgkin's disease: implications for different pathogenic mechanisms. *Histopathol* **26**:547

Kadin ME, Muramoto L, Said J (1988) Expression of T-cell antigens on Reed-Sternberg cells in a subset of patients with nodular sclerosing and mixed cellularity Hodgkin's disease. *Am J Pathol* **130**:345

Kanavaros P, Jiwa NM, van der Val PK, Walboomers JMM, Horstman A, Meijer CJLM (1993) Expression of Epstein-Barr virus latent gene products and related cellular activation and adhesion molecules in Hodgkin's disease and non-Hodgkin's lymphomas arising in patients without overt pre-existing immunodeficiency. *Hum Pathol* **24**:725

Kelly K, Cochran BH, Stiles CD, Leder P (1983) Cell specific regulation of the c-myc gene by lymphocyte mitogen and platelet-derived growth factor. *Cell* **35**:603

Koduru PRK, Susin M, Schulman P, Catell D, Goh JC, Krap L, Broome JD (1993) Phenotypic and genotypic characterisation of Hodgkin's disease. *Am J Hematol* **44**:117

Lear AL, Rowe MG, Kurilla S, Lee S, Henderson S, Kieff E, Rickinson AB (1992) The Epstein-Barr virus (EBV) nuclear antigen-1 BamHl F promoter is activated on entry of EBV-transformed B cells into the lytic cycle. *J Virol* **66**:7461

Mannick JB, Cohen JL, Birkenbach M, Marchini A, Kieff E (1991) The Epstein-Barr virus nuclear protein encoded by the leader of the EBNA RNAs is important in B-lymphocyte transformation. *J Virol* **65**:6826

Martin JM, Veis D, Korsmeyer SJ, Sugden B (1993) Latent membrane protein of Epstein-Barr virus induces cellular phenotypes independently of expression of bcl-2. *J Virol* **67**:5269

Miller CL, Longnecker R, Kieff E (1993) Epstein-Barr virus latent membrane protein 2A blocks calcium mobilisation in B lymphocytes. *J Virol* **67**:3087

Miyazaki I, Cheung RK, Dosch HM (1993) Viral interleukin-10 is critical for the induction of B- cell growth transformation by Epstein-Barr virus. *J Exp Med* **178**:439

Niedobitek G, Fahraeus R, Herbst H, Latza U, Ferszt A, Klein G (1992) The Epstein-Barr virus encoded membrane protein (LMP) induces phenotypic changes in epithelial cells. *Virchows Archiv (B cell Pathol)* **62**:55

O'Grady J, Stewart S, Lowrey J, Howie SEM, Krajewski AS (1994) CD40 expression in Hodgkin's disease. *Am J Pathol* **144**:21

Oka K, Mori N, Kojima M (1988) Anti-leu-3a antibody reactivity with Reed-Sternberg cells of Hodgkin's disease. *Arch Pathol Lab Med* **112**:139

Pallesen G, Hamilton Dutoit SJ, Rowe M, Young LS (1991) Expression of Epstein-Barr virus latent gene products in tumour cells of Hodgkin's disease. *Lancet* **337**:320

Pallesen G, Sandvej K, Hamilton Dutoit SJ, Rowe M, Young LS (1991) Activation of Epstein-Barr virus replication in Hodgkin and Reed-Sternberg cells. *Blood* **78**:1162

Rowe M, Rowe DT, Gregory CD, Young LS, Farrell PJ, Rupani H, Rickinson AB (1987) Differences in B cell growth phenotype reflect novel patterns of Epstein-Barr virus latent gene expression in Burkitt's lymphoma cells. *EMBO J* **6**:2743

Schmid C, Pan L, Diss T, Isaacson PG (1991) Expression of B-cell antigens by Hodgkin's and Reed-Sternberg cells. *Am J Pathol* **139**:701

Speck SH, Strominger JL (1989) Transcription of Epstein-Barr virus in latently infected, growth transformed lymphocytes. *Adv Vir Onc* **8**:133

Stewart JP, Rooney CM (1992) The interleukin-10 homolog encoded by Epstein-Barr virus enhances the reactivation of virus specific cytotoxic T- cell and HLA unrestricted killer cell responses. *Virology* **191**:773

Sugden, B (1992) EBV's open sesame. *TIBS* **17**:239

Szekely L, Selivanova G, Magnusson KP, Klein G, Wiman KG (1993) EBNA-5, an Epstein-Barr virus-encoded nuclear antigen, binds to the retinoblastoma and p53 proteins. *Proc Natl Acad Sci USA* **90**:5455

Vaux DL, Cory S, Adams JM (1988) bcl-2 gene promotes haemopoietic cell survival and co-operates with c-myc to immortalise pre-B cells. *Nature* **335**:440

Wang F, Gregory C, Sample C, Rowe M, Liebowitz D, Murray R, Rickinson A, Kieff E (1990) Epstein-Barr virus latent membrane protein (LMP-1) and nuclear proteins 2 and 3C are effectors of phenotypic changes in B lymphocytes: EBNA-2 and LMP-1 co-operatively induce CD23. *J Virol* **64**:2309

Weiss LM, Chen YY, Liu XF, Shibata D (1991) Epstein-Barr virus and Hodgkin's disease: a correlative *in situ* hybridisation and polymerase chain reaction study. *Am J Pathol* **139**:1259

THE EPIDEMIOLOGY OF EBV-ASSOCIATED HODGKIN'S DISEASE

Ruth F Jarrett and Alison A Armstrong

LRF Virus Centre, Department of Veterinary Pathology, University of Glasgow, Glasgow G61 1QH, UK

Our work is supported by the Leukaemia Research Fund.

INTRODUCTION

Preceding chapters provide evidence that the Epstein-Barr virus (EBV) is involved in the pathogenesis of Hodgkin's disease (HD). It is clear that EBV genomes or gene products cannot be detected within the affected tissues of all cases. We have designated those cases in which clonal EBV genomes are detectable within tumours or in which the Reed-Sternberg (RS) cells express EBV LMP-1 protein or EBER RNAs as EBV-associated (Armstrong *et al.*, 1992). Studies from a number of laboratories indicate that between 32 and 50% of cases in developed countries are EBV-associated using these criteria (Pallesen *et al.*, 1991; Weiss *et al.*, 1991; Herbst *et al.*, 1992; Delsol *et al.*, 1992; Armstrong *et al.*, 1994; Khan *et al.*, 1993).

Failure to detect EBV in all cases could be hypothetically explained in several ways. First, current assays used to detect EBV may not be sensitive enough to detect low copy number EBV genomes in RS cells. This is highly unlikely since the EBER *in situ* hybridisation assay, routinely used to detect EBV latent infection, is exquisitely sensitive by virtue of the high copy number of the EBV EBER transcripts (see Young and Niedobitek, this volume; Jarrett, this volume). Secondly, defective EBV genomes may be present in some cases and therefore escape detection. We have preliminary evidence for the existence of defective EBV genomes in a small minority of cases of HD; it is unlikely that this could explain the majority of negative results since there is a very good correlation between the results of various assays which detect DNA fragments from, or gene products encoded by, different parts of the EBV genome. For example, almost all cases positive for LMP-1 protein are

EBER RNA positive and, on Southern blot analysis, hybridise with probes from the EBV BamHI W and terminal repeat regions (Armstrong *et al.*, 1992). Thirdly, EBV may be a passenger virus in HD and of no importance in disease pathogenesis. Arguments against this include the clonality of the genomes, the presence of the virus in all tumour cells and the high level expression of the EBV LMP-1 protein (see Jarrett, this volume). Lastly, EBV may be associated with a distinct subgroup of HD cases.

Credence for this last possibility comes from epidemiological studies which suggest that HD is a heterogeneous condition which may have more than one aetiology (MacMahon, 1966; Correa and O'Conor, 1971). These studies are summarised by Freda Alexander in an earlier chapter. We reasoned that if EBV were more often associated with one particular subgroup of cases then this would strengthen the evidence for a causal link between the virus and those cases.

The age distribution of HD cases is bimodal and risk factors for disease development differ by age at diagnosis (MacMahon, 1966; McKinney *et al.*, 1989; Gutensohn, 1982; Alexander *et al.*, 1991). This has led to the two-disease hypothesis of HD which suggests that HD in young adults in developed countries has a different aetiology from HD in older adults. Paediatric HD is rare in developed countries but risk factors in this age group differ from those in young adults. On the basis of these differences MacMahon proposed in 1966 that HD was a heterogeneous condition with different aetiologies in different age groups (MacMahon, 1966). He defined the age groups 0-14 years, 15-34 years and ≥50 years and further suggested that HD in young adults was caused by an infectious agent.

In developing countries the first peak in the bimodal age incidence curve is observed in childhood and the young adult age incidence peak is lacking (MacMahon, 1966; Correa and O'Conor, 1971). By analogy with paralytic polio it has been proposed that HD in children in developing countries may have the same aetiology as HD in young adults in developed countries (see Alexander, this volume).

The distribution of the histological subtypes of HD varies by age (McKinney *et al.*, 1989; Glaser and Swartz, 1990). Nodular sclerosis HD (HDNS) largely accounts for the young adult age incidence peak observed in developed countries. Mixed cellularity (HDMC) and lymphocyte depleted HD (HDLD) are relatively more common in the older adult age group, in childhood, in developing countries at all ages, and in the context of HIV infection (MacMahon, 1966; Correa and O'Conor, 1971; McKinney *et al.*, 1989). Age and histological subtype are therefore confounding variables and must be considered together. Furthermore differences in epidemiological risk factors show a closer association with age at diagnosis than histological subtype, stressing the need to examine both variables. Distribution of cases by sex is also not uniform at all ages; male cases predominate in the paediatric and older adult age groups but not in the young adult group in developed countries in which a female excess has been reported (Spitz *et al.*, 1986; McKinney *et al.*, 1989; Glaser and Swartz, 1990).

EBV-ASSOCIATION - RELATIONSHIP WITH
HISTOLOGICAL SUBTYPE AND AGE

Adult Hodgkin's disease

Many investigators have analysed the association between EBV and histological subtype of HD. Most studies have found significant differences with HDMC cases being positive more often than HDNS cases (Pallesen et al., 1991; Weiss et al., 1991; Delsol et al., 1992; Hummel et al., 1992; Khan et al., 1993). Larger case series show positivity rates for HDMC between 56 and 90% whereas only 10-50% of HDNS cases have EBV-positive RS cells (Pallesen et al., 1991; Weiss et al., 1991; Herbst et al., 1992; Murray et al., 1992; Delsol et al., 1992; Khan et al., 1993). Relatively fewer cases of HDLD have been examined but the majority are EBV-associated. Lymphocyte predominance HD (HDLP) cases are infrequently EBV-positive.

We previously examined the relationship between age at diagnosis and EBV-positivity in HD using Southern blot analysis (Jarrett et al., 1991). Significant differences were found between different age groups with EBV genomes detected in the majority of paediatric and older adult cases ($\geq$50 years) but in the minority of cases in the young adult age incidence peak (15-34 years). These findings have generated controversy; some groups have reported similar results (Uhara et al., 1990; L Weiss, personal communication; I Lauder, personal communication) whereas others have failed to find differences in EBV-positivity by age (Vestlev et al., 1992; Khan et al., 1993). We therefore decided to regard our original study as hypothesis generating, and to test the hypothesis that there is a significant difference between the proportion of EBV-associated cases in young and older adults in a second study.

Study Design

There is no overlap between the cases included in the current study and those described in the hypothesis generating study. In April 1994, 176 adult cases were eligible for inclusion, of which 79 were non-selected. Cases were selected for older age in order to augment numbers in this age group. Sixty-two cases from an ongoing epidemiological study of young adult HD were included and there was therefore age selection in the 15-24 year age group. For this reason the 15-34 year age group has been split into 2 groups in the analysis.

We believe that there are sufficient data showing that HDLP is a distinct pathological entity to warrant a separate analysis of this histological subtype (Mason et al., 1994). Statistical analyses were therefore performed with and without inclusion of HDLP cases.

All cases were examined using the EBV EBER *in situ* hybridisation assay which we have found to be the most useful method for detecting EBV in HD. Briefly, sections of routinely processed, paraffin-embedded material were incubated with a biotinylated, antisense oligonucleotide probe and hybridisation detected using avidin-biotin complexes (Dako, UK) (Armstrong et al., 1992). Many cases were also analysed using antibodies (CS1-4) reactive with the EBV LMP-1 protein.

Results

The results are shown in Table 1 and Fig. 1. For comparison, the results of the analysis of paediatric cases from the UK are included in Fig. 1.

As in our previous study, we found a clear association between age at diagnosis and EBV-positivity with only 14% of young adult cases in the 15-24 year old age group being EBV-associated whereas the majority of older adult cases (>50 years) had EBV-positive RS cells. This difference is statistically significant, $p<0.001$ (see Table 1).

Table 1:   Analysis of EBV-association in Hodgkin's disease by age at diagnosis

Factors adjusted for	Inclusion of HDLP	15-24 years OR 95% CL	25-34 years OR 95% CL	≥50 years OR 95% CL	p value
None	Yes	1.00	2.09 0.57-7.66	5.75 2.43-13.59	<0.001
None	No	1.00	2.40 0.64-9.02	7.77 3.16-19.08	<0.001
HDNS	Yes	1.00	1.74 0.49-7.15	2.5 0.99-6.35	0.14
HDNS	No	1.00		3.1 1.1-8.74	0.05

The 15-24 year age group is taken as the reference group in the analysis. Abbreviations: HDLP, lymphocyte predominance Hodgkin's disease; HDNS, nodular sclerosis Hodgkin's disease; OR, odds ration; CL, confidence limits.

Analysis of the data by histological subtype revealed that HDMC cases were significantly more likely to be EBV-associated than HDNS cases ($p<0.001$), a result consistent with other studies. HDLP cases were only rarely positive (1/16 cases). There were only 9 HDLD cases included in the series but the majority (5/9) was EBV-associated.

Fig. 2 shows the results broken down by age and histological subtype. The most striking feature of these data is the lack of EBV-association in the young adult HDNS cases; only 1 out of 57 cases was positive. HDNS cases at other ages were more likely to be EBV-positive (see below). Statistically significant differences in EBV-association rates in different age groups were still observed following adjustment for the effects of histological subtype and sex (Table 1). The results therefore confirm those of our previous study. We believe that they provide clear evidence for an association between EBV and age at diagnosis and support the idea that HD is a heterogeneous condition.

Paediatric Hodgkin's disease

Paediatric HD is rare in developed countries but available results suggest that the proportion of EBV-associated cases is high in this age group (Weinreb *et al.*, 1992; Armstrong *et al.*, 1993). We have examined 19 cases of classical HD occurring in children aged <15 years in the UK; in 12 cases (63%) the RS cells harboured EBV. The relationship between EBV-association and age, within this age group, is discussed more fully below.

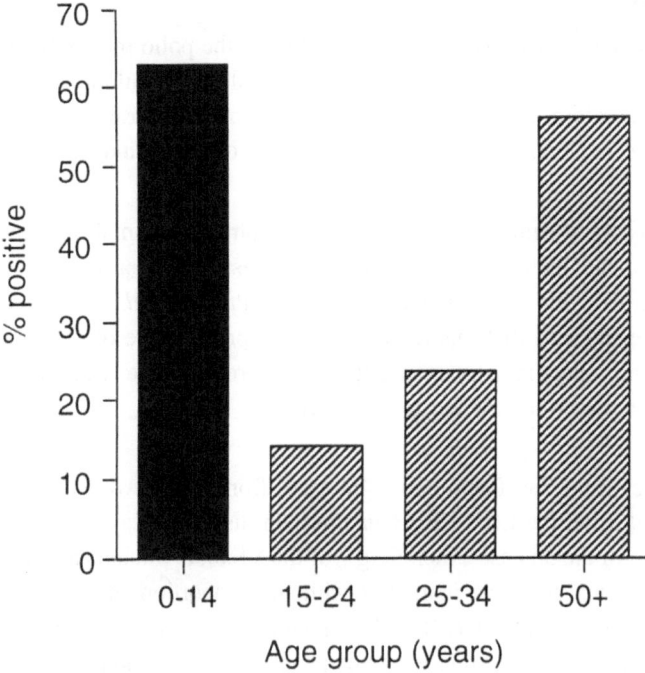

Fig. 1: EBV-association in Hodgkin's disease by age group. Results of analysis of paediatric cases are shown for comparison.

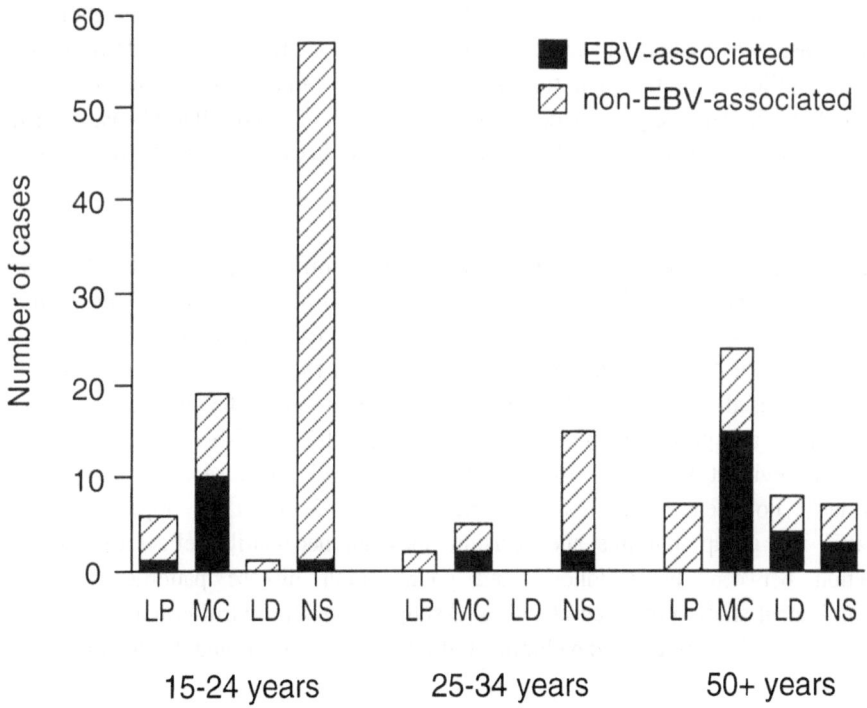

Fig. 2: EBV-association in Hodgkin's disease by age group and histological subtype. LP, lymphocyte predominance Hodgkin's disease; NS, nodular sclerosis Hodgkin's disease; MC, mixed cellularity Hodgkin's disease; LD, lymphocyte depleted Hodgkin's disease.

The polio model

As described above (see also Alexander, this volume), the polio model predicts that HD in young adults in developed countries has the same aetiology as childhood HD in developing countries. We therefore determined the EBV status of paediatric cases from developing countries; these were obtained from the Sao Paulo area of Brazil and from Saudi Arabia.

In the Sao Paulo area of Brazil the incidence of paediatric HD in males is highest in the age group 5-9 years, the incidence of acute lymphoblastic leukaemia is low, and a high incidence of non-Hodgkin's lymphoma is observed (Parkin *et al.*, 1988). This pattern of disease incidence suggests that this region may be representative of developing countries. In Saudi Arabia no cancer registry exists, but the features of acute lymphoblastic leukaemia are similar to those described for Western countries.

Twenty-five cases from Sao Paulo and eight cases from Saudi Arabia were examined; all were investigated using the EBV EBER *in situ* hybridisation assay. EBV was detected in the RS cells of 25 of the 33 cases, including 7 of the 8 Saudi Arabian cases and 18 of the 25 Brazilian cases. This contrasts markedly with the proportion of EBV-associated cases observed in young adults in the UK (Fig. 1), which is taken as representative of a developed country. These data provide strong evidence against the polio model of HD, at least with regard to the majority of cases.

A comparison of the proportion of EBV-associated paediatric cases in the UK and Brazil was then performed (Armstrong *et al.*, 1993). Although a higher proportion of the Brazilian cases was positive, the difference was not statistically significant. Differences did not persist following adjustment for the effects of age and subtype - differences were attributable to the relatively younger age and greater proportion of HDMC cases in Brazil. This confirms that there were slight differences in disease pattern between the two countries.

The most striking feature of the combined data from the different geographical locales was the association between EBV and age within this group of cases. In the age group 1-4 years, 6/7 cases were EBV-associated; the single negative case was a 3 year old patient with HDLP from Saudi Arabia. In the 5-9 year old age group 18/20 cases were EBV-associated but this decreased to 14/28 in the 10-14 year age group. The differences in EBV-association by age group are highly significant (p=0.004 by Fisher's exact test). This pattern was evident for cases from the UK and Brazil when separate analyses were performed. Fourteen years of age is conventionally taken as the upper limit of the paediatric age group, but this age division does not necessarily reflect the biological distinction between the childhood and young adult disease patterns. A recent epidemiological study, presented at this workshop by Seymore Grufferman, suggests that risk factors associated with the development of HD also change around the age of 10 years.

Other investigators have analysed the association between EBV and HD in developing countries (Ambinder *et al.*, 1993; Chang *et al.*, 1993). These studies also found a higher proportion of EBV-positive disease, rates in paediatric patients in Honduras and Peru were even higher than that reported above for Brazil. The latter difference probably reflects improved socio-economic conditions in the Sao Paulo area of Brazil.

Higher proportions of EBV-associated cases have also been reported in adults in developing countries (Chang *et al.*, 1993). It is not clear at the present time, in the absence of good cancer registry figures and population-based studies, whether this reflects an absolute increase in EBV-associated disease or simply the absence of the young adult age incidence peak seen in developed countries (see below).

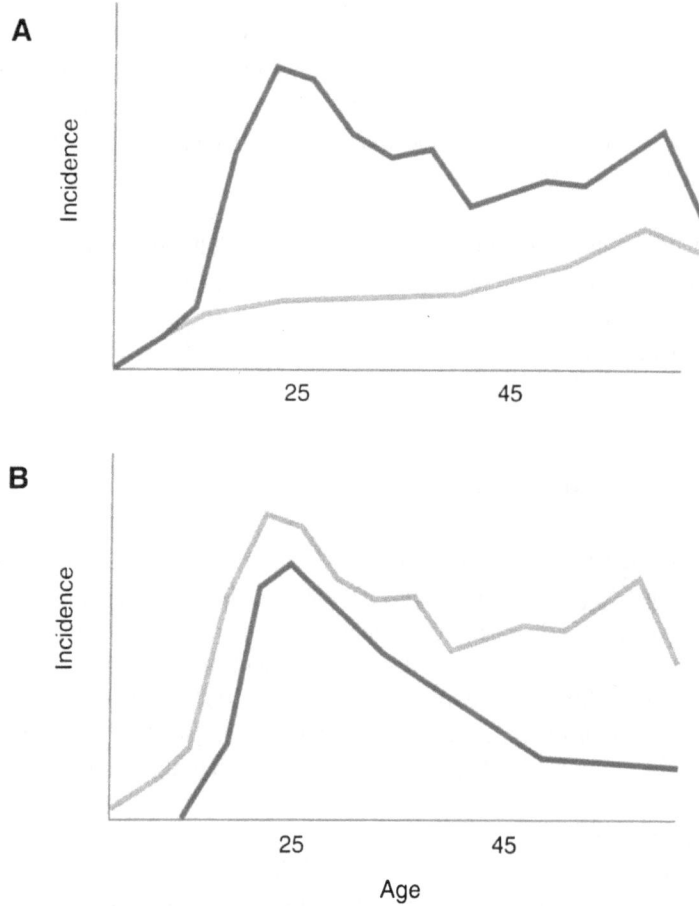

Fig. 3:   Speculation on the overall age distribution of EBV-associated and non-associated Hodgkin's disease cases. A. EBV-associated cases. Solid line, all Hodgkin's disease cases; shaded line, EBV-associated cases. B. Non-EBV-associated cases. Shaded line, all Hodgkin's disease cases; solid line, age distribution of cases which we speculate are related to another virus.

Conclusions

There are clearly differences in the proportions of EBV-associated HD cases in different age groups and in different histological subtypes. Our data show that differences by age group persist after adjustment for the effects of histological subtype. We believe that these results provide clear support for the hypothesis that HD is a heterogeneous condition with different aetiologies in different age groups. They do not support the polio model. We speculate that the age incidence of EBV-associated cases is as depicted in Fig. 3A. Although there is obviously some overlap between the disease patterns associated with the different age groups, we have identified two groups of cases in which the presence or absence of EBV is almost absolute; cases of classical (i.e. excluding HDLP) HD aged <10 years were virtually all EBV-associated whereas HDNS cases in the 15-24 year age range were almost exclusively EBV-negative.

It is not clear why some studies have not found this distribution of EBV-associated cases in relation to age. It is possible that case selection procedures, small case series and the use of mean ages may account for some differences. Despite these unexplained differences, a pattern does appear to be emerging. In age groups, geographical locales and immunodeficiency states which are associated with high rates of EBV-associated disease, there appears to be a lesser difference by histological subtype, i.e. a greater proportion of HDNS cases are EBV-positive (Ambinder et al., 1993; Chang et al., 1993; Gulley et al., 1994; Ambinder et al., 1992).

EBV-associated HD is found infrequently in young adult HDNS cases in developed countries. It is in this age group that there is most evidence for an infectious aetiology (see Alexander, this volume) and we therefore speculate that another virus is involved in the pathogenesis of this group of cases. We further speculate that the age incidence of HD related to this agent will be similar to that shown in Fig. 3B.

OTHER VIRUSES IN HODGKIN'S DISEASE

Epidemiological studies suggest that delayed exposure to a common virus may be important in the aetiology of young adult HD, the so-called late-host-response model (see Alexander, this volume). There are many reports describing seroepidemiological studies of HD but few molecular studies investigating involvement of known viruses; it is possible that negative results, although important, have not been documented. Experience gained from the investigation of the link between EBV and HD, emphasises the need for sensitive techniques and the examination of large case series (see Jarrett, this volume).

Human herpesvirus-6
Candidate agents are herpesviruses, other than EBV, since members of this family are common, have a widespread distribution and establish persistent infection. Much recent interest has focused on human herpesvirus-6 (HHV-6) which is a ubiquitous, lymphotropic virus first isolated in 1986 (Salahuddin et al., 1986). In all serological studies conducted to date HHV-6 antibody titres have been shown to be elevated in HD (Ablashi et al., 1988; Biberfeld et al., 1988; Clark et al., 1990; Torelli et al., 1991). In a case-control study including sera from 189 cases of HD, we showed that HD patients as a group have raised HHV-6 antibody levels. Furthermore, elevated titres were most marked in young adult cases, particularly those without siblings (Clark et al., 1990). Analysis of a separate data set suggests that non-EBV-associated cases are more likely to have high HHV-6 antibody levels than EBV-associated cases. These results suggest either that HHV-6 plays a role in the pathogenesis of HD, or that it is a marker for another risk factor.

Despite these data we have not detected HHV-6 genomic sequences in DNA samples from over 50 HD cases examined by Southern blot analysis (Jarrett *et al.*, 1988; Gledhill *et al.*, 1991). Similarly, Josephs *et al.* (1988) did not detect HHV-6 following the investigation of 8 cases. In contrast, in a PCR study Torelli and colleagues (1991) detected HHV-6 sequences in 3/25 cases and in 2 of these viral genomes were detectable by Southern blot analysis. There is no evidence, as yet, for HHV-6-infection of RS cells (Khan *et al.*, 1993), but there is a need for a sensitive *in situ* assay in order to eliminate a direct role for this virus. The involvement of HHV-6 in HD is described in greater detail by Levine *et al.* in a later chapter of this volume.

Antibodies to other viruses

The results of serological studies of the other human herpesviruses, *Herpes simplex* virus, *Varicella zoster* virus and cytomegalovirus (CMV) are less consistent, with some studies reporting elevated titres and others showing no significant differences between cases and controls (Catalano and Goldman, 1972; Henderson *et al.*, 1973; Langenhuysen *et al.*, 1974; Hesse *et al.*, 1977). We recently investigated serological responses to human herpesvirus-7 (HHV-7) in HD, however the cases did not have significantly different antibody titres from controls (unpublished results). No significant findings have emerged from studies of measles, rubella, adenovirus, parainfluenza virus, papovavirus and human T-cell leukaemia virus (Hesse *et al.*, 1977; Evans *et al.*, 1978; Evans and Gutensohn, 1984; our unpublished results).

Molecular studies

As mentioned above there are few documented molecular studies relating to the involvement of viruses, other than EBV, in HD. CMV probes have frequently been used as negative controls in Southern blot and *in situ* hybridisation studies investigating the role of EBV (Weiss *et al.*, 1987; Anagnostopoulos *et al.*, 1989; Coates *et al.*, 1991; Khan *et al.*, 1993). There is therefore good evidence that CMV does not play a direct role in HD. Similarly, Delsol *et al.* (1992) utilised human papillomavirus 11 and 18 probes as negative controls for *in situ* hybridisation studies. We recently analysed a series of >40 DNA samples from the affected lesions of HD for the presence of HHV-7 sequences, using Southern blot analysis and a panel of HHV-7 probes; no positive results were obtained (unpublished results with Zwi Berneman). Negative results were also obtained following the investigation of >20 cases using probes derived from adenovirus, lymphotropic papovavirus and SV40 (unpublished results).

SUMMARY

There is good evidence that a proportion, around 40%, of HD cases are associated with EBV. The distribution of EBV-associated is not random, strengthening the idea that EBV plays a role in disease pathogenesis in positive cases. The majority of studies have shown that cases of mixed cellularity disease are more likely to be EBV-associated than nodular sclerosis cases. Our data show a significant relationship between EBV and age at diagnosis. Paediatric cases, particularly those aged <10 years, and older adult cases are usually EBV-associated but young adult cases in the UK are rarely EBV-positive;

significant differences persist following adjustment for the effects of histological subtype. There is consistent evidence that the proportion of EBV-positive cases is higher in developing countries as compared to developed countries. Comparisons of EBV-association rates in children in developing countries with young adults in a developed country provide evidence which refutes the polio model of Hodgkin's disease, at least for the majority of cases. Overall, the data favour the multiple aetiology hypothesis of Hodgkin's disease which suggests that HD in different age groups has different aetiologies.

Epidemiological evidence for an infectious aetiology largely relates to cases in the young adult age incidence peak seen in developed countries. It is in this group of cases that there is least evidence for involvement of EBV; we therefore speculate that another virus is involved. HHV-6 antibody titres are elevated in HD cases compared to controls and titres are higher in young adult cases and in EBV-negative cases. At the present time there is no evidence for HHV-6 infection of RS cells and it remains to be seen whether this virus is simply a marker for another agent which plays a direct role. Evidence for involvement of other known viruses is currently lacking and we therefore suspect that a novel infectious agent is involved.

ACKNOWLEDGEMENTS

We should like to thank Freda Alexander for performing the statistical analysis and for helpful discussions. We also thank Alice Gallagher, Duncan Clark, June Freeland and Diane Gray for their contributions to the above data.

REFERENCES

Ablashi DV, Josephs SF, Hellman K, Nakamura S, Buchbinder A, Llana T, Lusso P, Kaplan M, Dahlberg J, Memon S, Imam F, Ablashi KL, Markham PD, Kramarsky B, Kreuger GRF, Biberfeld P, Salahuddin SZ (1988) Human B-lymphotropic virus (human herpesvirus-6). *J Virol Methods* **21**:29

Alexander FE, McKinney PA, Williams J, Ricketts TJ, Cartwright RA (1991) Epidemiological evidence for the "Two-disease hypothesis" in Hodgkin's disease. *Int J Epidemiol* **202**:354

Ambinder RF, Mann RB, Filipovich AH (1992) Association of EBV with Hodgkin's disease in patients with primary immunodeficiency and frequent detection of EBV in lymphoid tissue without neoplastic involvement in Wiskott-Aldrich syndrome: a survey of EBV in archival tissues from the immunodeficiency registry (ICR). *Vth International Symposium on Epstein-Barr Virus and Associated Diseases, Annecy, France* 76 (abstr)

Ambinder RF, Browning PJ, Lorenzana I, Leventhal BG, Cosenza H, Mann RB, MacMahon EM, Medina R, Cardona V, Grufferman S, Olshan A, Levin A, Petersen EA, Blattner W, Levine PH (1993) Epstein-Barr virus and childhood Hodgkin's disease in Honduras and the United States. *Blood* **81**:462

Anagnostopoulos I, Herbst H, Niedobitek G, Stein H (1989) Demonstration of monoclonal EBV genomes in Hodgkin's disease and KI-1-positive anaplastic large cell lymphoma by combined Southern blot and in situ hybridization. *Blood* **74**:810

Armstrong AA, Weiss LM, Gallagher A, Jones DB, Krajewski AS, Angus B, Brown G, Jack AS, Wilkins BS, Onions DE, Jarrett RF (1992) Criteria for the definition of Epstein-Barr virus association in Hodgkin's disease. *Leukemia* **6:**869

Armstrong AA, Alexander FE, Pinto Paes R, Morad NA, Gallagher A, Krajewski AS, Jones DB, Angus B, Adams J, Cartwright RA, Onions DE, Jarrett RF (1993) Association of Epstein Barr virus with paediatric Hodgkin's disease. *Am J Pathol* **142:**1683

Armstrong AA, Lennard A, Alexander FA, Angus BA, Proctor SJ, Onions DE, Jarrett RF (1994) Prognostic significance of Epstein-Barr virus association in Hodgkin's disease. *Eur J Cancer* **30:**1045

Biberfeld P, Petren A, Eklund A, Lindemalm C, Barkhem T, Ekman M, Ablashi D, Salahuddin Z (1988) Human herpesvirus-6 (HHV-6, HBLV) in sarcoidosis and lymphoproliferative disorders. *J Virol Methods* **21:**49

Catalano LW, Goldman JM (1972) Antibody to Herpesvirus hominis types 1 and 2 in patients with Hodgkin's disease and carcinoma of the nasopharynx. *Cancer* **29:**597

Chang KL, Albujar PF, Chen YY, Johnson RM, Weiss LM (1993) High prevalence of Epstein-Barr virus in the Reed-Sternberg cells of Hodgkin's disease occurring in Peru. *Blood* **81:**496

Clark DA, Alexander FE, McKinney P, Roberts BE, O'Brien C, Jarrett RF, Cartwright RA, Onions DE (1990) The seroepidemiology of human herpesvirus-6 (HHV-6) from a case control study of leukaemia and lymphoma. *Int J Cancer* **45:**829

Coates PJ, Slavin G, D'Ardenne AJ (1991) Persistence of Epstein-Barr virus in Reed-Sternberg cells throughout the course of Hodgkin's disease. *J Pathol* **164:**291

Correa P, O'Conor GT (1971) Epidemiologic patterns of Hodgkin's disease. *Int J Cancer* **8:**192

Delsol G, Brousset P, Chittal S, Rigal-Huguet F (1992) Correlation of the expression of Epstein-Barr virus latent membrane protein and in situ hybridization with biotinylated BamHI-W probes in Hodgkin's disease. *Am J Pathol* **140:**247

Evans AS, Carvalho RPS, Frost P, Jamra M, Pozzi DHB (1978) Epstein-Barr virus infections in Brazil. 2. Hodgkin's disease. *J Natl Cancer Inst* **61:**19

Evans AS, Gutensohn NM (1984) A population-based case control study of EBV and other viral antibodies among persons with Hodgkin's disease and their siblings. *Int J Cancer* **34:**149

Glaser SL, Swartz WG (1990) Time trends in Hodgkin's disease incidence. The role of diagnostic accuracy. *Cancer* **66:**2196

Gledhill S, Gallagher A, Jones DB, Krajewski AS, Alexander FE, Klee E, Wright DH, O'Brien C, Onions DE, Jarrett RF (1991) Viral involvement in Hodgkin's disease: detection of clonal type A Epstein-Barr viral genomes in tumour samples. *Br J Cancer* **64:**227

Gulley ML, Eagan PA, Quintanilla Martinez L, Picado AL, Smir BN, Childs C, Dunn CD, Craig FE, Williams JW, Jr., Banks PM (1994) Epstein-Barr virus DNA is abundant and monoclonal in the Reed-Sternberg cells of Hodgkin's disease: association with mixed cellularity subtype and Hispanic American ethnicity. *Blood* **83:**1595

Gutensohn NM (1982) Social class and age at diagnosis of Hodgkin's disease: new epidemiologic evidence for the 'two-disease hypothesis'. *Cancer Treat Rep* **66:**689

Henderson BE, Dworsky R, Menck H, Alena B, Henle W, Henle G, Terasaki P (1973) Case-control study of Hodgkin's disease. II. Herpesvirus group antibody titers and HL-A type. *J Natl Cancer Inst* **51:**1443

Herbst H, Steinbrecher E, Niedobitek G, Young LS, Brooks L, Muller LN, Stein H (1992) Distribution and phenotype of Epstein-Barr virus-harboring cells in Hodgkin's disease. *Blood* **80:**484

Hesse J, Levine PH, Ebbesen P, Connelly RR, Mordhorst CH (1977) A case control study on immunity to two Epstein-Barr virus-associated antigens, and to Herpes simplex virus and adenovirus in a population-based group of patients with Hodgkin's disease in Denmark, 1971-73. *Int J Cancer* **19:**49

Hummel M, Anagnostopoulos I, Dallenbach F, Korbjuhn P, Dimmler C, Stein H (1992) EBV infection patterns in Hodgkin's disease and normal lymphoid tissue: expression and cellular localization of EBV gene products. *Br J Haematol* **82**:689

Jarrett RF, Gledhill S, Qureshi F, Crae SH, Madhok R, Brown I, Evans I, Krajewski A, O'Brien CJ, Cartwright RA, Venables P, Onions DE (1988) Identification of human herpesvirus 6-specific DNA sequences in two patients with non-Hodgkin's lymphoma. *Leukemia* **2**:496

Jarrett RF, Gallagher A, Jones DB, Alexander FE, Krajewski AS, Kelsey A, Adams J, Angus B, Gledhill S, Wright DH, Cartwright RA, Onions DE (1991) Detection of Epstein-Barr virus genomes in Hodgkin's disease: relation to age. *J Clin Pathol* **44**:844

Josephs SF, Buchbinder A, Streicher HZ, Ablashi DV, Salahuddin SZ, Guo H, Wong-Staal F, Cossman J, Raffeld M, Sundeen J, Levine P, Biggar R, Krueger GRF, Fox RI, Gallo RC (1988) Detection of human B lymphotropic virus (human herpesvirus-6) sequences in B cell lymphoma tissues of three patients. *Leukemia* **2**:132

Khan G, Norton AJ, Slavin G (1993) Epstein-Barr virus in Hodgkin disease. Relation to age and subtype. *Cancer* **71**:3124

Langenhuysen MMAC, Cazemier T, Houwen B, Brouwers TM, Halie MR, The TH, Nieweg HO (1974) Antibodies to Epstein-Barr virus, cytomegalovirus, and Australia antigen in Hodgkin's disease. *Cancer* **34**:262

MacMahon B (1966) Epidemiology of Hodgkin's disease. *Cancer Res* **26**:1189

Mason DY, Banks PM, Chan J, Cleary ML, Delsol G, de Wolf Peeters C, Falini B, Gatter K, Grogan TM, Harris NL, et al (1994) Nodular lymphocyte predominance Hodgkin's disease. A distinct clinicopathological entity [editorial]. *Am J Surg Pathol* **18**:526

McKinney PA, Alexander FE, Ricketts TJ, Williams J, Cartwright RA (1989) A specialist leukaemia/lymphoma registry in the UK. Part 1: incidence and geographical distribution of Hodgkin's disease. *Br J Cancer* **60**:942

Murray PG, Young LS, Rowe M, Crocker J (1992) Immunohistochemical demonstration of the Epstein-Barr virus-encoded latent membrane protein in paraffin sections of Hodgkin's disease. *J Pathol* **166**:1

Pallesen G, Hamilton-Dutoit SJ, Rowe M, Young LS (1991) Expression of Epstein-Barr virus latent gene products in tumour cells of Hodgkin's disease. *Lancet* **337**:320

Parkin DH, Stiller CA, Draper GJ, Bieber CA, Terracini B, Young JL (1988) In *International incidence of childhood cancer* (Lyons: IARC Scientific Publications) p109

Salahuddin SZ, Ablashi DV, Markham PD, Josephs SF, Sturzenegger S, Kaplan M, Halligan G, Biberfeld P, Wong-Staal F, Kramarsky B, Gallo RC (1986) Isolation of a new virus, HBLV, in patients with lymphoproliferative disorders. *Science* **234**:596

Spitz MR, Sider JG, Johnson CC, Butler JJ, Pollack ES, Newell GR (1986) Ethnic patterns of Hodgkin's disease incidence among children and adolescents in the United States, 1973-82. *J Natl Cancer Inst* **76**:235

Torelli G, Marasca R, Luppi M, Selleri L, Ferrari S, Narni F, Mariano MT, Federico M, Ceccherini-Nelli L, Bendinelli M, Montagnani G, Montorsi M, Artusi T (1991) Human herpesvirus-6 in human lymphomas: identification of specific sequences in Hodgkin's lymphomas by polymerase chain reaction. *Blood* **77**:2251

Uhara H, Sato Y, Mukai K, Akao I, Matsuno Y, Furuya S, Hoshikawa T, Shimosato Y, Saida T (1990) Detection of Epstein-Barr virus DNA in Reed-Sternberg cells of Hodgkin's disease using the polymerase chain reaction and in situ hybridization. *Jpn J Cancer Res* **81**:272

Vestlev PM, Pallesen G, Sandvej K, Hamilton-Dutoit SJ, Bendtzen SM (1992) Prognosis of Hodgkin's disease is not influenced by Epstein-Barr virus latent membrane protein. *Int J Cancer* **51**:1

Weinreb M, Day PJR, Murray PG, Raafat F, Crocker J, Parkes SE, Coad NAG, Jones JT, Mann JR (1992) Epstein-Barr virus (EBV) and Hodgkin's disease in children: incidence of EBV latent membrane protein in malignant cells. *J Pathol* **168**:365

Weiss LM, Strickler JG, Warnke RA, Purtilo DT, Sklar J (1987) Epstein-Barr viral DNA in tissues of Hodgkin's disease. *Am J Pathol* **129**:86

Weiss LM, Chen Y, Liu X, Shibata D (1991) Epstein-Barr virus and Hodgkin's disease. A correlative in situ hybridization and polymerase chain reaction study. *Am J Pathol* **139**:1259

HODGKIN'S DISEASE IN HUMAN IMMUNODEFICIENCY VIRUS INFECTED INDIVIDUALS

Stephen J Hamilton-Dutoit

Laboratory of Immunopathology, Institute of Pathology,
Aarhus University Hospital, Aarhus, Denmark

HIV-ASSOCIATED HODGKIN'S DISEASE

The first cases of Hodgkin's disease (HD) arising in patients with, or at risk for, human immunodeficiency virus (HIV) infection were recognised at an early stage in the current HIV epidemic (Robert and Schneiderman, 1984; Ioachim *et al.,* 1984). Subsequently, several hundred cases of HIV-associated HD (HIV-HD) have been reported (for references see Rubio, 1994). Whether the overall incidence of HD is increased in HIV-infected individuals is still, however, controversial. Whereas the HIV-associated malignancies Kaposi's sarcoma, primary lymphoma of the brain, high grade non-Hodgkin's B-cell lymphoma, and invasive cervical carcinoma are accepted as acquired immunodeficiency syndrome (AIDS)-defining conditions, HIV-HD is not included in the current US Centers for Disease Control case definition of AIDS (Centers for Disease Control. 1985, 1987 and 1992).

Nevertheless, it is clear that the clinico-pathological features and natural history of HD in HIV-infected patients differ considerably from those normally found in cases in the non-infected population. Study of HIV-HD is of value, because it not only throws light on the possible mechanisms involved in the development of neoplasia in HIV infection, but also provides clues to the aetiology and pathogenesis of HD in the general population.

EPIDEMIOLOGY OF HIV-ASSOCIATED HODGKIN'S DISEASE

Although numerous individual case studies and small patient series have raised the possibility of an increased incidence of HD in HIV-seropositive patients, definitive proof of this is not available (Biggar and Rabkin, 1992). Studies using data from population-based cancer registries have found no consistent excess incidence of tumours other than Kaposi's sarcoma and non-Hodgkin's lymphoma among single young American males, a surrogate group for homosexual men (Biggar *et al.*, 1989; Rabkin *et al.*, 1991).

There are, however, difficulties involved in using such registries to identify HIV-related neoplasms, since the age-specific incidence rates for HD are high in those age groups most at risk for HIV infection. Thus, cases of HD would be expected to occur in the virus-infected population in the absence of a causal relationship between HIV and HD. Moreover, even if such a relationship were to exist, the number of expected excess tumours would be small. These two factors mean that large population groups must be followed for long periods in order to establish an eventual slight excess cancer risk. In addition, an increased incidence of HD in risk groups other than homosexual men may be overlooked because of the difficulties involved in finding suitable surrogates for these groups.

An alternative approach involves studying the occurrence of cancer in defined groups of HIV-infected patients. Two recent studies have found an increased incidence of HD in cohorts of homosexual men from the San Francisco area. Hessol *et al.* (1992) reported a five-fold excess of HD in HIV-positive compared with HIV-negative homosexual men in the San Francisco City Clinic Cohort, although these data were based upon a total of only eight cases of HD. In a larger study, Reynolds *et al.* (1993) linked data from the California Tumor Registry and the San Francisco AIDS Surveillance Registry and found an elevated incidence of HD that appeared to be increasing with time, with an 18-fold excess for the latest period assessed (1986-1987). These authors also reported that HIV-HD appeared to occur more frequently among non-whites than did other HIV-associated malignancies. Although this finding was based upon a relatively small number of tumours, it raises the interesting possibility that there may be racial and/or genetic differences that predispose to the development of HD in HIV-infection.

These cohort studies can be criticised on the grounds that they have overlapping data sets (their results are, therefore, not independent), that their conclusions are based on relatively low numbers of HD cases, and that they did not undertake systematic histological review in order to exclude misclassification of non-Hodgkin's lymphomas as a confounding factor (Biggar and Rabkin, 1992). Nevertheless, they do suggest that there is a real, and possibly increasing, excess incidence of HD among American homosexual men.

Studies from northern Europe have generally reflected the picture initially found in North America, with HD being reported in low numbers among homosexual males (Pedersen *et al.*, 1991). In contrast, a quite different pattern has emerged in parts of southern Europe, particularly in Spain (Serrano *et al.*, 1990; Rubio, 1994) and Italy (Tirelli *et al.*, 1992), and to a lesser extent in France (Andrieu *et al.*, 1992). In these countries, there is an apparent increase in the incidence of HD in HIV-seropositive individuals. This is suggested primarily by a reduction in the ratio of non-Hodgkin's lymphoma to HD (NHL:HD) reported among HIV-associated malignancies. Thus, whereas NHL:HD rates of between 5:1 and 30:1 are found in representative American series (Ioachim et al., 1985; Di Carlo et al., 1986; Lowenthal et al., 1988; Knowles et al., 1988), corresponding rates of 3:1 or less are found in France and Italy (Carbone et al., 1991; Andrieu et al., 1993), and a rate of under 2:1 is reported from Spain (Rubio, 1994).

It has been suggested that the differences between NHL:HD ratios in southern Europe and in the USA result from a relatively greater increased incidence of non-Hodgkin's lymphomas in American patients (Hessol et al., 1992), rather than from an increased incidence of HD among European HIV-infection risk groups. The evidence for this is, however, poor. Final proof that HD is increased in incidence in HIV-infected patients in southern Europe (both absolutely, and relative to HIV-infected populations in the USA) will depend on the results from cohort studies. However, the large numbers of HIV-HD cases now reported in Europe appear to make this likely.

A possible clue to explain the observed geographical variation in NHL:HD rates may be found in the relative distribution of the different HIV-infection risk groups in these regions. Thus, intravenous drug abuse is the predominant risk behaviour for HIV-infection in Spain (Rubio, 1994) and Italy (Tirelli et al., 1992) and is an important risk factor in France (Andrieu et al., 1993). In contrast, male homosexuals continue to make up the main HIV-risk group in both North America and northern Europe. Interestingly, the NHL:HD ratio is significantly lower in intravenous drug abusers compared with homosexuals in Spanish patients (Rubio, 1994), and similar findings are reported from France (Andrieu et al., 1993) and Italy (Carbone et al., 1991).

These data provide at least circumstantial evidence that the site and route of HIV infection may be important factors in deciding the types of malignancy found in HIV-infected patients. Although this raises the possibility that blood-borne agents transmitted during intravenous drug abuse may be involved in the pathogenesis of HIV-HD, alternative explanations such as differences in the genetic make-up of the infected population, or geographical variation in risk factors may also be involved.

Studies from outside southern Europe have not shown so clear an association between intravenous drug abuse and development of HIV-HD. Ahmed et al. (1987) found an excess of HD cases among New York State prisoners who were intravenous drug addicts, compared with the general population (Ahmed et al., 1987). However, they also found a marked increase in the number of non-Hodgkin's lymphomas developing in the same population, with an NHL:HD ratio of 4:1. Similarly, cases of HIV-HD in the United States are more likely to occur in homosexual men than they are in intravenous drug abusers. Between 15% and 30% of U.S. HIV-HD cases occurred in intravenous drug addicts (Lowenthal et al., 1987; Knowles et al., 1988; Ree et al., 1990) compared with 80% or more in patient series from Spain and Italy (Serrano et al., 1990; Tirelli et al., 1992). Since homosexuals make up by far the largest HIV risk group in North America, this may obscure any underlying association of HIV-HD with intravenous drug abuse in this region.

However, Reynolds *et al.* (1993) found no evidence that cancer type among intravenous drug abusers differed from that among other risk groups in the San Francisco Bay area.

Some evidence that a parenteral route of HIV-infection does not in itself favour the development of HD rather than non-Hodgkin's lymphoma comes from studies of HIV-seropositive haemophiliacs. Ragni *et al.* (1993) studied the incidence of malignancies in 3041 U.S. haemophiliacs followed between 1978 and 1989, of whom 1295 (56.6%) were HIV-seropositive. Fourteen developed non-Hodgkin's lymphomas, a 36.5 fold greater risk in the HIV-seropositive compared with HIV-seronegative patients. In contrast, no cases of HD were reported in the HIV-seropositive group. In a second study, Rabkin *et al.* (1992) studied 1701 U.S. haemophiliacs, of whom 1065 (63%) were HIV-seropositive, and found an up to 38-fold relative increase in non-Hodgkin's lymphoma compared with the HIV-sero-negative population. Although, in this series, cases of HD were found among the HIV-seropositive cohort with a 6.6-fold increased risk, this was not statistically significant. Thus, American HIV-infected haemophiliacs appear to develop similar types of lymphoma to those reported in homosexual risk groups. It will be interesting to see whether this holds true for HIV-infected haemophiliacs from southern Europe.

CLINICAL FEATURES OF HIV-ASSOCIATED HODGKIN'S DISEASE

In large series of HIV-HD, almost all reported cases have occurred in men, with a median age range of 27-38 years (Lowenthal *et al*, 1987; Knowles *et al*, 1988; Ree *et al*, 1990; Serrano *et al*, 1990; Tirelli *et al*, 1992; Andrieu *et al.*, 1993; Rubio, 1994), these features reflecting the composition of the predominant HIV-risk groups - male homosexuals and intravenous drug abusers. In addition, however, these series suggest a particular clinical profile in HIV-infected compared with non-infected HD patients. Thus, the great majority of the former have B symptoms, have advanced stage (III or IV) disease at presentation (often with bone marrow involvement), and have clinically aggressive tumours which show a poor therapeutic response (often complicated by opportunistic infections), with a relatively shorter survival time. Another feature noted by some investigators is the relative infrequency of mediastinal disease compared with sporadic HD (Andrieu *et al.*, 1993).

Interestingly, most patients with HIV-HD have only moderately decreased CD4 cell counts, and only very few have had a previous AIDS-defining illness (Serrano *et al*, 1990; Tirelli *et al*, 1992; Andrieu *et al.*, 1993; Reynolds *et al.*, 1993; Rubio, 1994). This is in sharp contrast to the findings in HIV-infected patients with large cell non-Hodgkin's lymphomas, but resembles the pattern reported with AIDS-associated Burkitt's lymphoma (Pedersen *et al.*, 1991), and suggests that HIV-HD development is not directly related to the severity of the HIV-associated immunodeficiency. On the other hand, the development of HD in an HIV-infected individual is a poor prognostic sign, patients showing a subsequent high rate of progression to AIDS (Serrano *et al*, 1990; Tirelli *et al*, 1992; Andrieu *et al.*, 1993; Rubio, 1994) and a clinical course closely parallel to that seen in AIDS patients (Ames *et al.*, 1991). In particular, HIV-HD patients have a high incidence of opportunistic infections, some of which may presumably be attributed to the additional immunodeficiency associated with the use of antineoplastic treatment.

HISTOLOGICAL FEATURES

Reports from both Europe and the United Sates have found a predominance of mixed cellularity and lymphocyte depletion subtypes among cases of HIV-HD, a reversal of the situation found in sporadic HD (Serrano *et al*, 1990; Ree *et al.*, 1991; Tirelli *et al*, 1992; Andrieu *et al.*, 1993; Hessol *et al.*, 1993; Rubio, 1994). A few of these cases (especially in the lymphocyte depletion subtype) may be attributed to misdiagnosis of non-Hodgkin's lymphoma. However, this cannot account for the majority of cases, and there appears to be a genuine shift towards more histologically aggressive subtypes in HIV-HD. An additional observation, emphasised by one report, is the presence of increased numbers of non-lymphoid cells (particularly fibrohistiocytoid stromal cells) in HIV-HD (Ree *et al.*, 1990).

Whether the morphological changes reported in HIV-HD can be attributed entirely to an altered host response to the neoplastic cells in immunodeficient HIV-infected patients, or whether they reflect a more fundamental difference in the pathogenesis of HD in HIV-infected compared with non-infected patients remains to be established.

ASSOCIATION WITH EPSTEIN-BARR VIRUS

There is now considerable evidence of an association between Epstein-Barr virus (EBV) infection and HD (reviewed in Pallesen *et al.*, (1993) and elsewhere in this volume). One of the earliest studies to demonstrate that EBV was actually present in the Hodgkin and Reed-Sternberg (H-RS) cells in viral-positive HD cases was performed on HIV-infected patients (Uccini *et al.*, 1989 and 1990). In their initial study, these authors used Southern blotting analysis to detect EBV genomes in 10 of 12 HIV-HD cases, and confirmed the presence of EBV in H-RS cells in 4 of 7 cases using DNA *in situ* hybridisation. Furthermore, in 4 of 4 cases, analysis of the configuration of the EBV terminal region indicated that a clonal viral infection was present (Uccini *et al.*, 1989).

Subsequently, more sensitive *in situ* hybridisation methods for detecting EBV in tissue sections have been developed based upon the demonstration of EBV encoded small RNAs (EBERs) (Hamilton-Dutoit and Pallesen, 1994). Using this technique to examine 11 cases of HIV-HD, our group has demonstrated the consistent presence of EBV genomes in the H-RS cells (Hamilton Dutoit *et al.*, 1993). Similar results have been reported by Herndier *et al.*, (1993). The results of EBER *in situ* hybridisation confirm an earlier immunohistological study on 16 HIV-HD cases that found consistent expression of the EBV encoded oncogene latent membrane protein-1 in the great majority of H-RS cells (Audouin *et al.*, 1992). These studies show that EBV is almost invariably present in H-RS cells in HIV-HD (thus suggesting that infection occurs at an early stage of tumour development), and that the virus is transcriptionally active, expressing viral oncogene products that may play an important role in the pathogenesis of the neoplasm.

EBV-positive HIV-HD cases show a so called latency II (Kerr *et al.*, 1992) pattern of EBV gene expression (Boiocchi *et al.*, 1993) characterised by the absence of EBNA-2, similar to that found in sporadic HD (Pallesen *et al.*, 1993). There is no evidence of replicative EBV gene expression (Hamilton-Dutoit, unpublished observation).

At least two groups have analysed the EBV subtype (as defined by the organisation of EBNA-2 gene locus) found in HIV-HD (Boyle *et al.*, 1992; De Re *et al.*, 1993), the results from these few cases indicating an approximate equal distribution of EBV types 1 and 2. These findings are similar to those reported in EBV-positive AIDS-related non-Hodgkin's lymphomas (Boyle *et al.*, 1991), but contrast with the situation found in cases of sporadic EBV-positive HD which show almost exclusively type-1 EBV infection (Jarrett *et al.*, 1991). Whether this merely reflects the change in the proportion of EBV type-1 and type-2 infections in immunocompromised individuals compared with the normal population (Boyle *et al.*, 1991), or whether there is an additional increase in the tumourigenicity of the type-2 virus in immunodeficiency states is not yet clear.

Although the overall number of cases of HIV-HD analysed for EBV remains relatively small, the data described above suggest that infection with the virus is a fundamental characteristic of these lymphomas. The almost invariable association of EBV with HIV-HD contrasts with the findings in AIDS-related non-Hodgkin's lymphomas in which the association with EBV varies with lymphoma type. Thus, while primary CNS AIDS lymphomas show a similarly high rate of EBV infection, the virus is found in 77% of systemic large cell lymphomas and in only 34% of Burkitt-type lymphomas (Hamilton-Dutoit *et al.*, 1993). Similarly, the association with EBV is much clearer in HIV-HD than it is in cases of sporadic HD occurring in developed countries, only about one-half of which carry EBV in their H-RS cells (Pallesen *et al.*, 1993).

SIMILARITIES BETWEEN HODGKIN'S DISEASE IN HIV-INFECTED PATIENTS AND OTHER SPECIAL EPIDEMIOLOGICAL GROUPS

There are interesting parallels between the type of HD found in association with HIV-infection and that occurring in other special epidemiological groups. For example, several studies have shown that HD in developing countries is clinically more aggressive, presents with more advanced stage disease, shows a higher rate of bone marrow involvement and a lower rate of mediastinal tumour, has an increased incidence of mixed cellularity subtype, and is more likely to be associated with EBV infection, compared with cases occurring in more developed countries (Correa and O'Conor, 1971; Cohen *et al.*, 1980; Levy, 1988; Glaser, 1990; Riyat, 1992; Ramadas *et al.*, 1994); all these features are also characteristic of HIV-HD.

Similarly, HD occurring in young children (especially under 10 years of age) is frequently associated with EBV infection (Armstrong *et al.*, 1993) suggesting that it may represent a separate disease entity to sporadic HD in adults.

Finally, there is evidence for geographical and ethnic variation in the association between EBV and HD with, for example, a relatively increased frequency of viral infection in HD cases in Hispanic groups (Ambinder *et al.*, 1993; Chang *et al.*, 1993; Gulley *et al.*, 1994).

Again, there may be parallels with the apparent variation in HD incidence described among different HIV-seropositive ethnic populations.

As yet, there is only circumstantial evidence linking the type of HD found in HIV-infected patients and that described in other special epidemiological groups. However, these similarities raise the possibility that common geographical, genetic or environmental (infective?) co-factors may be present in these various populations, and suggest possible fruitful areas for future research.

CONCLUSION

There is increasing evidence of an association between HIV-infection and the development of HD. However, this association is much weaker than that described with non-Hodgkin's lymphoma, and the development of HD in an HIV-infected patient is not presently regarded as an AIDS-defining disease. A much clearer pattern is seen in certain geographical regions such as southern Europe, where cases of HIV-HD are numerous, and the NHL:HD ratio in HIV-infected individuals is relatively low. This geographical variation may reflect the greater importance of intravenous drug abuse as a risk factor for HIV infection in this area.

Regardless of whether the incidence of HD is or is not increased in HIV-infection, it is clear that those cases developing in HIV-infected patients show clinicopathological features and biological behaviour distinct from sporadic HD. Furthermore, HIV-HD usually occurs in patients without previous AIDS-defining events, but is a poor prognostic sign predicting the onset of AIDS in many cases. These factors alone suggest that the development of HD should be considered as an AIDS-defining illness in HIV-infected individuals.

The pathogenesis of HIV-HD remains unclear. There is little evidence that immunodeficiency *per se* predisposes to HD development (for example in organ transplant recipients (Penn, 1984)), although occasional cases are reported in patients with congenital immunodeficiency (Frizzera *et al.,* 1980). Thus, other pathogenetic factors must be sought. Cases of HIV-HD are characterised by clinically aggressive disease, advanced stage, an excess of mixed cellularity and lymphocyte depleted subtype cases. and a consistent association with EBV infection. This may partly reflect a shift in the features of HD caused by the imposition of added immunodeficiency. However, this seems an inadequate explanation, particularly since patients with HIV-HD are not usually severely immunodeficient, and sporadic HD is often associated with a degree of impaired immunity. An alternative explanation is that HIV (or a co-factor spread with HIV) preferentially predisposes to the development of a type of clinically aggressive, EBV-associated HD. Thus, this type of tumour would account for the increased HD incidence in association with HIV infection, the number of cases of EBV-negative, nodular sclerosing subtype remaining at levels comparable to those found in the general population.

A similar process may account for the distinctive forms of HD reported in developing countries, in certain ethnic groups, in the very young, and possibly in the elderly. The presence of common genetic, geographic, or environmental factors in HIV-infected populations, and in these special epidemiological groups may define one particular pathogenetic type of HD. The definition of this type, and the identification of the factors involved in its development, would be important steps towards a better understanding of the pathogenesis of HD as a whole.

REFERENCES

Ahmed T, Wormser GP, Stahl RE, Mamtani R, Cimino J, Glasser M, Mittelamn A, Friedland M, Arlin Z (1987) Malignant lymphomas in a population at risk for acquired immune deficiency syndrome. *Cancer* **60**:719

Ambinder RF, Browning PJ, Lorenzana I, Leventhal BG, Cosenza H, Mann RB, MacMahon EM, Medina R, Cardona V, Grufferman S, Olshan A, Levin A, Petersen EA, Blattner W, Levine PH (1993) Epstein-Barr virus and childhood Hodgkin's disease in Honduras and the United States. *Blood* **81**:462

Ames ED, Metroka CE, Goldberg AF (1993) Hodgkin's disease and HIV. *Ann Intern Med* **118**:313

Andrieu JM, Roithman S, Tournai JM, Levy R, Desablens B, le Maignan C, Gastaut JA, Brice P, Raphael M, Taillan B (1989) Hodgkin's disease during HIV1 infection: the French registry experience. *Ann Oncol* **4**:635

Armstrong AA, Alexander FE, Pinto Paes R, Morad NA, Gallagher A, Krajewski AS, Jones DB, Angus B, Adams J, Cartwright RA, Onions DE, Jarrett RF (1993) Association of Epstein-Barr virus with pediatric Hodgkin's disease. *Am J Pathol* **142**:1683

Audouin J, Diebold J, Pallesen G (1992) Frequent expression of Epstein-Barr virus latent membrane protein-1 in tumour cells of Hodgkin's disease in HIV-positive patients. *J Pathol* **167**:381

Biggar RJ, Rabkin CS (1992) The epidemiology of acquired immunodeficiency syndrome-related lymphomas. *Curr Opin Oncol* **4**:883

Boyle MJ, Sewell WA, Sculley T, Apolloni A, Turner JJ, Swanson CE, Penny R, Cooper DA (1991) Subtypes of Epstein-Barr virus in human immunodeficiency virus-associated non-Hodgkin's lymphoma. *Blood* **78**:3004

Boyle MJ, Vasak E, Tschuchnigg M, Turner JJ, Sculley T, Penny R, Cooper DA, Tindall B, Sewell WA (1993) Subtypes of Epstein-Barr virus (EBV) in Hodgkin's disease: association between B-type EBV and immunocompromise. *Blood* **81**:468

Carbone A, Tirelli U, Vaccher E, Volpe R, Gloghini A, Bertola G, De Re V, Rossi C, Boiocchi M, Monfardini S (1992) A clinicopathologic study of lymphoid neoplasia associated with human immunodeficiency virus infection in Italy. *Cancer* **68**:842

Centers for Disease Control (1985) Revision of the case definition of acquired immunodeficiency syndrome for national reporting - United States. *MMWR* **34**:373

Centers for Disease Control (1987) Revision of the CDC surveillance case definition for acquired immunodeficiency syndrome. *MMWR* **36 (Suppl 1S)**:1S

Centers for Disease Control (1992) Revised classification system for HIV infection and expanded surveillance case definition for AIDS among adolescents and adults. *MMWR* **41**:1

Chang KL, Albujar PF, Chen YY, Johnson RM, Weiss LM (1993) High prevalence of Epstein-Barr virus in the Reed-Sternberg cells of Hodgkin's disease occurring in Peru. *Blood* **81**:496

Cohen C, Hamilton DG (1980) Epidemiologic and histologic patterns in Hodgkin's disease. *Cancer* **46**:186

Correa P, O'Conor GT (1971) Epidemiologic patterns in Hodgkin's disease. *Int J Cancer* **8**:192

De Re V, Boiocchi M, de Vita S, Dolcetti R, Gloghini A, Uccini S, Baroni C, Scarpa A, Cattoretti G, Carbone A (1993) Subtypes of Epstein-Barr virus in HIV-1 associated and HIV-1 unrelated Hodgkin's disease cases. *Blood* **81**:468

Di Carlo EF, Amberson JB, Metroka CE, Ballard P, Morre A, Mouradian JA (1986) Malignant lymphomas and the acquired immunodeficiency syndrome. *Arch Pathol Lab Med* **110**:1012

Glaser SL (1991) Hodgkin's disease in black populations: a review of the epidemiologic literature. *Semin Oncol* **17**:643

Gulley ML, Eagan PA, Quintanilla-Martinez L, Picado AL, Smir BN, Childs C, Dunn CD, Craig FE, Williams JW, Banks P (1994) Epstein-Barr virus DNA is abundant and monoclonal in the Reed-Sternberg cells of Hodgkin's disease: association with mixed cellularity subtype and Hispanic American ethnicity. *Blood* **83**:1595

Hamilton-Dutoit SJ, Audouin J, Diebold J, Lisse I, Pedersen C, Oksenhendler E, Pallesen G (1993) *In situ* demonstration of Epstein-Barr virus small RNAs (EBER 1) in acquired immunodeficiency syndrome-related lymphomas: correlation with tumour morphology and primary site. *Blood* **82**:619

Hamilton-Dutoit SJ, Pallesen G (1994) Detection of Epstein-Barr virus small RNAs in routine paraffin sections using non-isotopic RNA/RNA *in situ* hybridisation. *Histopathol* **25**:101

Herndier BG, Sanchez HC, Chang KL, Chen YY, Weiss LM (1993) High prevalence of Epstein-Barr virus in the Reed-Sternberg cells of HIV-associated Hodgkin's disease. *Am J Pathol* **142**:1073

Hessol NA, Katz MH, Liu JY, Buchbinder SP, Rubino CJ, Holmberg SD (1992) Increased incidence of Hodgkin's disease in homosexual men with HIV infection. *Ann Intern Med* **117**:309

Hessol NA, Katz MH, Buchbinder SP (1993) Hodgkin's disease and HIV. *Ann Intern Med* **118**:313

Ioachim HL, Cooper MC, Hellman GC (1984) Hodgkin's disease and the acquired immunodeficiency syndrome. *Ann Intern Med* **101**:876

Ioachim HL, Cooper MC, Hellman GC (1985) Lymphomas in men at high risk for acquired immunodeficiency syndrome (AIDS). *Cancer* **56**:2831

Jarrett RF, Gallagher A, Jones DB, Alexander FE, Krajewski AS, Kelsey A, Adams J, Angus B, Gledhill S, Wright DH, Cartwright RA, Onions DE (1991) Detection of Epstein-Barr virus genomes in Hodgkin's disease: relation to age. *J Clin Pathol* **44**:844

Knowles DM, Chamulak GA, Subar M, Burke JS, Dugan M, Wernz J, Slywotzky C, Pelicci PG, Dalla-Favera R, Raphael B (1988) Lymphoid neoplasia associated with the acquired immunodeficiency syndrome (AIDS). *Ann Int* Med **108**:744

Kerr BM, Lear AL, Rowe M, Croom-Carter D, Young LS, Rookes SM, Gallimore PH, Rickinson AB (1992) Three transcriptionally distinct forms of Epstein-Barr virus latency in somatic cell hybrids: cell phenotype dependence of virus promoter usage. *Virology* **187**:189

Levy LM (1988) Hodgkin's disease in black Zimbabweans. *Cancer* **61**:189

Lowenthal DA, Straus DJ, Campbell SW, Gold JWM, Clarkson BD, Koziner B (1988) AIDS-related lymphoid neoplasia. The Memorial Hospital experience. *Cancer* **61**:2325

Pallesen G, Hamilton-Dutoit SJ, Zhou X (1993) The association of Epstein-Barr virus (EBV) with T cell lymphoproliferations and Hodgkin's disease: two new developments in the EBV field. *Adv Cancer Res* **62**:179

Pedersen C, Gerstoft J, Lundgren JD, Skinhoj P, Bottschauw J, Geisler C, Hamilton-Dutoit SJ, Thorsen S, Lisse I, Ralfkiaer E, Pallesen G (1991) HIV-associated lymphoma: histopathology and association with Epstein-Barr virus genome related to clinical, immunologic, and prognostic features. *Eur J Cancer* **27**:1416

Penn I (1984) Allograft transplant cancer registry. In *Immune Deficiency and Cancer*. Purtilo DT, ed (New York: Plenum) p281

Rabkin CS, Biggar RJ, Horm JW (1991) Increasing incidence of cancers associated with the human immunodeficiency virus epidemic. *Int J Cancer* **47**:692

Rabkin CS, Hilgartner MW, Hedberg KW, Aledort LM, Hatzakis A, Eichinger S, Eyster ME, White GC, Kessler C, Lederman MM, Moerloose P, Bray G, Cohen A, Andes WA, Manco-Johnson M, Schramm W, Kroner B, Blattner W, Goedert J (1992) Incidence of lymphomas and other cancers in HIV-infected and HIV-uninfected patients with hemophilia. *JAMA* **267**:1090

Ragni MV, Belle SH, Jaffe RA, Duerstein SL, Bass DC, McMillan CW, Lovrien EW, Aledort LM, Kisker CT, Stabler SP, Hoots WK, Hilgartner MW, Cox-Gill J, Buchanan GR, Sanders N, Brettler DB, Barron LE, Goldsmith JC, Ewenstein B, Smith KJ, Green D, Addiego JE, Kingsley LA (1993) Acquired immunodeficiency syndrome associated non-Hodgkin's lymphomas and other malignancies in patients with hemophilia. *Blood* **81**:1889

Ramadas K, Sankaranarayanan R, Nair MK, Nair B, Padmanabhan TK (1994) Adult Hodgkin's disease in Kerala. *Cancer* **73**:2213

Ree HJ, Strauchen JA, Khan AA, Gold J, Crowley JP, Kahn H, Zalusky R (1990) Human immunodeficiency virus-associated Hodgkin's disease. *Cancer* **67**:1614

Reynolds P, Saunders LD, Layefsky ME, Lemp GF (1993) The spectrum of acquired immunodeficiency syndrome (AIDS)-associated malignancies in San Francisco, 1980-1987. *Am J Epidemiol* **137**:19

Riyat MS (1992) Hodgkin's disease in Kenya. *Cancer* **69**:1047

Robert NJ, Schneiderman H (1984) Hodgkin's disease and the acquired immunodeficiency syndrome. *Ann Intern Med* **101**:142

Rubio R (1994) Hodgkin's disease associated with human immunodeficiency virus infection. *Cancer* **73**:2400

Serraino D, Carbone A, Franceschi S, Tirelli U (1993) Increased frequency of lymphocyte depletion and mixed cellularity subtypes of Hodgkin's disease in HIV-infected patients. *Eur J Cancer* **29A**:1948

Serrano M, Bellas C, Campo E, Ribera J, Martin C, Rubio R, Ruiz C, Ocana I, Buzon L, Yebra M, Font M, Martinez MA (1990) Hodgkin's disease in patients with antibodies to human immunodeficiency virus. *Cancer* **65**:2248

Tirelli U, Errante D, Vaccher E, Repetto L, Rizzardini G, Spina M, Gastaldi R, Bertola G, Serraino D, Carbone A (1992) Hodgkin's disease in 92 patients with HIV infection: the Italian experience. *Ann Oncol* **3 (Suppl 4)**:69

Uccini S, Monardo F, Ruco LP, Baroni CD, Faggioni A, Agliano AM, Gradilone A, Manzari V, Vago L, Costanzi G, Carbone A, Boiocchi M, De Re V (1989) High frequency of Epstein-Barr virus genome in HIV-positive patients with Hodgkin's disease. *Lancet* i:1458

Uccini S, Monardo F, Stoppacciaro A, Gradilone A, Agliano AM, Faggioni A, Manzari V, Vago L, Costanzi G, Ruco LP, Baroni CD (1990) High frequency of Epstein-Barr virus genome detection in Hodgkin's disease of HIV-positive patients. *Int J Cancer* **46**:581

HODGKIN'S DISEASE AND HUMAN HERPESVIRUS-6: A MODEL FOR STUDIES OF NEW AETIOLOGICAL AGENTS

Paul H Levine,[1] Mark Manak,[2] Linda Jagodzinski[2]

[1]Viral Epidemiology Branch, National Cancer Institute, Bethesda, Md. 20892;
[2]Biotech Research Laboratories, Rockville, Md. 20707, USA

INTRODUCTION

Although clinical and epidemiological evidence support the role of infectious agents in the aetiology of Hodgkin's disease (HD), it has not been possible to identify a single pathogen that fits with the epidemiology of this disease. Epstein-Barr virus (EBV) has been linked to HD by a number of serological (Levine *et al.*, 1970; Johannson *et al.*, 1970; Levine *et al.*, 1971; Hesse *et al.*, 1977; Mueller *et al.*, 1989) and virological (Weiss *et al.*, 1989; Bignon *et al.*, 1990; Jarrett *et al.*, 1991; Pallesen *et al.*, 1991; Ambinder *et al.*, 1993) studies with the strongest EBV-association identified in children and older adults (Jarrett *et al.*, 1991; Ambinder *et al.*, 1993), and this virus remains a serious aetiological candidate for a significant proportion of HD cases. Human herpesvirus-6 (HHV-6), first described in 1986 (Salahuddin *et al.*, 1986), has also been linked to HD by serological (Ablashi *et al.*, 1988; Biberfeld *et al.*, 1988; Clark *et al.*, 1990; Iyengar *et al.*, 1991; Levine *et al.*, 1992a) and virological (Torelli *et al.*, 1991; Gompels *et al.*, 1993) studies. This report will summarise the data suggesting an aetiological association, review relevant studies to place these findings in perspective, and discuss possible approaches to the identification of other candidate aetiological agents based on the studies of these two herpesviruses.

SEROLOGICAL STUDIES

In 1988, the association of HHV-6 with HD was first suggested by Ablashi *et al.* and Biberfeld *et al.* using available samples from patients with HD, a variety of other malignant and non-malignant diseases, and healthy individuals. Subsequently, Clark *et al.* (1990) studied 98 HD patients, as part of a case-control study of 477 patients with haematological malignancies, and noted a higher proportion of HD cases with elevated anti-HHV-6 antibody titres compared to controls (11% had antibody titres greater than 1:640 compared to 2% in the matched controls). A serological association between HD and HHV-6 was strengthened in collaborative studies with Dr. Gary Pearson and his colleagues at Georgetown University using monoclonal antibodies directed against an HHV-6 early antigen (Iyengar *et al.*, 1991). Using sera from healthy donors and from patients with a variety of diseases, the sera being selected by previously performed IFA tests as having elevated VCA antibody titres, 60% of HD sera had elevated antibodies to the immunoaffinity purified HHV-6 early antigen in comparison to 30% of the healthy individuals, 29% of the patients with heterophile negative infectious mononucleosis, and 9.5% of patients with chronic fatigue syndrome. These three groups had similar mean titres to the immunoaffinity purified HHV-6 late antigen (African Burkitt's lymphoma patients in this study had the highest antibody levels against both early and late HHV-6 antigens).

All of the above studies used samples from treated as well as untreated HD patients and since our early studies (Levine *et al.*, 1970 and 1971) had shown that EBV antibodies were higher in treated than untreated HD patients, we evaluated serum samples from a population-based group of Danish HD patients which distinguished between pre- and post-treatment samples (Levine *et al.*, 1992a). We confirmed the elevation of anti-HHV-6 antibody titres using two assays, one an ELISA assay utilising purified viral antigen and the other an immunofluorescent assay using HHV-6 infected HSB-2 cells, which apparently contain a number of virus-associated antigens. In this study, we found no significant difference between the untreated cases and our controls, but there were marked changes associated with treatment and clinical course. In the patients who did not relapse, antibodies steadily declined over time although there was a transient initial post-therapy increase observed with the IFA test. In the relapsing patients, however, a steady increase was observed in the IFA assay, and a delayed rise was also noted in the ELISA assay (Fig. 1).

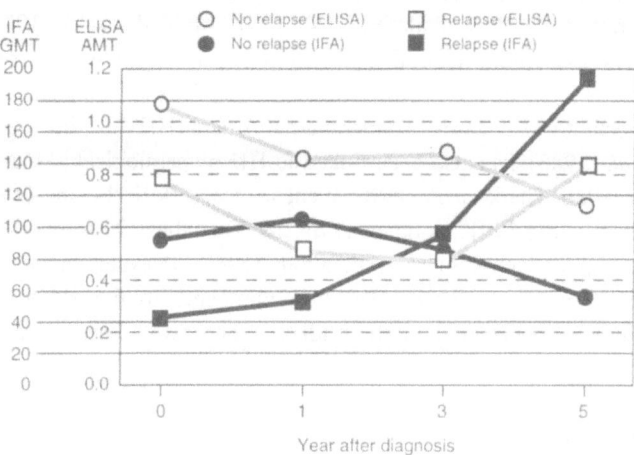

Fig. 1:   Changes in HHV-6 antibody titres in patients with Hodgkin's disease, comparing those in long-term remission with those having relapses.

VIRUS DETECTION

Since elevated viral antibodies can occur for a number of reasons, including viral reactivation due to immune dysregulation, a phenomenon commonly associated with HD, detection of virus in the involved tissue is generally accepted as the primary criterion for a virus potentially being aetiologically related to the neoplasms. A viral aetiology for HD had been suggested in earlier studies reporting the presence of EBV sequences in many of the cases examined (Weiss *et al.*, 1989; Bignon *et al.*, 1990; Jarrett *et al.*, 1991; Pallesen *et al.*, 1991; Ambinder *et al.*, 1993). The use of *in situ* hybridisation methods revealed intense concentrations of EBV nucleic acids in positive tissues localised in virtually all Reed-Sternberg (RS) cells and variants, but not in the lymphocytes, histiocytes, plasma cells, or eosinophils (Weiss *et al.*, 1989). Although the RS cells and their variants represent only a small fraction of the total cells in involved tissues, these cells are presumed to represent the neoplastic element in HD. The specific association of EBV with these cells, therefore, provides a strong argument for the role of this virus in HD. HHV-6 has been detected in HD biopsies by several techniques but only on relatively rare occasions. In the first two studies using polymerase chain reaction (PCR) and Southern blot (Jarrett *et al.*, 1988; Josephs *et al.*, 1988) no evidence of HHV-6 was detected in 37 consecutive HD cases. Torelli *et al.* (1991) were the first to report HHV-6 sequences in three cases within a series of 25 HD patients. All three positive cases were shown to belong to a particular subtype called nodular sclerosis-lymphocyte depletion. Subsequent samples, however, including those of the same histological subtype, did not have evidence of HHV-6 (Torelli, personal communication). Studies by Gompels *et al.* (1993) have shown an association of HHV-6 with another two cases of HD. More recently, Maeda *et al.* (1993) used *in situ* hybridisation to demonstrate the presence of HHV-6 sequences in the lymph node tissues of a HD patient. The HHV-6 sequences, however, were localised in the macrophages and lymphocytes, but not in the RS cells. Therefore, unlike EBV, there has still been no linkage of HHV-6 with the RS cells in any reported case, raising the question of the relevance of the role of HHV-6 in HD.

Having been successful in obtaining apparent HHV-6 identification in pathologically involved tissues from patients with sinus histiocytosis and massive lymphadenopathy (Rosai-Dorfman disease) and non-Hodgkin's lymphoma (Levine *et al.*, 1992b; Shen *et al.*, 1993), we attempted to determine whether similar patterns of infection by HHV-6 could be observed in HD cases, as had been reported by Maeda *et al.* (1993). Using *in situ* hybridisation techniques we examined 15 cases of HD, that were available from concurrent studies of EBV, for the presence of HHV-6 sequences.

In our initial studies, we used digoxigenin labelled pZVH14 (8.7 kb insert) probes using *in situ* hybridisation conditions as described previously (Levine *et al.*, 1992b; Shen *et al.*, 1993). These conditions were very similar to those used by Maeda *et al.*, (1993) but using a different portion of the viral genome as probe. Control hybridisations with a Chinese non-Hodgkin's lymphoma tissue known to be positive for HHV-6 sequences from a previous study again gave positive hybridisation results within the tumour tissues (Fig. 2). We were unable, however, to detect any significant patterns of specific hybridisation of HHV-6 in any of the 15 HD tissues examined. These experiments were again repeated with a larger HHV-6 probe, pZVB70 (23 kb insert) which contains several large open reading frames and part of the repeat sequence which is found at each end of the HHV-6 genome. Control hybridisations with formalin-fixed, paraffin-embedded cultured cells infected with HHV-6 gave strong specific signals, while uninfected HSB-2 cells were negative (Fig. 3). Again, these hybridisations failed to detect significant virus levels in any of the HD cases. It should be noted that in several of the cases (6 of 15), an occasional isolated cell appeared to show specific hybridisation with the HHV-6 probes (Fig. 4). This hybridisation, however, was not within RS cells, but may represent rare infiltrating lymphocytes infected with HHV-6 within these tissues. Because of the very low incidence of positive cells, we consider all of these tissues to be essentially negative for HHV-6 infection.

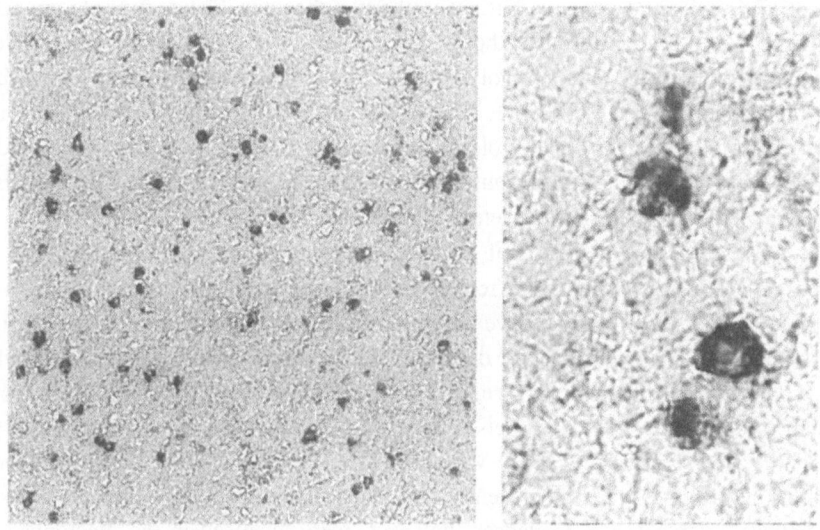

Fig. 2:   *In situ* hybridisation of paraffin-embedded formalin fixed tissue from a non-Hodgkin's lymphoma patient hybridised with digoxigenin-labelled HHV-6 probes (pZVH14). The panel on the left shows diffuse scattering of positive cells throughout the tissue (20 x magnification). The panel on the right is a close-up view of a positively staining cell (100 x magnification).

Although our *in situ* studies have not been able to confirm the association of HHV-6 with HD, it can not be ruled out that these tumour cells may contain very low levels of HHV-6 sequences below the levels of limit of detection in our assay. From comparisons with slot blot data of infected cultured cells, we estimate that cells containing 100-200 copies of the viral genome per cell could be readily detected in our assays. Thus, cells containing actively replicating virus should have been detected had they existed. However, quiescent cells in which virus was integrated, but not replicating could be missed. The application of more sensitive assays such as the recently developed *in situ* PCR, may shed some additional light on this issue.

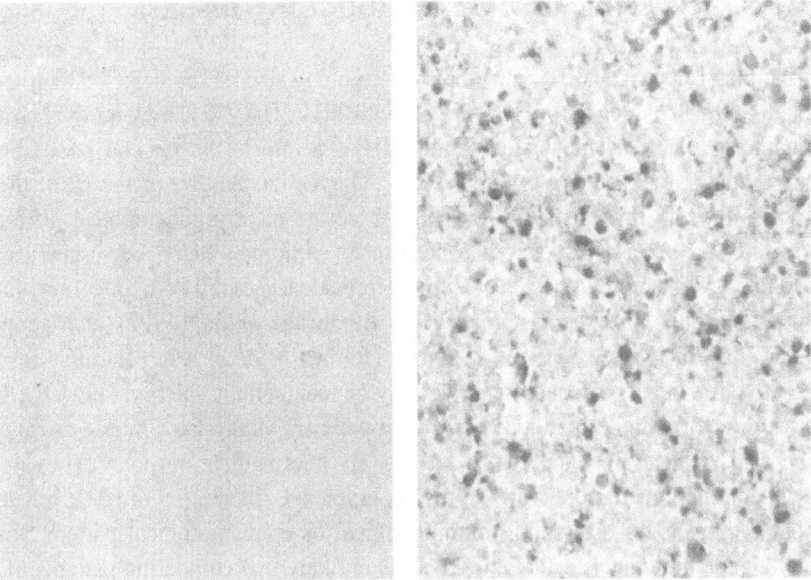

Fig. 3:   Controls for HHV-6 *in situ* hybridisation. The panel on the left shows hybridisation of a paraffin-embedded pellet of cultured uninfected HSB-2 cells fixed in formalin and hybridised with digoxigenin-labelled HHV-6 probes (pZVB70). The panel on the right shows hybridisation of a similarly prepared paraffin-embedded pellet of a mixture of HHV-6-infected and uninfected HSB-2 cells (1:10) fixed in formalin and hybridised with digoxigenin-labelled HHV-6 probes (pZVB70) (20 x magnification).

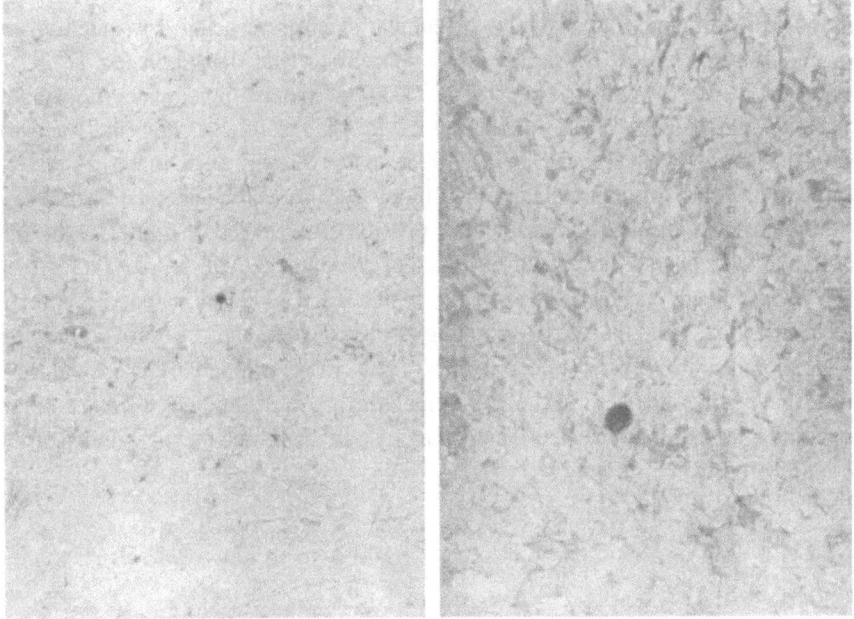

Fig. 4:   *In situ* hybridisation of Hodgkin's lymphoma tissues. Paraffin-embedded formalin-fixed tissue of a Hodgkin's lymphoma case hybridised with digoxigenin-labelled HHV-6 probes. Left panel is at 20 x magnification and right panel is at 40 x magnification.

HHV-6 AND THE EPIDEMIOLOGY OF HODGKIN'S DISEASE

The principles relevant to EBV and the epidemiology of HD, as described earlier in this symposium, are also relevant to HHV-6. As with EBV, HHV-6 is a ubiquitous virus and in many countries, such as the United States, HHV-6 infects the general population at an earlier age than EBV (Levine *et al.*, 1992c). Our understanding of the epidemiology of HHV-6 is relatively limited, however, since there are clearly geographical differences in age of infection and host response that need to be explored. Independent studies indicate that certain Asian populations have a lower prevalence of HHV-6 infection than other racial/ethnic groups (Levine *et al.*, 1992d; Buchwald *et al.*, 1992) and geographical differences have also been reported in Africa (Ranger *et al.*, 1991). Immune response to HHV-6 infection also is affected by geography or racial/ethnic factors, infected Ghanaians having significantly higher antibody titres than infected Malaysian Chinese or Indians, age matched U.S. Caucasians apparently having an intermediate response (Levine *et al.*, 1992d). Although much more work on the geographical distribution of HHV-6 needs to be done, since there is no discernible immunological or epidemiological pattern associating EBV with epidemiological features of HD, it is unlikely that comparing patterns of HHV-6 infection with patterns of HD will be greatly rewarding. As noted elsewhere in this symposium, it is apparent that HD is more than one biological entity with more than one aetiological explanation. The biological characteristics of HHV-6 as well as experience with a number of virus-induced diseases, makes it plausible to consider HHV-6 as an occasional trigger of HD. In addition to roseola (Yamanishi *et al.*, 1988), HHV-6 can cause hepatitis, encephalitis, pneumonia, and a number of other syndromes (Dubedat and Kappagoda, 1989; Steeper *et al.*, 1990; Pruksananonda *et al.*, 1992; Knox and Carrigan, 1994; Yoshikawa *et al.*, 1992; Drobyski *et al.*, 1994; for a recent review, see Levine, in press). There are numerous examples of clinical syndromes being associated with more than one virus, including Kawasaki disease (Keim *et al.*, 1977; Okano *et al.*, 1989; Hagiwara *et al.*, 1992; Whitby *et al.*, 1991), HTLV-associated myelopathy/tropical spastic paraparesis (Gessain *et al.*, 1985; Osame *et al.*, 1987; Jacobson *et al.*, 1993) and roseola (Yamanishi *et al.*, 1988; Tanaka *et al.*, 1994), and HD may well represent an unusual response to a variety of agents determined by as yet unspecified factors, such as genetic susceptibility. To further investigate the possible role of HHV-6 in HD, it may be important to understand those factors affecting the outcome of primary infection with HHV-6. An analogy can be drawn to EBV which often produces inapparent infection in children, infectious mononucleosis in young adults, and probably Burkitt's lymphoma in children coping with repeated malarial infections. EBV-associated HD in children may prove to have a co-factor substituting for malaria in determining a different oncogenic process, and attention to diverse co-factors may help to determine the plausibility of HHV-6 as an aetiological agent in HD.

CONCLUSIONS

What can we learn from our studies of HHV-6 (and EBV) thus far that can help us in regard to the search for other viruses?

First, as noted elsewhere in this symposium we need to focus on specific subgroups of HD for virological studies. EBV-associated HD has particular characteristics (mixed cellularity, age <10 and >45) and the unusual histological subtype associated with HHV-6 by Torelli *et al.* (1991) may be an important lead. Another group that deserves particular attention is the young adult population. This is the group with the most evidence for geographical clustering (Cole *et al.*, 1988; see other chapters) and the least evidence for an EBV-association (Jarrett *et al.*, 1991; Ambinder *et al.*, 1993). In addition, the histological pattern of nodular sclerosis suggests a strong immune response and therefore an antibody response to the triggering agent should be identifiable. In general, considering the strong antibody response to the candidate agents identified thus far (EBV and HHV-6) and the histological evidence of immune reactivity in the non-EBV-associated subtypes, such as nodular sclerosis and lymphocyte predominance, it would appear that a "stealth agent" that does not produce antibody is unlikely. Therefore, by appropriate selection of subgroups for seroepidemiological studies, it may be possible to target particular candidate agents for specific subgroups. Secondly, there must be an emphasis on comparing several techniques for virus detection in paraffin blocks of HHV-6-containing tissues. Finally, it is reasonable to believe that HHV-6 can, on occasion, be the primary agent triggering HD, but the evidence for its playing an important role in the pathogenesis of this disease is still lacking.

REFERENCES

Ablashi DV, Josephs SF, Buchbinder A, Hellman K, Nakamura S, Llana T, Lusso P, Kaplan M, Dahlberg J, Memon S, Imam F, Ablashi KL, Markham PD, Kramarsky B, Krueger GRF, Biberfeld P, Wong-Staal F, Salahuddin SZ, Gallo RC (1988) Human B-lymphotropic virus (human herpes virus-6). *J Virol Methods* **21**:29

Ambinder RF, Browning PJ, Lorenzana I, Leventhal BG, Cosenza H, Mann RB, MacMahon EME, Medina R, Cardona V, Grufferman S, Olshan A, Levin A, Petersen EA, Blattner W, Levine PH (1993) Epstein-Barr virus and childhood Hodgkin's disease in Honduras and the United States. *Blood* **81**:462

Biberfeld P, Petrén A-L, Eklund A, Lindemalm C, Barkhem T, Ekman M, Ablashi D, Salahuddin Z (1988) Human herpesvirus-6 (HHV-6, HBLV) in sarcoidosis and lymphoproliferative disorders. *J Virol Methods* **21**:49

Bignon Y-J, Bernard D, Curé H, Fonck Y, Pauchard J, Travade P, Legros M, Dastugue B, Plagne R (1990) Detection of Epstein-Barr viral genomes in lymph nodes of Hodgkin's disease patients. *Mol Carcinogenesis* **3**:9

Buchwald D, Hooton TM, Ashley RL (1992) Prevalence of herpesvirus, human T-lymphotropic virus type I, and treponemal infections in Southeast Asian refugees. *J Med Virol* **38**:195

Clark DA, Alexander FE, McKinney PA, Roberts BE, O'Brien C, Jarrett RF, Cartwright RA, Onions DE (1990) The seroepidemiology of human herpesvirus-6 (HHV-6) from a case-control study of leukaemia and lymphoma. *Int J Cancer* **45**:829

Cole P, MacMahon B, Aisenberg A (1968) Mortality from Hodgkin's disease in the U.S. *Lancet* **ii**:1371

Drobyski WR, Knox KK, Majewski D, Carrigan DR (1994) Fatal encephalitis due to variant B human herpesvirus-6 infection in a bone marrow transplant recipient. *N Engl J Med* **19**:1356

Dubedat S, Kappagoda N (1989) Hepatitis due to human herpesvirus-6. *Lancet* **ii**:1463

Gessain A, Barin F, Vernant JC, Gout O, Maurs L, Calendar A, de Thé G (1985) Antibodies to human T-lymphotropic virus type-I in patients with tropical spastic paraparesis. *Lancet* **ii**:407

Gompels UA, Carrigan DR, Carss AL, Arno J (1993) Two groups of human herpesvirus 6 identified by sequence analyses of laboratory strains and variants from Hodgkin's lymphoma and bone marrow transplant patients. *J gen Virol* **74**:613

Hagiwara K, Komura H, Kishi F, Kaji T, Yoshida T (1992) Isolation of human herpesvirus-6 from an infant with Kawasaki disease (letter). *Eur J Pediatr* **151**:867

Hesse J, Levine PH, Ebbesen P, Connelly RR, Mordhorst CH (1977) A case-control study on immunity to two Epstein-Barr virus-associated antigens, and to herpes simplex virus and adenovirus in a population-based group of patients with Hodgkin's disease in Denmark, 1971-1973. *Int J Cancer* **19**:49

Iyengar S, Levine PH, Ablashi D, Neequaye J, Pearson GR (1991) Sero-epidemiological investigations on human herpesvirus 6 (HHV-6) infections using a newly developed early antigen assay. *Int J Cancer* **49**:551

Jacobson S, Lehky T, Nishimura M, Robinson S, McFarlane DE, Dhibjalbut S (1993) Isolation of HTLV-II from a patient with chronic, progressive neurological disease clinically indistinguishable from HTLV-I-associated myelopathy/tropical spastic paraparesis. *Ann Neurol* **33**:392

Jarrett RF, Gledhill S, Qureshi F, Crae SH, Madhok R, Brown I, Evans I, Krajewski A, O'Brien CJ, Cartwright RA, Venables P, Onions DE (1988) Identification of human herpesvirus 6-specific DNA sequences in two patients with non-Hodgkin's lymphoma. *Leukemia* **2**:496

Jarrett RF, Gallagher A, Jones DB, Alexander FE, Krajewski AS, Kelsey A, Adams J, Angus B, Gledhill S, Wright DH, Cartwright RA, Onions DE (1991) Detection of Epstein-Barr virus genomes in Hodgkin's disease: relation to age. *J Clin Pathol* **44**:844

Johannson B, Klein G, Henle W, Henle G (1970) Epstein-Barr virus (EBV) associated antibody patterns in malignant lymphoma and leukemia. I. Hodgkin's disease. *Int J Cancer* **6**:450

Josephs SF, Buchbinder A, Streicher HZ, Ablashi DV, Salahuddin SZ, Guo HG, Wong-Staal F, Cossman J, Raffeld M, Sundeen J, Levine P, Biggar R, Krueger GRF, Fox RI, Gallo RC (1988) Detection of human B-lymphotropic virus (human herpesvirus 6) sequences in B-cell lymphoma tissues of three patients. *Leukemia* **2**:132

Keim DE, Keller EW, Hirsch MS (1977) Mucocutaneous lymph-node syndrome and parainfluenza 2 virus infection. *Lancet* **ii**:303

Knox KK, Carrigan DR (1994) Disseminated active HHV-6 infections in patients with AIDS. *Lancet* **343**:577

Levine PH, Ablashi DV, Berard CW, Carbone PP, Waggoner DE, Malan L (1970) Elevated antibody titers to herpes-type virus in Hodgkin's disease. *Proc Am Assoc Cancer Res* **11**:49

Levine PH, Ablashi DV, Berard CW, Carbone PP, Waggoner DE, Malan L (1971) Elevated antibody titers to Epstein-Barr virus in Hodgkin's disease. *Cancer* **27**:416

Levine PH, Ebbesen P, Ablashi DV, Saxinger WC, Nordentoft A, Connelly RR (1992a) Antibodies to human herpes virus-6 and clinical course in patients with Hodgkin's disease. *Int J Cancer* **51**:53

Levine PH, Jahan N, Murari P, Manak M, Jaffe ES (1992b) Detection of human herpesvirus 6 in tissues involved by sinus histiocytosis with massive lymphadenopathy Rosai-Dorfman disease. *J Infect Dis* **166**:291

Levine PH, Jarrett R, Clark DA (1992c) The epidemiology of human herpesvirus-6. In *Human Herpesvirus-6: Epidemiology, Molecular Biology, and Clinical Pathology.* Ablashi DV, Krueger GRF, Salahuddin SZ, eds (Amsterdam: Elsevier Science Publishers) p9

Levine PH, Neequaye J, Yadav M, Connelly R (1992d) Geographic/ethnic differences in human herpesvirus-6 antibody patterns. *Microbiol Immunol* **36**:169

Levine PH: Human herpesvirus-6 and human herpesvirus-7. In *Viral Infections of Humans*, fourth edition. Evans A, Kaslowe R, eds (in press)

Maeda A, Sata T, Enzan H, Tanaka K, Wakiguchi H, Kurashige T, Yamanishi K, Kurata T (1993) The evidence of human herpesvirus 6 infection in the lymph nodes of Hodgkin's disease. *Virchows Arch A Pathol Anat Histopathol* **423**:71

Müeller N, Evans A, Harris NL, Comstock GW, Jellum E, Magnus K, Orentreich N, Polk BF, Vogelman J (1989) Hodgkin's disease and Epstein-Barr virus. Altered antibody pattern before diagnosis. *N Engl J Med* **320**:689

Okano M, Luka J, Thiele GM, Sakiyama Y, Matsumoto S, Purtilo DT (1989) Human herpesvirus-6 infection and Kawasaki disease. *J Clin Microbiol* **27**:2379

Osame M, Matsumoto M, Usuku K, Izumo S, Ijichi N, Amitani H, Tara M, Igata A (1987) Chronic progressive myelopathy associated with elevated antibodies to human T-lymphotropic virus type I and adult T-cell leukemia-like cells. *Ann Neurol* **21**:117

Pallesen G, Hamilton-Dutoit SJ, Rowe M, Young LS (1991) Expression of Epstein-Barr virus latent gene products in tumour cells of Hodgkin's disease. *Lancet* **337**:320

Pruksananonda P, Hall CB, Insel RA, McIntyre K, Pellett PE, Long CE, Schnabel KC, Pincus PH, Stamey FR, Dambaugh TR, Stewart JA (1992) Primary human herpesvirus 6 infection in young children. *N Engl J Med* **326**:1445

Ranger S, Patillaud S, Denis F, Himmich A, Sangare A, M'Boup S, Ztoua-N'gaporo A, Prince-David M, Chout R, Cevallos R, Agut H (1991) Seroepidemiology of human herpesvirus-6 in pregnant women from different parts of the world. *J Med Virol* **34**:194

Salahuddin SZ, Ablashi DV, Markham PD, Josephs SF, Sturzenegger S, Kaplan M, Halligan G, Biberfeld P, Wong-Staal F, Kramarsky B, Gallo RC (1986) Isolation of a new virus, HBLV, in patients with lymphoproliferative disorders. *Science* **234**:596

Shen YY, Huang AM, Jahan N, Manak M, Jaffe ES, Levine PH (1993) *In situ* hybridisation detection of human herpesvirus-6 in biopsy specimens from Chinese patients with non-Hodgkin's lymphoma. *Arch Pathol Lab Med* **117**:502

Steeper TA, Horwitz CA, Ablashi DV, Salahuddin SZ, Saxinger C, Saltzman R, Schwartz B (1990) The spectrum of clinical and laboratory findings resulting from human herpesvirus-6 (HHV-6) in patients with mononucleosis-like illnesses not resulting from Epstein-Barr virus or cytomegalovirus. *Am J Clin Pathol* **93**:776

Tanaka K, Kondo T, Torigoe S, Okada S, Mukai T, Yamanishi K (1994) Human herpesvirus 7: another causal agent for roseola (exanthem subitum). *J Pediatr* **125**:1

Torelli G, Marasca R, Luppi M, Selleri L, Ferrari S, Narni F, Mariano MT, Federico M, Ceccherini-Nelli L, Bondinelli M, Montagnani G, Montorsi M, Artusi T (1991) Human herpes virus-6 in human lymphomas: identification of specific sequences in Hodgkin's lymphoma by polymerase chain reaction. *Blood* **77**:2251

Weiss LM, Movahed LA, Estmir RA (1989) Detection of Epstein-Barr viral genomes in Reed-Sternberg cells of Hodgkin's disease. *N Engl J Med* **320**:502

Whitby D, Hoad JG, Tizard EJ, Dillon MJ, Weber JN, Weiss RA, Schulz TF (1991) Isolation of measles virus from a child with Kawasaki disease. *Lancet* **338**:1215

Yamanishi K, Okuno T, Shiraki K, Takahashi M, Kondo T, Asano Y, Kurata T (1988) Identification of human herpes virus-6 as a causal agent for exanthem subitum. *Lancet* **i**:1065

Yoshikawa T, Nakashima T, Suga S, Asano Y, Yazaki T, Kimura H, Morishima T, Kondo K, Yamanishi K (1992) Human herpesvirus-6 DNA in cerebrospinal fluid of a child with exanthem subitum and meningoencephalitis. *Pediatrics* **89**:888

THE REED-STERNBERG CELL AND THE CD30 ANTIGEN

Horst Dürkop[1], Ute Latza[1], Brunangelo Falini[2], Gianpaolo Nadali[3],
Giovanni Pizzolo[3], and Harald Stein[1]

[1]Institute of Pathology, Klinikum Steglitz, Free University of Berlin, Germany;
[2]Institute of Hematology, Perugia University, Italy;
[3]Department of Hematology, Policlinico Borgo Roma, Verona University,
  School of Medicine, Italy

The tumour cells of Hodgkin's disease (HD), called Reed-Sternberg and Hodgkin cells (H-RS cells) are, in conjunction with an admixture of non-malignant cells of various types, pathognomonic for HD. In most cases, the H-RS cells account for less than 5% and in some cases for less than 1% of the total tumour mass. In classical cases, the H-RS cells are large cells with bi-lobed or poly-lobed nuclei so that the cells appear binucleated or multinucleated. It is possible that, in some cases, *bona fide* multinucleation does actually occur. The nuclear membrane is thick and sharply defined, and there is a very large, variously shaped, acidophilic central nucleolus surrounded by a clear halo. In the most typical example of the H-RS cell, the two nuclear lobes face each other. The non-malignant counterpart of these cells is still a matter of debate. In most cases they react with the lymphoid surface markers CD30 (Schwab *et al.*, 1982; Stein *et al.*, 1985) and CD70 and in some cases with CDw75, CD3, T-cell receptor ß chain (TcRß) (Dallenbach and Stein, 1989), J-chain (Stein *et al.*, 1986) and CD20. These data, as well as the detection of immunoglobulin and T-cell receptor (TcR) rearrangements by Southern blot in some instances (Herbst *et al.*, 1989), provide evidence for a lymphoid origin of these tumour cells. In particular, Dallenbach and Stein (1989) were able to detect TcRß on the H-RS cells of 24 of 65 cases and CD20 on the H-RS cells of 12 of 65 cases. Recently, support of these data came from Tamaru *et al.* (1994) who reported that in 67% of HD cases with B-cell-antigen positive H-RS cells, clonal VDJ rearrangements of the immunoglobulin heavy chain (IgH) gene were detectable; no VDJ rearrangements were detected in cases with T-cell antigen-positive H-RS cells. Furthermore, somatic mutations of the V_H segment could be detected in 6 of 10 sequenced HD cases with rearranged IgH genes. These data lead to the conclusion that the H-RS cells of cases with rearranged IgH genes are B-cell-related and correspond in their differentiation stage either to naive pre-germinal centre B-cells or to germinal centre/post-germinal centre-derived memory B-cells. However, Daus and co-workers (1994) reported the absence of IgH and TcR rearrangement. They obtained their data by a single cell polymerase chain reaction (PCR) which was performed on viable single cells isolated from cell suspensions of HD lymph nodes. These data point towards a non-lymphoid, e.g., macrophage, origin of H-RS cells.

For the immunohistological detection and verification of the H-RS cells the following markers proved to be useful: CD30, CD15, CD20, CDw75, and lectins such as peanut agglutinin (Ree et al., 1989) or Bauhinia purpurea (Sarker et al., 1992). The expression pattern of the majority of these markers is not restricted to H-RS cells. CD15 also stains granulocytes, the malignant cells in some non-Hodgkin's lymphomas (NHLs), many adenocarcinomas, and urothelial carcinomas (Perkins and Kjeldsberg, 1993). However, in 15-30% of cases of HD, H-RS cells may not stain with CD15 (Perkins and Kjeldsberg, 1993). CD20 and CDw75 are mainly expressed on H-RS cells in cases of the lymphocyte predominant subtype and are only occasionally present in the other subtypes (Perkins and Kjeldsberg, 1993). Peanut agglutinin has been used as a marker for H-RS cells. In one study, staining in a characteristic paranuclear and membranous pattern was observed in 86% of cases (Ree et al., 1989); in other studies as few as 44% of all investigated cases revealed peanut agglutinin positive H-RS cells (Sarker et al., 1992). Sarker et al. (1992), reported that Bauhinia purpurea stained the H-RS cells of 97% of all cases of HD investigated. Peanut agglutinin and Bauhinia purpurea also stain histiocytes, macrophages, granulocytes, and plasma cells (Sarker et al., 1992; Perkins and Kjeldsberg, 1993). Recently, Bilbe et al. (1992) reported on the molecular characterisation of a novel intermediate filament-associated protein, called Restin, which is highly expressed in H-RS cells and in the tumour cells of anaplastic large cell lymphomas (ALCL) (Delabie et al., 1992). Delabie and co-workers (1992) showed that Restin was expressed in the H-RS cells of 28 of a series of 41 cases of HD. In particular, the H-RS cells of the paragranuloma and mixed cellularity subtypes of HD were not labelled by the Restin antibodies.

Among all other markers mentioned, the CD30 antigen proved to be most restricted in its expression to H-RS cells (Fig. 1) (Schwab et al., 1982; Stein et al., 1985). In normal lymphoid tissue the CD30 antigen is only expressed by a few extrafollicular, activated B and T blasts and the B blasts located at the rim of germinal centres (Fig. 2) (Schwab et al., 1982; Stein et al., 1985). Because of the resemblance in morphology and tissue distribution between normal CD30[+] blasts and H-RS cells, these normal CD30[+] blasts appear to be likely candidates for the non-malignant counterpart of H-RS cells. CD30 is an inducable molecule; it becomes detectable on 15-30% of all peripheral lymphocytes after in vitro stimulation with phytohaemagglutinin (PHA), concanavalin A, phorbol myristate acetate, anti-CD3 and S. aureus (Stein et al., 1985; Ellis et al., 1993). The peripheral blood lymphocytes induced to express CD30 have been shown to be a subset of cells derived from CD45R0[+] T-cell precursors (Ellis et al., 1993). The strongest inducers of CD30 expression are the human T-cell lymphotropic virus I and II (HTLV-I and HTLV-II) and the Epstein-Barr virus (EBV). The CD30 antigen has therefore been characterised as an activation marker (Andreesen et al., 1984; Stein et al., 1985). Interestingly, CD30 expression was detectable on the tumour cells of ALCL. This was the first indication that HD could be related to this particular group of lymphomas (Stein *et al.*, 1985). Before its description, most cases of this tumour had been diagnosed as malignant histiocytosis or anaplastic carcinoma because of the bizarre morphology of tumour cells. In most of the studied series (Stein *et al.*, 1985; Schwarting *et al.*, 1989; Ralfkiaer *et al.*, 1987; Norton *et al.*, 1987; Miettinen, 1992; Carbone *et al.*, 1992), all cases of ALCL were consistently CD30[+]. This prompted some groups to imprecisely designate these malignancies as "Ki-1 lymphoma" or "CD30[+] lymphoma".

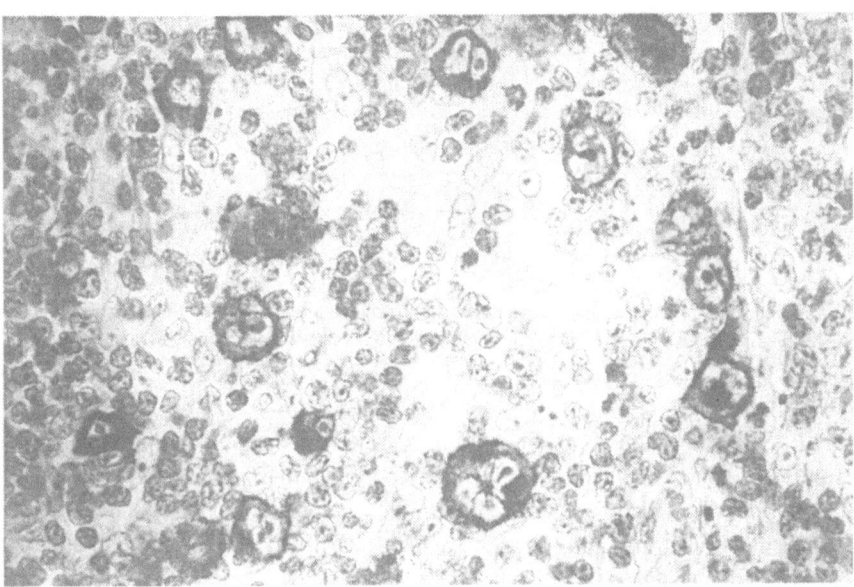

Fig. 1:   Immunostaining of a lymph node involved by Hodgkin's disease with the CD30 monoclonal antibody Ber-H2. The strong and consistent staining of Hodgkin and Reed-Sternberg cells is typical.

Additionally, CD30 is expressed on the tumour cells of some large cell NHLs and on the large blasts of about 25% of all T-cell NHL. No CD30 expression is detectable on the tumour cells of low grade malignant B-cell NHL, e.g., B-cell chronic lymphatic leukaemia, centroblastic-centrocytic NHL, mantle cell lymphoma, and MALT lymphoma. CD30 is consistently absent on the tumour cells of B and T-cell lymphoblastic (precursor) NHL. Thus, immunohistological detection of CD30 on tumour cells excludes more than 60% of all NHL from the histopathological differential diagnosis. Furthermore, CD30 expression in NHLs other than ALCL is mostly weak, often only being detectable in frozen sections (Stein *et al.*, 1985; Stein and Dallenbach, 1992). In nine series (Schwarting *et al.*, 1989; Schwab *et al.*, 1982; Stein *et al.*, 1985; Ralfkiaer *et al.*, 1987; Norton *et al.*, 1987; Miettinen, 1992; Carbone *et al.*, 1992; Hall *et al.*, 1988; van der Putte *et al.*, 1988) of morphologically and immunohistologically characterised NHL, CD30 expression was seen in 15% of the non-ALCL B-cell lymphomas and 27% of the non-ALCL T-cell lymphomas.

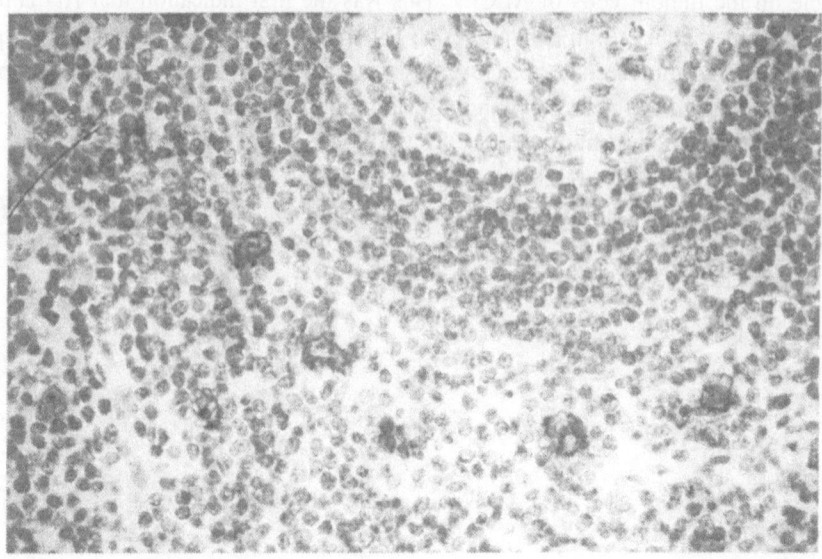

Fig. 2:   Immunostaining of a tonsil with the CD30 monoclonal antibody Ber-H2 after microwave treatment of the tissue. CD30 expression is detected in a few lymphoid blasts which are located at the rim of the follicle.

Certainly, CD30 expression is rarely found in non-haematolymphoid neoplasms. Pallesen and Hamilton-Dutoit (1989) reported CD30 positivity in 8 of 10 embryonal carcinomas and in the embryonal elements of four mixed germ cell tumours. This has been confirmed by other groups and is the only non-lymphoid neoplasm with a weak but reproducible pattern of CD30 expression (Latza *et al.*, 1995).

In all other reported cases of CD30 positivity of non-lymphoid malignancies the CD30 detection was non-reproducible or inconsistent. Thus, Schwarting and co-workers (1989) reported a weak Ber-H2 reactivity on paraffin sections of pancreatic carcinomas, salivary gland carcinomas, malignant melanomas, and breast carcinomas. When the staining was repeated using the Ber-H2 and Ki-1 monoclonal antibodies on frozen material, confirmation of these results was not produced. Mechtersheimer and Möller (1990) described the expression of the CD30 antigen in several mesenchymal tumours such as leiomyomas, leiomyosarcomas, rhabdomyosarcomas, aggressive fibromatoses, fibrosarcomas, synovial sarcomas, one giant cell tumour of tendon sheath, malignant fibrous histiocytomas, osteosarcomas, Ewing's sarcomas, malignant schwannomas, in the Schwann cell compartment of ganglioneuromas and in the myoepithelial compartment of fibroadenomas. These results were obtained using frozen material and the monoclonal antibody Ki-1. Furthermore, it was not possible for the above-mentioned authors to reproduce these staining results on the same tissue samples in our laboratory. Andreesen and colleagues (1989) reported Ki-1 reactivity on human blood monocytes which had been cultured on hydrophobic Teflon foils for more than 8 days. In our hands, the same staining pattern was available with an unrelated, subgroup-matched, murine IgG3 (unpublished data). No reactivity of these stimulated monocytes was revealed when stained with another murine CD30 monoclonal antibody (Ber-H2) belonging to the IgG1 subclass. Moreover, monocytes and their putative derivates, e.g., macrophages and interdigitating cells, are consistently CD30-negative in frozen and paraffin-embedded tissue sections.

The situation became more complex by the findings of Hansen and co-workers (1989 and 1990) who described an intracellular molecule which reacts with only the Ki-1 monoclonal antibody but not with other CD30 antibodies, such as Ber-H2, HRS-1, HRS-2 and HRS-3. In contrast to the transmembrane CD30 antigen, the intracellular Ki-1 antigen is smaller (57 kD versus 120 kD), as revealed by sodium dodecylsulfate polyacrylamide gel electrophoresis (SDS-PAGE), and shows protein kinase activity (Hansen *et al.*, 1990). This antigen, designated Ki-1/57, is located in the cytoplasm and in association with the nuclear envelope, chromatin structures, and the nucleoli (Rohde *et al.*, 1992). Possible cross-reactivity with this intracellular Ki-1/57 antigen should be considered and could account for the immunohistological reaction pattern of the Ki-1 antibody and the discrepancies found in the reaction patterns of other CD30 antibodies.

Because of the restricted *in vivo* reaction pattern of CD30 monoclonal antibodies, Falini and colleagues (1992a) were encouraged to investigate the therapeutic potential of anti-CD30 reagents. First, they showed the *in vivo* targeting of H-RS cells with the CD30 monoclonal antibody Ber-H2 in 4 patients. Biopsy samples, removed 24-72 h after injection of the Ber-H2 monoclonal antibody, were incubated with the secondary rabbit anti-mouse immunoglobulin followed by APAAP complexes (Falini *et al.*, 1992b; Cordell *et al.*, 1984). All morphologically recognisable H-RS cells, including those encased by bundles of fibrous tissues in cases of nodular sclerosing HD, were specifically labelled *in vivo* by the injected murine antibody. In all cases, *in vivo* labelling was highly selective in that no other cell constituents, apart from H-RS cells, were stained by the secondary antibody in frozen sections.

These results prompted Falini and co-workers to treat 9 patients suffering from refractory HD with an immunotoxin consisting of a CD30 (Ber-H2) monoclonal antibody covalently linked to saporin (Falini *et al.*, 1992a and 1993). The immunotoxin was applied in doses ranging from 0.4 mg/kg to 0.8 mg/kg body weight. The side effects included fever, fatigue, signs of mild capillary leak syndrome, slight thrombocytopaenia and an increase in liver enzymes of up to 5 times. All clinical signs and laboratory changes reversed within 7-10 days. Computer tomographic images, performed 7 and 30 days after the infusion of immunotoxins, showed a rapid reduction in size of the tumour masses, ranging from 50% to more than 75% in 5 of the 9 patients (Figs. 3A and 3B). Two patients also stated a complete relief from systemic symptoms. The duration of the response ranged from 2 to 4 months. In general, no tumour growth at new sites was observed but regrowth at all original sites occurred. Most patients developed an immune response to the antibody and the toxin moieties.

Due to the restricted expression of the CD30 antigen on H-RS cells, CD30 antibodies were investigated as potential tools for immunoscintigraphy in HD for improving staging or restaging (Falini *et al.*, 1992b; da Costa *et al.*, 1992). In 6 patients, Falini and co-workers (1992b) could only detect, on average, 50% of the sites previously documented to be clinically or pathologically involved by HD. In contrast, da Costa and colleagues (1992) found that it was possible to detect nodal, splenic, bone marrow, and muscular involvements by radioactively labelled anti-CD30 antibodies. Furthermore, this group reported that involvement of several of these sites had initially been revealed by immunoscintigraphy.

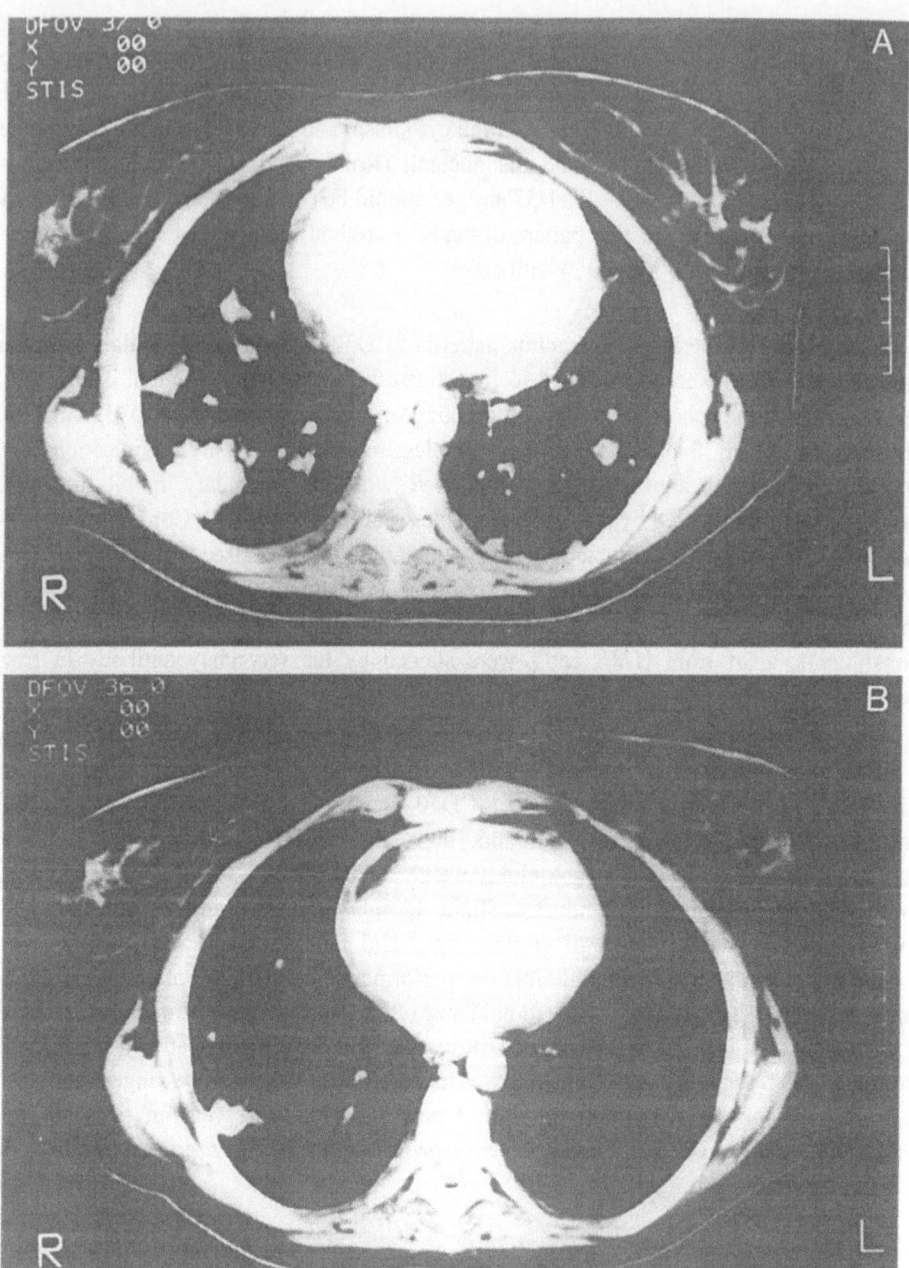

Figs. 3A and 3B: CT lung scans of a patient with refractory Hodgkin's disease before (A) and 7 days after treatment (B) with an anti-CD30 (Ber-H2) immunotoxin.

The biochemical and molecular characterisation of the CD30 antigen reveals that its mature form is a transmembrane protein of about 120 kD (Froese *et al.*, 1987). This mature form develops from a precursor of approximately 84 kD during its passage through the Golgi complex (Froese *et al.*, 1987). The molecular weight shift from 84,000 to 120,000 is almost completely explainable through glycosylation (Nawrocki *et al.*, 1988). Nawrocki and colleagues (1988) found that the 84 kD core protein migrates, after processing by high mannose N-linked glycosylation, with an apparent molecular weight of 90,000. This protein is then further processed to the mature 120 kD protein; this is accomplished by O-linked glycosylation, the addition of sialic acid residues, and by conversion of N-linked oligosaccharides from the high mannose to the complex type (Nawrocki *et al.*, 1988). Pulse chase experiments and treatment of CD30[+] cells with the ionophore monensin demonstrated that the processing of the 90 kD precursor to the mature 120 kD protein occurs during the passage from the trans-cisternae of the Golgi complex to the cell membrane (Fig. 4) (Froese *et al.*, 1987).

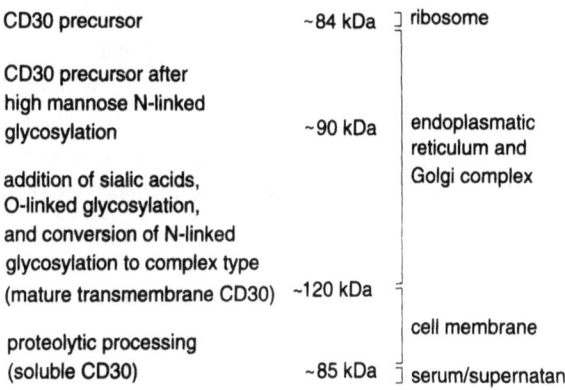

Fig. 4:   Biosynthesis of the CD30 antigen as outlined in Froese *et al.* (1987), Nawrocki *et al.* (1988) and Josimovic-Alasevic *et al.* (1989).

The soluble form of the CD30 antigen (sCD30) is shed by CD30[+] cells, and therefore CD30[+] malignancies can be monitored by measuring the sCD30 serum level (Josimovic-Alasevic *et al.*, 1989; Gause *et al.*, 1991; Nadali *et al.*, 1994). Probably because of proteolytic processing, the sCD30 has a smaller molecular weight (85,000) than the membrane bound CD30 (120,000) as revealed by SDS-PAGE (Josimovic-Alasevic *et al.*, 1989). However, the mechanism of CD30 release remains to be investigated, although it is already known that CD30 shedding occurs as an active process from viable CD30[+] cells and is not merely due to release from dying or dead cells.

The fluid phase CD30 can be detected by a sensitive enzyme-linked immunosorbent assay (ELISA) (Josimovic-Alasevic *et al.*, 1989). Apart from CD30[+] malignancies, sCD30 is only detectable in sera of patients with acute infectious mononucleosis (Pfreundschuh *et al.*, 1990), HIV infection (Pizzolo *et al.*, 1990), and virus B- and C-related chronic hepatitis (Pizzolo *et al.*, 1994). sCD30 is absent from the sera of healthy blood donors (Josimovic-Alasevic *et al.*, 1989; Nadali *et al.*, 1994). Data regarding the rate of detection of sCD30 in patients with untreated HD are controversial. The reported results depend mainly on the sensitivity of the assay which was used. In a prospective study of 90 untreated patients with newly diagnosed HD, Gause and co-workers (1992) reported detectable sCD30 in the serum of 20 patients (22%). With this assay it was only possible to detect sCD30 in patients with B symptoms. The highest sCD30 levels were found by Gause *et al.* (1991) in the sera of patients with stage IVB disease. The percentage of patients with sCD30[+] sera increased slightly with the stage of HD (IIB: 33%; IIIB: 58%; IVB: 53%). No detectable sCD30 could be found in serum from 4 patients suffering from lymphocyte predominant HD. However, 6 of 49 patients (12%) with nodular sclerosis HD were sCD30[+], 9 of 31 patients (29%) with mixed cellularity disease were sCD30[+], and all 5 patients with the lymphocyte depletion subtype exhibited positive sCD30 serum levels (Gause *et al.*, 1991).

In contrast, Vinante *et al.* (1991) and Nadali *et al.* (1994) reported that detectable levels of sCD30 are found in the serum of the majority of untreated HD patients at the time of diagnosis. In particular, Nadali and co-workers, who used a modified, second generation ELISA which was more sensitive than that used by Gause and co-workers (1991), demonstrated that 87% of patients with HD had elevated sCD30 serum levels at the time of diagnosis (Fig. 5). Using this more sensitive ELISA, Nadali *et al.* (1994) were able to detect elevated serum levels of sCD30 in patients with and without B symptoms. They confirmed that the sCD30 level increased with the stage of HD; the level was significantly higher in patients with stage III and IV, compared to patients with stage I and II disease. The serum levels of sCD30 also correlated with the presence of B symptoms and bulky disease.

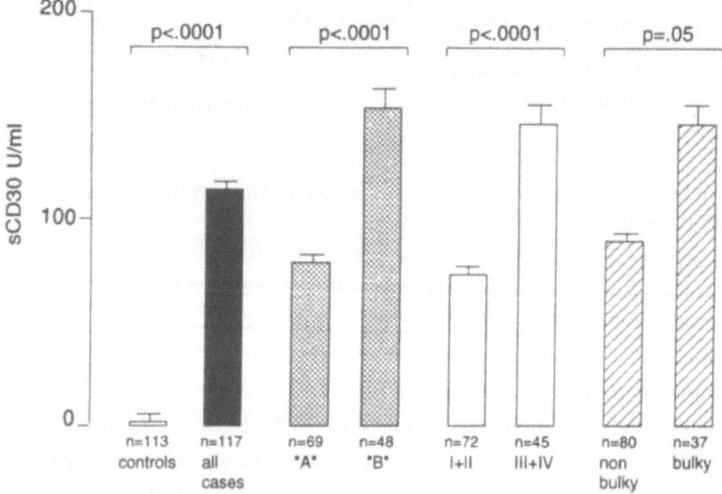

Fig. 5:   Level of sCD30 in sera from 117 patients with Hodgkin's disease at the time of diagnosis. The correlation between the serum concentration of sCD30 and presence of Bsymptoms, and that between the sCD30 serum level and stage of the disease are highly significant

In patients with untreated HD, it would seem that only the sCD30 serum level, and not the soluble CD8 (sCD8) and soluble CD25 (sCD25) serum levels, correlates strictly with disease activity. Thus it is favoured in the follow-up of patients in remission (Gause *et al.*, 1992).

Elevated sCD30 levels are also detectable in patients with the following CD30[+] malignant tumours: ALCL (Pfreundschuh et al., 1990), angioimmunoblastic lymphadenopathy-like T-cell lymphoma (Pizzolo et al., 1990), HTLV-I[+] T-cell leukaemia/lymphoma (Pfreundschuh et al., 1990), and embryonal carcinomas (Latza et al., 1995).

The molecular characterisation of the CD30 antigen was performed by immunoscreening a λgt 11 cDNA library, constructed from the HUT-102 cell line (Gazdar et al., 1980), with the CD30-reactive monoclonal antibodies Ki-1 and Ber-H2 (Dürkop et al., 1992). This approach yielded one clone with Ber-H2 reactivity. This clone was used for colony hybridisation of another HUT-102 cDNA library, which was cloned in the pCDM8 vector. Five partially overlapping clones were obtained. The extreme 3' end of the mRNA was subsequently amplified by PCR. The open reading frame extends from nucleotide 231 to 2015 of the cDNA sequence and consists of two similar domains sharing 77% homology (nucleotides 381 to 742 and 906 to 1270). There are two polyadenylation sites in the 3' untranslated region which are preceded by the unusual poly(A) signal sequences TGTAAA and AATAAT respectively. In Northern blot analysis of poly(A)[+] RNA from various human cell lines, a major mRNA species of about 3,800 nucleotides, and a minor species of about 2,600 nucleotides could be detected. Only the larger transcript of 3,800 nucleotides was detected with a cDNA fragment which covered the extreme 3' untranslated region of the CD30 cDNA (Fig. 6). The sizes of the two mRNA species of CD30 correspond with the positions of the two polyadenylation sites. In accordance with the known antigenic distribution, the cell lines SU-DHL-1 (Morgan et al., 1989), L591 (Diehl et al., 1982), L540 (Diehl et al., 1981), L428 (Schaadt et al., 1980), Ho (Jones et al., 1989), Co (Jones et al., 1985), and HUT-102 are positive for CD30 mRNA, whilst no CD30 transcripts are detectable in peripheral blood lymphocytes and in the cell lines HPB-ALL (Boylston et al., 1985), MDA-MB-231 (Cailleau et al., 1974) and U-937 (Sundström and Nilsson, 1976).

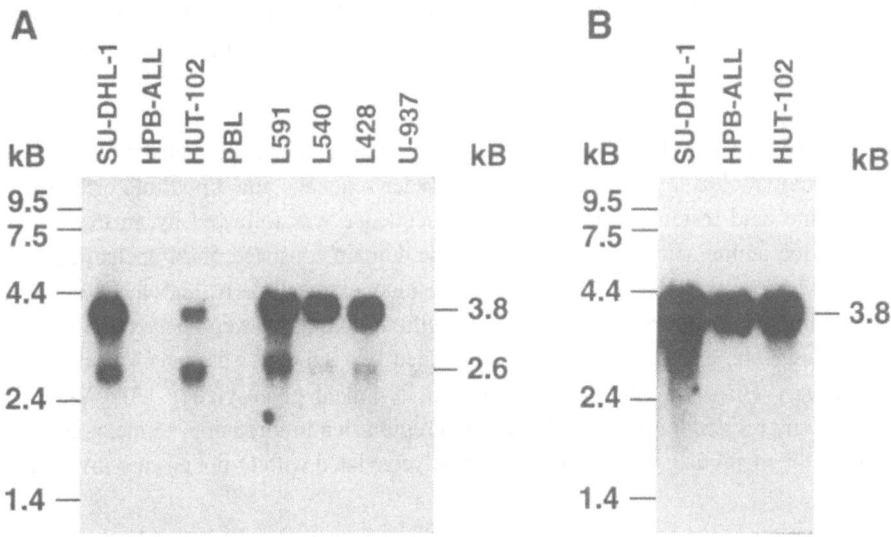

Fig. 6:   Northern blot analysis of poly(A)+ RNA (2 μg per lane) from various cell lines, and peripheral blood lymphocytes hybridised with a ^{32}P-labelled CD30 cDNA probe containing the entire CD30 coding region (A) and a CD30 cDNA probe containing the extreme 3' non-translated region of the CD30 mRNA (B).

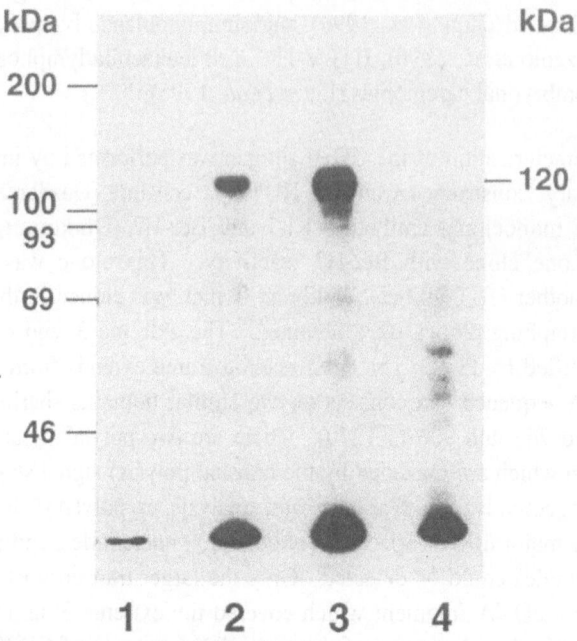

Fig. 7:   Immunoprecipitation of CD30, with the monoclonal antibody Ber-H2, from [125]I- labelled COS
            cells transfected with a CD30 cDNA construct containing the entire CD30 coding region (lane 2),
            HUT-102 cell line (lane 3), mock transfected COS cells (lane 4), and an immunoprecipitation of
            CD30-transfected COS cells with an unrelated monoclonal antibody (lane 1). Molecular size
            markers are indicated (left) as well as a specifically precipitated protein (right).

Immunoprecipitation of recombinant CD30 antigen with the monoclonal antibody Ber-H2 after ^{125}I surface-labelling of CD30 cDNA-transfected COS cells, yielded a single band on SDS-PAGE which corresponded to the 120 kD native CD30 antigen precipitated from HUT-102 cells (Fig. 7).

Translation of the CD30 DNA sequence into the protein sequence predicted a typical type I transmembrane molecule with a hydrophobic leader sequence and a possible cleavage site behind amino acid residue 18 (Fig. 8). This sequence was followed by an extracellular domain of 365 amino acids. The transmembrane domain consisted of 24 uncharged amino acids, flanked by one charged amino acid residue extracellularly and three positively charged amino acids intracellularly. The intracellular domain was composed of 188 amino acid residues. The extracellular domain contained two subunits sharing 64% identity on 129 residues. There are two putative sites for N-linked glycosylation (Asn-X-Ser/Thr, where X is any residue but Pro and Asp), and a region rich in threonine, serine, and proline adjacent to the membrane which is likely to be glycosylated with O-linked carbohydrates.

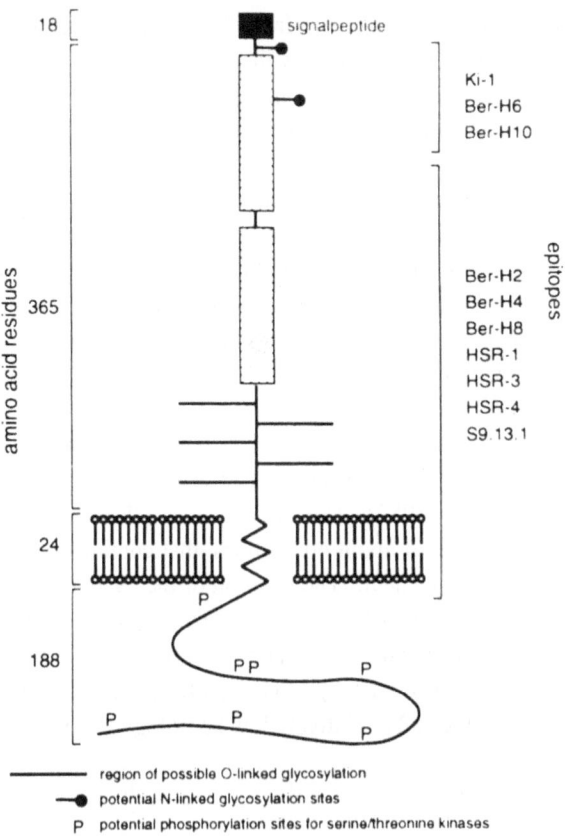

18 ⌈ signalpeptide

Ki-1
Ber-H6
Ber-H10

amino acid residues

365

epitopes

Ber-H2
Ber-H4
Ber-H8
HSR-1
HSR-3
HSR-4
S9.13.1

24

P

188

P P

P

P

P

P

———— region of possible O-linked glycosylation

—● potential N-linked glycosylation sites

P potential phosphorylation sites for serine/threonine kinases

Fig. 8: Schematic view of the CD30 antigen. N-terminal is a hydrophobic leader sequence which consists of 18 amino acid residues and is removed after entering the export pathway. The extracellular domain contains 365 amino acid residues and is organised in two subunits with 64% homology to each other on 129 amino acid residues. There are two sites for N-linked glycosylation in the extracellular domain. Sites for O-linked glycosylation are situated mainly in the hinge region of the CD30 molecule. The epitopes of the Ki-1, Ber-H6, and Ber-H10 monoclonal antibodies map to the N-terminal end up to residue 93. The epitopes of all other investigated CD30 monoclonal antibodies (Ber-H2, Ber-H4, Ber-H8, HRS-1, HRS-3, HRS-4, and S9.13.1) cluster between amino acid residue 112 and 412. The transmembrane domain contains 24 amino acids. In the intracellular domain there are seven potential phosphorylation sites for serine and threonine kinases.

There are potential phosphorylation sites for serine/threonine kinases in the intracellular domain. These data are in consensus with reports on the phosphorylation of the mature CD30 molecule which have demonstrated that the CD30 antigen is mainly phosphorylated on serine and to a small extent on threonine in the L540 cell line (Hansen *et al.*, 1990). In keeping with sequence data, revealing possible sites for tyrosine kinase to be only in the extracellular domain of the CD30 molecule, no phosphorylation on tyrosine residues was detected after *in vitro* phosphorylation of lysates of the L540 cell line. Furthermore, in Western immunoblotting experiments there was no reactivity of immunoprecipitated CD30 antigen with anti-phosphotyrosine monoclonal antibodies (Dürkop *et al.*, 1992).

The intracellular domain of the CD30 molecule contains the conserved sequence for ATP-binding Gly-X-Gly-X-X-Gly, found in many protein kinases, with hydrophobic residues occupying positions one and seven upstream from the first conserved glycine (residue 537). However, there are no conserved features with the consensus sequence of the catalytic kinase domain (Hanks *et al.*, 1988). This is in agreement with Hansen *et al.* (1990), who could not demonstrate protein kinase activity of the membrane-associated CD30 molecule.

Comparison of the predicted amino acid sequence, with the NBRF and SWISSPROT database using the search logarithm FASTA (Pearson and Lipman, 1988), indicates that the CD30 polypeptide is similar to the transmembrane growth factor receptors of the tumour necrosis factor receptor/nerve growth factor receptor (TNFR/NGFR) superfamily (Mallett and Barclay, 1991). Most markedly the first, and to a minor extent the second, cysteine-rich subunit of the extracellular domain shares common features with the comparable regions of the human tumour necrosis factor receptors (TNFR1, TNFR2) (Schall *et al.*, 1990; Heller *et al.*, 1990; Loetscher *et al.*, 1990; Kohno *et al.*, 1990; Engelmann *et al.*, 1990; Smith *et al.*, 1990) and with the human low-affinity nerve growth factor receptor (NGFR) (Fig. 9) (Johnson *et al.*, 1986). The CD30 extracellular domain can be divided into six internal cysteine-rich motifs, of about 40 residues, that contain six cysteines, except for motifs 1B and 3B, which are truncated. The position of the cysteines are highly conserved in both subdomains. The homology of the other members of the NGFR/TNFR superfamily with CD30 decreases in the following order: TNFR2, TNFR1, CD40 (Stamenkovic *et al.*, 1989), and FAS/APO-1 (Itoh *et al.*, 1991). The NGFR, 4-1BB (Kwon *et al.*, 1989), and CD27 (Camerini *et al.*, 1991) reveal a relatively low homology to CD30 (Fig. 10).

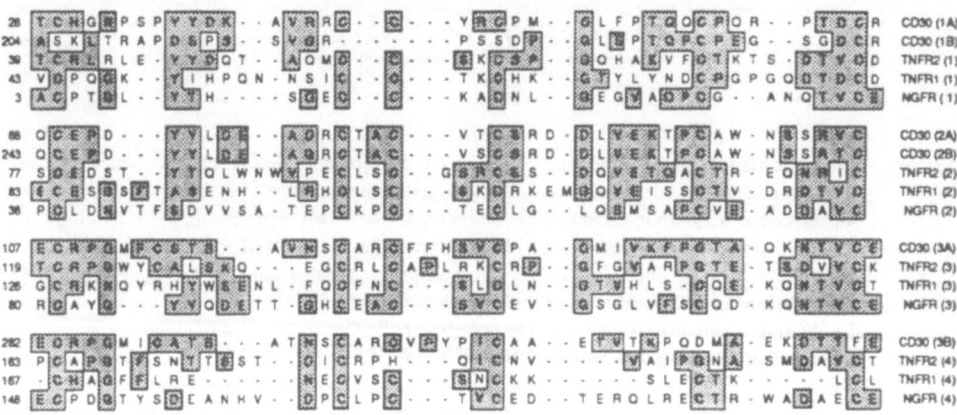

Fig. 9: Consensus alignment of the cysteine-rich motifs of CD30, TNFR1, TNFR2 and NGFR. Boxed residues reflect those common to CD30 and at least one other family member. Amino acid positions are indicated at the beginning of each line. Sequences were aligned using a combination of the FASTA and ALIGN programmes. For maximum alignment, gaps were introduced.

The CD30 gene was localised to chromosome 1p36 by the hybridisation of metaphases of PHA-stimulated peripheral blood lymphocytes (Fonatsch *et al.*, 1992). Thus, the CD30 gene is closely linked with other members of the TNFR/NGFR superfamily, such as the human TNFR2 and human OX40 genes which are also located at 1p36 (Kemper *et al.*, 1991; Latza *et al.*, 1994).

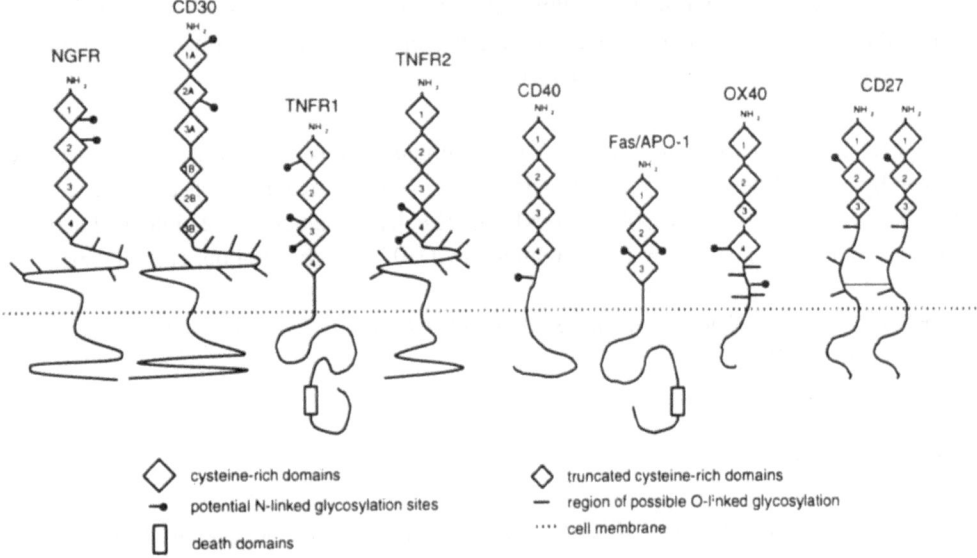

NGFR CD30 TNFR2 CD40 OX40 CD27 TNFR1 Fas/APO-1

◇ cysteine-rich domains
→● potential N-linked glycosylation sites
▯ death domains

◇ truncated cysteine-rich domains
— region of possible O-linked glycosylation
···· cell membrane

Fig. 10:   Schematic view of some members of the TNFR/NGFR superfamily. TNFR1 and TNFR2 refer to the two receptors that bind both TNF-α and TNF-ß and are also referred to by their approximate molecular mass, i.e., 55 and 75 kD, respectively. Cysteine-rich motifs are indicated by diamonds. Cysteine-rich motifs which are too small to form a normal TNFR/NGFR motif are indicated by small diamonds. The Fig. is based on Mallett and Barclay (1991).

Smith *et al.,* (1993) identified a ligand for CD30 (CD30L). Furthermore, they showed that the CD30L induced proliferation of activated T-cells and the HD-derived cell line HDLM-2, whilst in the ALCL-derived cell line, Karpas 299, it induced cytostasis and cytolysis (see Gruss and Dower, this volume). However, the specific response to CD30 stimulation, including cell death or proliferation, depends upon cell type, stage of differentiation, transformation status, and the presence of other stimuli. These data therefore unequivocally characterise the CD30 antigen as a cytokine receptor. Recently, Bowen and co-workers (1993) reported that stimulation of the CD30 antigen by the monoclonal antibody C10 induces inhibition of cytotoxicity of CD30[+] YT cells on Raji cells by 70% to 80% (Yodoi *et al.,* 1985; Epstein *et al.,* 1966). The authors demonstrated a down-regulation of the CD28 molecule on C10-stimulated YT cells. Because the ligand of CD28, the CD80 molecule (B7 antigen or BB1 antigen) (Linsley *et al.,* 1990), is expressed on Raji cells, diminished conjugation of YT cells to Raji cells due to the observed down-regulation of CD28 might explain the loss of cytotoxicity. Furthermore, CD30 stimulation by the monoclonal antibody C10 induces an increase of CD25 (IL-2R p55) expression and a decrease of CD45 expression. On the other hand, most CD30[+] cells showed increased expression of CD30 after *in vitro* stimulation with IL-2 (Ellis *et al.,* 1993). Signalling through the CD30 pathway would seem to participate in the regulation of the CD28/CD80 and IL-2R/IL-2 pathways. Furthermore, because CD30 regulates the CD28/CD80 system, CD30 could also be involved in the control of the CD40/CD40L signal, T-cell proliferation and B-cell maturation induced by T-cell cytokines (Clark and Ledbetter, 1994). Thus, the CD30 molecule seems to transmit information which is essential for the immune response.

Only little is known about the intracellular transduction of the CD30-mediated signal. Ellis and colleagues (1993) reported that cross-linking of the membrane-bound CD30 induced Ca^{2+} in TcR^+, but not in TcR^-, Jurkat T-cells.

Initial results on the regulation of the CD30 antigen have revealed that its expression is driven by a TATA-less promoter. Furthermore, there are upstream consensus sequences for the transcription factors SP1, AP-1, and NFkB. The determination of the biological activity of these transcription factors in the regulation of CD30 is the object of our current work.

GenBank Accession Number: the accession number for the CD30 cDNA sequence reported in this paper is M83554.

REFERENCES

Andreesen R, Brugger W, Löhr GW, Bross KJ (1989) Human macrophages can express the Hodgkin's cell-associated antigen Ki-1 (CD30). *Am J Pathol* **134**:187

Andreesen R, Osterholz J, Löhr GW, Bross KJ (1984) A Hodgkin cell-specific antigen is expressed on a subset of auto- and alloactivated T (helper) lymphoblasts. *Blood* **63**:1299

Bilbe G, Delabie J, Brüggen J, Richener H, Asselbergs FAM, Cerletti N, Sorg C, Odink K, Tarcsay L, Wiesendanger W, de Wolf-Peeters C, Shipman R (1992) Restin: a novel intermediate filament-associated protein highly expressed in the Reed-Sternberg cells of Hodgkin's disease. *EMBO J* **11**:2103

Bowen MA, Olsen KJ, Cheng L, Avila D, Podack ER (1993) Functional effects of CD30 on a large granular lymphoma cell line, YT. *J Immunol* **151**:5896

Boylston AW, Cosford P (1985) Growth of normal human T lymphocytes induced by monoclonal antibody to the T cell antigen receptor. *Eur J Immunol* **15**:738

Cailleau R, Young R, Olive M, Reeves WJ (1974) Breast tumour cell lines from pleural effusions. *J Natl Cancer Inst* **53**:661

Camerini D, Walz G, Loenen WAM, Borst J, Seed B (1991) The T cell activation antigen CD27 is a member of the NGF/TNF receptor gene family. *J Immunol* **147**:3165

Carbone A, Gloghini A, Volpe R (1992) Paraffin section immunohistochemistry in the diagnosis of Hodgkin's disease and anaplastic large cell (CD30+) lymphomas. *Virchows Arch A Pathol Anat* **420**:527

Clark EA, Ledbetter JA (1994) How B and T cells talk to each other. *Nature* **367**:425

Cordell JL, Falini B, Erber WN, Gosh AK, Abdulazir Z, MacDonald S, Pulford KA, Stein H, Mason DY (1984) Immunoenzymatic labeling of monoclonal antibodies using immune complexes of alkaline phosphatase and monoclonal anti-alkaline phosphatase (APAAP complexes). *J Histochem Cytochem* **32**:219

da Costa L, Carde P, Lumbroso JD, Ricard M, Pfreundschuh M, Bosq J, Manil L, Diehl V, Parmentier C (1992) Immunoscintigraphy in Hodgkin's disease and anaplastic large cell lymphomas: results in 18 patients using the iodine radiolabelled monoclonal antibody HRS-3. *Ann Oncol* **3 (Suppl 4)**:53

Dallenbach FE, Stein H (1989) Expression of T-cell-receptor ß chain in Reed-Sternberg cells. *Lancet* **ii**:828

Daus H, Trümper L, Roth J, Jacobs G, von Bonin F, Gause A, Pfreundschuh M (1994) Molecular analysis of single Reed-Sternberg cells of Hodgkin's disease: absence of IGH and TCR rearrangements: p53 and N-RAS mutations. *J Cancer Res Clin Oncol* **120 (Suppl)**:R56

Delabie J, Shipman R, Brüggen J, de Strooper B, van Leuven F, Tarcsay L, Cerletti N, Odink K, Diehl V, Bilbe G, de Wolf-Peeters C (1992) Expression of the novel intermediate filament-associated protein restin in Hodgkin's disease and anaplastic large-cell lymphoma. *Blood* **80**:2891

Diehl V, Kirchner HH, Burrichter H, Stein H, Fonatsch C, Gerdes J, Schaadt M, Heit W, Uchanska-Ziegler B, Ziegler A, Heintz F, Sueno K (1982) Characteristics of Hodgkin's disease-derived cell lines. *Cancer Treat Rep* **66**:615

Diehl V, Kirchner HH, Schaadt M, Fonatsch C, Stein H, Gerdes J, Boie, C (1981) Hodgkin's disease, establishment and characterisation of four *in vitro* cell lines. *J Cancer Res Clin Oncol* **101**:111

Dürkop H, Latza U, Hummel M, Eitelbach M, Seed B, Stein H (1992) Molecular cloning and expression of a new member of the nerve growth factor receptor family that is characteristic for Hodgkin's disease. *Cell* **68**:421

Ellis TM, Simms PE, Slivnick DJ, Jäck HM, Fisher RI (1993) CD30 is a signal-transducing molecule that defines a subset of human activated CD45RO+ T cells. *J Immunol* **151**:2380

Engelmann H, Novick D, Wallach D (1990) Two tumour necrosis factor-binding proteins purified from human urine. Evidence for immunological cross-reactivity with cell surface tumour necrosis factor receptors. *J Biol Chem* **265**:1531

Epstein MA, Achong BG, Barr YM, Zajac B, Henle G, Henle W (1966) Morphological and virological investigations on cultured Burkitt tumour lymphoblasts (strain Raji). *J Natl Cancer Inst* **37**:547

Falini B, Bolognesi A, Flenghi L, Tazzari PL, Broe MK, Stein H, Dürkop H, Aversa F, Corneli P, Pizzolo G, Barbabietola G, Sabattini E, Pileri S, Martelli MF, Stirpe F (1992a) Response of refractory Hodgkin's disease to monoclonal anti-CD30 immunotoxin. *Lancet* **339**:1195

Falini B, Flenghi L, Fedeli L, Broe MK, Bonino C, Stein H, Dürkop H, Bigerna B, Barbabietola G, Venturi S, Aversa F, Pizzolo G, Bartoli A, Pileri S, Sabattini E, Palumbo R, Martelli MF (1992b) *In vivo* targeting of Hodgkin and Reed-Sternberg cells of Hodgkin's disease with monoclonal antibody Ber-H2 (CD30): immunohistological evidence. *Br J Haematol* **82**:38

Falini B, Flenghi L, Pasqualucci L, Bolognesi A, Tazzari PL, Pileri S, Stein H, Martelli MF, Stirpe F (1993) Therapy of refractory Hodgkin's disease with anti-CD30/saporin immunotoxin. *Tissue Antigens* **42**:243.

Fonatsch C, Latza U, Dürkop H, Rieder H, Stein H (1992) Assignment of the human CD30 (Ki-1) gene to 1p36. *Genomics* **14**:825

Froese P, Lemke H, Gerdes J, Havensteen B, Schwarting R, Hansen H, Stein H (1987) Biochemical characterisation and biosynthesis of the Ki-1 antigen in Hodgkin-derived and virus-transformed human B and T lymphoid cell lines. *J Immunol* **139**:2081

Gause A, Jung W, Schmits R, Tschiersch A, Scholz R, Pohl C, Hasenclever D, Diehl V, Pfreundschuh M (1992) Soluble CD8, CD25 and CD30 antigens as prognostic markers in patients with untreated Hodgkin's lymphoma. *Ann Oncol* **3 (Suppl 4)**:49

Gause A, Pohl C, Tschiersch A, da Costa L, Jung W, Diehl V, Hasenclever D, Pfreundschuh M (1991) Clinical significance of soluble CD30 antigen in the sera of patients with untreated Hodgkin's disease. *Blood* **77**:1983

Gazdar AF, Carney DN, Bunn PA, Russell EK, Jaffe ES, Schechter GP, Guccion JG (1980) Mitogen requirements for the *in vitro* propagation of cutaneous T-cell lymphomas. *Blood* **55**:409

Hall PA, D'Ardenne AJ, Stansfeld AG (1988) Paraffin section immunohistochemistry. I. Non-Hodgkin's lymphoma. *Histopathol* **13**:149

Hanks SK, Quinn AM, Hunter T (1988) The protein kinase family: conserved features and deduced phylogeny of the catalytic domains. *Science* **241**:42

Hansen H, Bredfeldt G, Havsteen B, Lemke H (1990) Protein kinase activity of the intracellular but not of the membrane-associated form of the Ki-1 antigen (CD30). *Res Immunol* **141**:13

Hansen H, Lemke H, Bredfeldt G, Könnecke I, Havsteen B (1989) The Hodgkin-associated Ki-1 antigen exists in an intracellular and a membrane-bound form. *Biol Chem Hoppe-Seyler* **370**:409

Heller RA, Song K, Onasch M, Fischer WH, Chang D, Ringold, GM (1990) Complementary DNA cloning of a receptor for tumour necrosis factor and demonstration of a shed form of the receptor. *Proc Natl Acad Sci USA* **87**:6151

Herbst H, Tippelmann G, Anagnostopoulos I, Gerdes J, Schwarting R, Boehm T, Pileri S, Jones DB, Stein H (1989) Immunoglobulin and T-cell receptor gene rearrangements in Hodgkin's disease and Ki-1-positive anaplastic large cell lymphoma: dissociation between phenotype and genotype. *Leuk Res* **13**:103

Itoh N, Yonehara S, Ishii A, Yonehara M, Mizushima SI, Sameshima M, Hase A, Seto Y, Nagata S (1991) The polypeptide encoded by the cDNA for human cell surface antigen Fas can mediate apoptosis. *Cell* **66**:33

Johnson D, Lanahan A, Buck R, Sehgal A, Morgan C, Mercer E, Bothwell M, Chao M (1986) Expression and structure of the human NGF receptor. *Cell* **47**:545

Jones DB, Furley AJW, Gerdes J, Greaves MF, Stein H, Wright DH (1989) Phenotypic and genotypic analysis of two cell lines derived from Hodgkin's disease tissue biopsies. In *New Aspects in the Diagnosis and Treatment of Hodgkin's disease*. Diehl V, Pfreundschuh M, Loeffler M, eds (Berlin: Springer) p62

Jones DB, Scott CS, Wright DH, Stein H, Beverley PLC, Payne SV, Crawford DH (1985) Phenotypic analysis of an established cell line derived from a patient with Hodgkin's disease (HD). *Hematol Oncol* **3**:133

Josimovic-Alasevic O, Dürkop H, Schwarting R, Backé E, Stein H, Diamantstein T (1989) Ki-1 (CD30) antigen is released by Ki-1-positive tumour cells *in vitro* and *in vivo*. I. Partial characterisation of soluble Ki-1 antigen and detection of the antigen in cell culture supernatants and in serum by an enzyme-linked immunosorbent assay. *Eur J Immunol* **19**:157

Kemper O, Derré J, Cherif D, Engelmann H, Wallach D, Berger R (1991) The gene for the type II (p75) tumour necrosis factor receptor (TNF-RII) is localized on band 1p36.2-p36.3. *Hum Genet* **87**:623

Kohno T, Brewer MT, Baker SL, Schwartz PE, King MW, Hale KK, Squires CH, Thompson RC, Vannice JL (1990) A second tumour necrosis factor receptor gene product can shed a naturally occurring tumour necrosis factor inhibitor. *Proc Natl Acad Sci USA* **87**:8331

Kwon BS, Weissman SM (1989) cDNA sequences of two inducible T-cell genes. *Proc Natl Acad Sci USA* **86**:1963

Latza U, Dürkop H, Schnittger S, Ringeling J, Eitelbach F, Hummel M, Fonatsch C, Stein H (1994) The human OX40 homolog: cDNA structure, expression and chromosomal assignment of the ACT35 antigen. *Eur J Immunol* **24**:677

Latza U, Foss HD, Dürkop H, Eitelbach F, Dieckmann KP, Loy V, Unger M, Stein H (1995) Cellular and soluble CD30 antigen in embryonal carcinoma and during embryogenesis. *Am J Pathol* **146**:463

Linsley PS, Clark EA, Ledbetter JA (1990) T-cell antigen CD28 mediates adhesion with B cells by interacting with activation antigen B7/BB-1. *Proc Natl Acad Sci USA* **87**:5031

Loetscher H, Pan YCE, Lahm HW, Gentz R, Brockhaus M, Tabuchi H, Lesslauer W (1990) Molecular cloning and expression of the human 55 kd tumour necrosis factor receptor. *Cell* **61**:351

Mallett S, Barclay AN (1991) A new superfamily of cell surface proteins related to the nerve growth factor receptor. *Immunol Today* **12**:220

Mechtersheimer G, Möller P (1990) Expression of Ki-1 antigen (CD30) in mesenchymal tumours. *Cancer* **66**: 1732

Miettinen M (1992) CD30 distribution. Immunohistochemical study on formaldehyde-fixed, paraffin-embedded Hodgkin's and non-Hodgkin's lymphomas. *Arch Pathol Lab Med* **116**:1197

Morgan R, Smith SD, Hech K, Christy V, Mellenti JD, Warnke R, Cleary ML (1989) Lack of involvement of the c-*fms* and N-*myc* genes by chromosomal translocation t(2:5)(p23:q35) common to malignancies with features of so-called malignant histiocytosis. *Blood* **73**:2155

Nadali G, Vinante F, Ambrosetti A, Todeschini G, Veneri D, Zanotti R, Meneghini V, Ricetti MM, Benedetti F, Vassanelli A, Perona G, Chilosi M, Menestrina F, Fiacchini M, Stein H, Pizzolo G (1994) Serum levels of soluble CD30 are elevated in the majority of untreated patients with Hodgkin's disease and correlate with clinical features and prognosis. *J Clin Oncol* **12**:793

Nawrocki JF, Kirsten ES, Fisher RI (1988) Biochemical and structural properties of a Hodgkin's disease-related membrane protein. *J Immunol* **141**:672

Norton AJ, Isaacson PG (1987) Detailed phenotypic analysis of B-cell lymphoma using a panel of antibodies reactive in routinely fixed wax-embedded tissue. *Am J Pathol* **128**: 225

Pallesen G, Hamilton-Dutoit SJ (1988) Ki-1 (CD30) antigen is regularly expressed by tumour cells of embryonal carcinoma. *Am J Pathol* **133**:446

Pearson WR, Lipman DJ (1988) Improved tools for biological sequence comparison. *Proc Natl Acad Sci USA* **85**:2444

Perkins SL, Kjeldsberg CR (1993) Immunophenotyping of lymphomas and leukemias in paraffin-embedded tissues. *Am J Clin Pathol* **99**:362

Pfreundschuh M, Pohl C, Berenbeck C, Schroeder J, Jung W, Schmits R, Tschiersch A, Diehl V, Gause A (1990) Detection of a soluble form of the CD30 antigen in the sera of patients with lymphoma, adult T-cell leukemia, and infectious mononucleosis. *Int J Cancer* **45**:869

Pizzolo G, Stein H, Josimovic-Alasevic O, Vinante F, Zanotti R, Chilosi M, Feller AC, Diamantstein T (1990) Increased serum levels of soluble IL-2 receptor, CD30 and CD8 molecules, and gamma-interferon in angioimmunoblastic lymphadenopathy: possible pathogenetic role of immunoactivation mechanisms. *Br J Haematol* **75**:485

Pizzolo G, Vinante F, Morosato L, Nadali G, Chilosi M, Gandini G, Sinicco A, Raiteri R, Semenzato G, Stein H, Perona G (1994) High serum level of the soluble form of CD30 molecule in the early phase of HIV-1 infection as an independent predictor of disease progression. *AIDS* **8**:741

Ralfkiaer E, Bosq J, Gatter KC, Schwarting R, Gerdes J, Stein H, Mason DY (1987) Expression of a Hodgkin and Reed-Sternberg cell associated antigen (Ki-1) in cutaneous lymphoid infiltrates. *Arch Dermatol Res* **279**:285

Ree HJ, Neimann RS, Martin AW, Dallenbach F, Stein H (1989) Paraffin section markers for Reed-Sternberg cells: a comparative study of peanut agglutinin, Leu-M1, LN-2, and Ber-H2. *Cancer* **63**:2030

Rohde D, Hansen H, Hafner M, Lange H, Mielke V, Hansmann ML, Lemke H (1992) Cellular localisations and processing of the two molecular forms of the Hodgkin-associated Ki-1 (CD30) antigen. The protein kinase Ki-1/57 occurs in the nucleus. *Am J Pathol* **140**:473

Sarker AB, Akagi T, Jeon HJ, Miyake K, Murakami I, Yoshino T, Takahashi K, Nose S (1992) Bauhinia purpurea-a new paraffin section marker for Reed-Sternberg cells of Hodgkin's disease. A comparison with Leu-M1 (CD15), LN2 (CD74), peanut agglutinin, and Ber-H2 (CD30). *Am J Pathol* **141**:19

Schaadt M, Diehl V, Stein H, Fonatsch C, Kirchner HH (1980) Two neoplastic cell lines with unique features derived from Hodgkin's disease. *Int J Cancer* **26**:723

Schall TJ, Lewis M, Koller KJ, Lee A, Rice GC, Wong GHW, Gatanaga T, Granger GA, Lentz R, Raab H, Kohr W, Goeddel DV (1990) Molecular cloning and expression of a receptor for human tumour necrosis factor. *Cell* **61**:361

Schwab U, Stein H, Gerdes J, Lemke H, Kirchner H, Schaadt M, Diehl V (1982) Production of a monoclonal antibody specific for Hodgkin and Sternberg-Reed cells of Hodgkin's disease and a subset of normal lymphoid cells. *Nature* **299**:65

Schwarting R, Gerdes J, Dürkop H, Falini B, Pileri S, Stein H (1989) BER-H2: a new anti-Ki-1 (CD30) monoclonal antibody directed at a formol-resistant epitope. *Blood* **74**:1678

Smith C, Davis T, Anderson D, Solam L, Beckmann MP, Jerzy R, Dower SK, Cosman D, Goodwin RG (1990) A receptor for tumour necrosis factor defines an unusual family of cellular and viral proteins. *Science* **248**:1019

Smith CA, Gruss HJ, Davis T, Anderson D, Farrah T, Baker E, Sutherland GR, Brannan CI, Copeland NG, Jenkins NA, Grabstein KH, Gliniak B, McAlister IB, Fanslow W, Anderson M, Falk B, Gimpel S, Gillis S, Din WS, Goodwin RG, Armitage RJ (1993) CD30 antigen, a marker for Hodgkin's lymphoma, is a receptor whose ligand defines an emerging family of cytokines with homology to TNF. *Cell* **73**:1349

Stamenkovic I, Clark EA, Seed B (1989) A B-lymphocyte activation molecule related to the nerve growth factor receptor and induced by cytokines in carcinomas. *EMBO J* **8**:1403

Stein H, Dallenbach F (1992) Diffuse large cell lymphomas of B and T cell type. In *Neoplastic Hematology.* Knowles DM, ed (Baltimore: William & Wilkins) p675

Stein H, Hansmann ML, Lennert K, Brandtzaeg P, Gatter KC, Mason DY (1986) Reed-Sternberg cells in lymphocyte-predominant Hodgkin's disease of nodular subtype contain J chain. *Am J Pathol* **86**:292

Stein H, Mason DY, Gerdes J, O'Connor N, Wainscoat J, Pallesen G, Gatter K, Falini B, Delsol G, Lemke H, Schwarting R, Lennert K (1985) The expression of the Hodgkin's disease associated antigen Ki-1 in reactive and neoplastic lymphoid tissue: evidence that Reed-Sternberg cells and histiocytic malignancies are derived from activated lymphoid cells. *Blood* **66**:848

Sundström C, Nilsson K (1976) Establishment and characterisation of a human histiocytic lymphoma cell line (U-937). *Int J Cancer* **17**:565

Tamaru J, Hummel M, Zemlin M, Kalvelage B, Stein H (1994) Hodgkin's disease with a B cell phenotype often shows a VDJ rearrangement and somatic mutations in the V_H genes. *Blood* **84**:708

van der Putte SCJ, Toonstra J, van Wichen DF, van Unnik JAM, van Vloten WA (1988) The expression of the Hodgkin's disease-associated antigen Ki-1 in cutaneous infiltrates. *Acta Derm Venereol (Stockholm)* **68**:202

Vinante F, Morosato L, de Sabata D, Pizzolo G (1991) Soluble molecules in lymphoproliferative disorders. *Leukemia* **5 (Suppl 1)**:18

Yodoi J, Teshigawara K, Nikaido T, Fukui K, Noma T, Honjo T, Takigawa M, Sasaki M, Minato N, Tsudo M, Uchiyama T, Maeda M (1985) TCGF (IL 2)-receptor inducing factor(s). I. Regulation of IL 2 receptor on a natural killer-like cell line (YT cells). *J Immunol* **134**:1623

CD30 LIGAND: CLONING, CHARACTERISATION AND BIOLOGICAL ACTIVITIES

Hans-Jürgen Gruss[1,2] and Steven K Dower[1]

[1]Department of Biochemistry, Immunex Research and Development Corporation, Seattle, WA 98101, USA;
[2]Department of Medical Oncology and Applied Molecular Biology, Virchow Klinikum, Robert-Rössle Cancer Center and Max Delbruck Center for Molecular Medicine, D-13122, Berlin, Germany

ABSTRACT

Hodgkin's disease (HD) is characterised by the presence of a small number (< 1% of total tumour mass) of the typical, presumed malignant Hodgkin and Reed-Sternberg (H-RS) cells in a hyperplastic background of normal, reactive lymphocytes, plasma cells, histiocytes, neutrophils, eosinophils, and stromal cells. The histopathological presentation and characteristic clinical features of HD correlate with an unbalanced production of multiple cytokines. H-RS cells produce various growth factors, cytokine receptors and activation antigens, implying a role for growth factors in the pathophysiology of HD. HD may therefore be characterised as a tumour of cytokine producing cells. The CD30 antigen has been described as a marker for cultured and primary H-RS cells, and found to be overexpressed in HD and some large cell anaplastic non-Hodgkin's lymphoma cases. The molecular cloning of the CD30 antigen revealed that CD30 is a member of the tumour necrosis factor/nerve growth factor receptor superfamily. The cloning of the cognate for CD30, currently termed CD30 ligand, confirmed that the CD30 antigen functions as a cytokine receptor. Recombinant CD30 ligand is a type-II membrane-associated protein with pleiotropic biological activities for different $CD30^+$ lymphoma types, but also normal immune system cells, predominantly T-cells. CD30L belongs to the emerging tumour necrosis factor ligand superfamily by virtue of homology to CD27L, CD40L, TNF and other members. These findings suggest that CD30-CD30L interactions could have a pathophysiological role in HD and large cell anaplastic lymphomas, and could also be involved in the activation and function of the immune system, particularly T-cell responses.

INTRODUCTION

Lymphomas, malignant neoplasms of the lymphoid tissues, are classified into two groups, Hodgkin's disease (HD) and non-Hodgkin's lymphomas (NHLs). HD contains a low proportion of the presumed malignant Hodgkin and Reed-Sternberg (H-RS) cells (usually less than 1% of the total tumour mass). The diagnosis of HD is based on the histopathological presence of the typical H-RS cells in a normal reactive milieu with lymphocytes, plasma cells, histiocytes, neutrophils, eosinophils and stromal cells (Kaplan, 1980). Almost every cell lineage and type has been described as the origin and normal counterpart of the H-RS cells (Drexler, 1992 and 1993).

Clinical presentation of HD includes constitutional "B" symptoms with fever, night sweats and weight loss, elevations of acute phase proteins in the serum, generalised itching, presence of mild thrombocytosis, and impaired immune functions. Certain histopathological features such as sclerosis, polykaryon formation, T-cell accumulation and rosetting around H-RS cells, T-cell activation, plasmacytosis, and eosinophilia are common features of active HD. Clinical and histopathological findings can be related to an abnormal or unbalanced secretion/production of cytokines (Gruss et al., 1992) and thus HD can be considered to be a tumour of cytokine producing cells (Gruss et al., 1994a).

A small number of HD-derived cell lines have been established and claimed to be derived from the in vivo H-RS cells (Drexler, 1993). Antibodies against CD30 were generated against L-428 cultured H-RS cells (Schwab et al., 1982). Using immunostaining techniques the H-RS cells, except those occurring in the lymphocyte-predominant subform, are commonly stained by antibodies reacting with CD30 and CD15 antigens, but also other less specific activation markers such as CD25, CD71 and HLA-DR can be detected (Drexler, 1993). Recent cDNA cloning studies showed that CD30 is a member of the tumour necrosis factor (TNF)/nerve growth factor (NGF) receptor superfamily (Dürkop et al., 1992; see Dürkop et al., this volume). The subsequently cloned membrane-bound CD30 ligand (CD30L) belongs to the TNF ligand superfamily and confirms that CD30 acts as a cytokine receptor (Smith et al., 1993). CD30L has proliferative effects on some HD-derived cell lines and a T-ALL cell line, but has strong antiproliferative effects on CD30[+] large cell anaplastic lymphoma (LCAL) cell lines (Gruss et al., 1994b and 1994c). CD30L has co-stimulatory activity for peripheral blood T (PBT) cells (Smith et al., 1993). The characteristics of the human CD30 and CD30L molecules are summarised in Table 1.

This overview will discuss the recent cloning and characterisation of the CD30 ligand. Further, it will summarise preliminary data on the in vitro activities of CD30L and its potential involvement in the pathology of CD30[+] lymphomas, particularly HD.

CD30 IS A LYMPHOID ACTIVATION ANTIGEN, BUT WAS ORIGINALLY IDENTIFIED AS A CELL SURFACE ANTIGEN ON H-RS CELLS

HD-derived cell lines were used to develop monoclonal antibodies that could be used to visualise H-RS cells in tissue sections. CD30 antigen was originally identified as a cell surface antigen on primary and cultured H-RS cells by the monoclonal antibody Ki-1 (Schwab *et al.*, 1982). Subsequently, monoclonal antibodies with similar reactivity were obtained, and have been assigned to the CD30 cluster. CD30 antibodies detect a phosphorylated membrane glycoprotein of 120 kD, which derives from a non-phosphorylated 84 kD apoprotein (Froese *et al.*, 1987; see Dürkop *et al.*, this volume). Subsequent studies have demonstrated that the CD30 antigen is neither cell lineage-restricted nor specific for H-RS cells (Drexler *et al.*, 1989; Drexler, 1992 and 1993). CD30 has also been detected on mitogen- or antigen-activated T and B-cells, Epstein-Barr virus (EBV)-transformed B-cells and human T-cell lymphotropic virus (HTLV)-I or -II transformed lymphocytes (Herbst *et al.*, 1993). Also a small population of lymphoid cells, of mainly parafollicular localisation in reactive lymph nodes and tonsils, has been reported to be CD30[+] (Schwab *et al.*, 1982; Stein *et al.*, 1982). With the exception of the lymphocyte predominant form (only ~ 32% positivity), the CD30 antigen is present on the H-RS cells of 84 - 91% of HD cases (Drexler, 1992). CD30 antibodies also identify a new entity of NHL with anaplastic morphology (CD30[+] large cell anaplastic lymphomas/LCAL) (Stein *et al.*, 1985). Quantitative detection of a soluble 85 kD form of CD30 in sera from HD patients, has been shown to be a marker for disease activity with prognostic significance (Nadali *et al.*, 1994).

Recently, a cDNA encoding CD30 was cloned (Dürkop *et al.*, 1992; see Dürkop *et al.*, this volume). Sequence analysis revealed that CD30 is a member of the TNF/NGF receptor superfamily by virtue of homology in the extracellular domain, with characteristic cysteine-rich motifs. This receptor family also includes the low affinity NGF receptor, two distinct receptors for TNF and lymphotoxin (LT)-α (p60 TNFR-I and p80 TNFR-II), TNFR-related protein (receptor for LT-α/β_2 heterotrimer), CD27, CD40, 4-1BB, OX40, CD95 (FAS/APO-1) and several viral open reading frames encoding soluble secreted homologues (Smith *et al.*, 1994). The CD30 protein has 595 amino acids and can be translated from two mRNA species of 2.6 kb and 3.8 kb, which utilise different poly-adenylation signals (Dürkop *et al.*, 1992). The CD30 molecule was predicted to be a molecule with receptor function.

CLONING OF A COGNATE FOR THE CD30 ANTIGEN

Sequence homology between TNF (TNF-α), LT-α (TNF-β), and CD40 ligand (CD40L) suggested that the potential cognates for the TNF receptor superfamily members might be membrane-bound cytokines/ligands and form a parallel ligand family (Farrah and Smith, 1992). In order to identify a cognate for CD30, cell lines were screened by flow cytometry for cell surface expression using a CD30-Fc fusion-protein consisting of the extracellular portion of human CD30 linked to the Fc domains of human immunoglobulin G_1 (IgG$_1$) heavy chain (Smith *et al.*, 1993). The chimeric, soluble CD30 receptor was transiently expressed in CV-1/EBNA cells and purified by protein A chromatography. CD30-Fc was used to show, by flow cytometry, that activated murine T-cell clones and human PBT cells express membrane-associated CD30L (Fig. 1). The murine CD30L cDNA was isolated using a direct expression cloning strategy performed on a library prepared from activated 7B9 cells, a T-cell clone (Smith *et al.*, 1993). The cDNA clone corresponding to murine CD30L encodes a 239 amino acid protein with characteristic features of a type-II membrane protein. The predicted protein lacks a signal peptide, contains an internal 21-residue hydrophobic transmembrane domain and a 172-residue C-terminal extracellular domain containing six potential N-linked glycosylation sites. The human CD30L cDNA clone was isolated by cross-hybridisation, using the murine cDNA clone as a probe, from a λgt10 library constructed from PBT mRNA after 12 hours stimulation. The human CD30L cDNA encodes a 234-residue type-II membrane protein with 72% amino acid homology to the murine ligand and five potential N-linked glycosylation sites. The schematic organisation of the human CD30L protein is shown in Fig. 2.

Table 1: Characteristics of the human CD30 and CD30L molecules.

Characteristic	CD30	CD30L
Molecular weight	120 kD	26-40 kD
Length (amino acids)	595	234
Chromosomal location	1p36	9q33
Form a) membrane-bound b) soluble	type I 88 kD	type II ?
Expression a) immune system	activated T-cells and clones NK cell clones virus-transformed B-cells*	activated T-cells and clones stimulated monocytes/macrophages granulocytes subset of B-cells Burkitt-like lymphoma cells
b) malignancies	Hodgkin and Reed-Sternberg cells subgroup of large cell anaplastic lymphoma cells other blastoid lymphoma cells	
Function	cytokine receptor	cytokine (membrane-bound)
Family	TNF/NGF receptor superfamily	TNF ligand superfamily

*Has also been reported to be expressed on stimulated (SAC) tonsil B-cells (Stein et al., 1985; Schwarting et al., 1987)

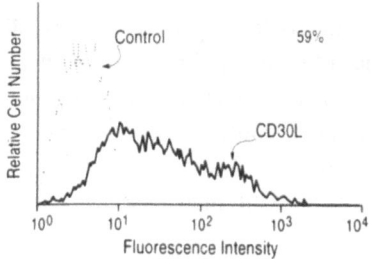

Fig. 1: Expression of membrane-bound CD30L on activated peripheral blood T-cells using flow cytometry. Peripheral blood T-cells were stimulated with PMA and ionomycin for 16 hours and then stained with biotinylated CD30-Fc followed by phycoerythrin labelling. Analysis was performed on a FACSscan and results presented as a histogram. The percentage of CD30L+ cells is indicated.

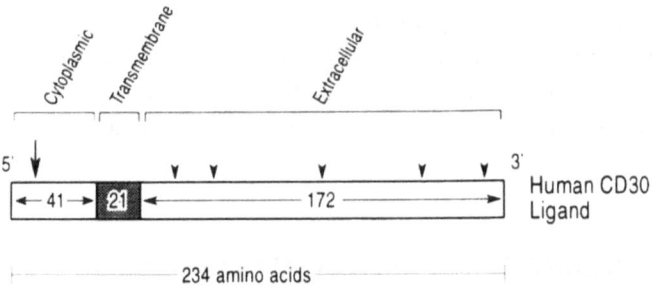

Fig. 2: Schematic organisation of the human CD30L protein, a type-II membrane protein. Arrow, start codon; arrowheads, five potential N-linked glycosylation sites.

CD30L shows sequence homology with other members of the TNF ligand family of cytokines. This family currently includes TNF (TNF- α), LT-α (TNF-β), LT-β, CD27L, CD30L, CD40L, 4-1BBL, OX40L and CD95L (FASL) (Smith *et al.*, 1994). Sequence homology between CD30L and other members of the TNF ligand superfamily is limited to the C-terminal (receptor-binding) region of the extracellular domains. CD30L exhibits the lowest sequence similarity with 12-18% identity with various other family members. For comparison, human TNF and LT-α share 36% identity, human TNF and CD40L share 27% identity, and human LT-α and CD40L share 25% identity. CD30L possesses the two sequence motifs that are strictly conserved across the family: the L-X-W motif and G-φ-Y-φ-φ-X-X-Q-φ motif. TNF and LT-α have been shown to fold into a β-sandwich (Eck *et al.*, 1989; Jones *et al.*, 1990). Based on the sequence homology and possible conserved disulphide pairing, CD30L could fold into a similar tertiary structure.

MOLECULAR AND BIOCHEMICAL CHARACTERISATION OF CD30L

Recombinant murine and human CD30L, expressed as membrane-bound forms on CV-1/EBNA cells, displayed a single class of binding sites for CD30-Fc fusion protein (K_D of ~ 3nM). Similar affinities are seen with native CD30L. CD30 ligands were analysed by surface radioiodination and immunoprecipitation with CD30-Fc. Under reducing conditions, the dominant species for both human and murine CD30L is seen as a diffuse band migrating at 40 kD. The predicted molecular weight for the CD30L protein is 26 kD, thus the cell surface form may contain up to 14 kD of oligosaccharides, a reasonable amount, based on the presence of several N-linked glycosylation sites (Fig. 2). Disulphide-linked dimers of human CD30L appear under non-reducing conditions, and even higher oligomers are seen for the murine CD30L. Reducing conditions can convert most, but not all, of these oligomers to monomers. These results raise the possibility that disulphide-linked oligomeric forms of native CD30L may exist.

The mouse chromosomal location of CD30L was determined by interspecific backcross analysis. The mapping results indicate that CD30L is located in the proximal region of mouse chromosome 4 linked to the loci of major urinary protein, hexabrachion, ribonucleotide reductase-M2 pseudogene 1, and brown. No recombinations were found between hexabrachion and CD30L, suggesting that these loci are within 1.7 cM of each other. To determine the chromosomal location of human CD30L, 25 metaphases from healthy males were examined by fluorescence *in situ* hybridisation with a biotinylated human CD30L probe. Of these metaphases, signals were found on one or both chromatids of chromosome 9 in the region 9q32-34. Sixty-seven per cent of this signal was at 9q33. No obvious disease-marker association was found for the human or mouse CD30L chromosomal locus.

CD30L EXPRESSION OF IMMUNE SYSTEM CELLS AND MALIGNANT HEMATOPOIETIC TUMOUR CELL LINES

The CD30 monoclonal antibody Ki-1 was first reported to stain specifically a cell surface antigen on primary and cultured H-RS cells (Schwab *et al.*, 1982; Stein *et al.*, 1982). Subsequent studies clearly demonstrated that the CD30 antigen is neither cell lineage-restricted nor specific for H-RS cells (Drexler, 1982). The antigen is present on normal T-cells and B-cells upon activation with various stimuli or after viral transformation (Stein *et al.*, 1985; Schwarting *et al.*, 1989). Controversial data exist regarding CD30 expression on interferon (IFN)-γ activated blood monocytes/macrophages; some reports claim that CD30 is a "lymphoid cell-specific activation marker" (Andreesen *et al.*, 1989; Gruss *et al.*, 1994b; Herbst *et al.*, 1993). It was of interest to determine the pattern of CD30L expression on lymphoid cells. Northern blot analysis revealed that human PBT cells induced with either calcium ionophore, anti-CD3, PMA or PHA, uninduced tonsil T-cells, lipopolysaccharide (LPS)-induced or IFN-γ stimulated monocytes/macrophages, peripheral blood and tonsil B-cells and unstimulated neutrophils all express a single mRNA transcript for CD30L. Interleukin (IL)-7 treated PBT cells, uninduced Jurkat cells, LPS-activated THP-1 monocyte line and granulocyte macrophage colony-stimulating factor (GM-CSF) stimulated monocytes did not express CD30L mRNA. Similar expression of CD30L by activated T-cells and monocytes/macrophages was seen in the murine system. CD30L expression has been found in CD4[+] and CD8[+] subsets, and also in all CD45 isoforms (Gruss *et al.*, 1995). Thus, CD30L expression seems broadly expressed on induced T-cells and monocytes/macrophages, and also granulocytes and a subset of B-cells. CD30L is a membrane bound protein, but whether it, like TNF, also exists in an alternative soluble secreted or cleaved form is presently unclear.

CD30L mRNA expression was also studied in a panel of 105 continuous human leukaemia-lymphoma cell lines (10 pre B-cell lines, 26 B-cell lines, 8 plasma cell lines, 19 T-cell lines,

7 HD-derived and malignant histiocytosis cell lines, 14 myeloid cell lines, 17 monocytic cell lines and 4 megakaryocytic-erythroid cell lines). Six of the 26 B-cell lines expressed CD30L mRNA. CD30L transcripts were found in four Burkitt's lymphoma, one Burkitt-type acute lymphoblastic leukaemia and one NHL cell line (Gruss *et al.*, 1994b). All HD-derived cell lines were CD30L mRNA and surface protein expression negative (Gruss *et al.*, 1994c). So far, no cell line has been identified that expresses both CD30L and CD30 mRNA or protein. In contrast, PBT cells and T-cell clones can express both CD30 and CD30L. Thus, CD30L expression appears restricted to a specific type of neoplastic haematopoietic tumour cell, namely Burkitt-type cells. The functional relevance of this expression remains to be determined.

BIOLOGICAL ACTIVITIES OF RECOMBINANT CD30L ON IMMUNE SYSTEM CELLS

The members of the TNF ligand family have been reported to co-stimulate the proliferation of PBT cells (Smith *et al.*, 1993). We were interested in studying the response of purified human PBT cells to culture for 5 days in the presence of immobilised CD3 antibody and a titration of fixed CV-1/EBNA cells transfected with either vector alone or human full length CD30L. In contrast with control cells transfected with vector alone, CV-1/EBNA cells expressing CD30L co-stimulated proliferation of T-cells in a dose-dependent manner. This enhanced proliferation could be blocked by the inclusion of 10μg/mL CD30-Fc. CD30L was unable to induce proliferation of resting T-cells secondary to the lack of CD30 surface expression. We generated two new CD30 monoclonal antibodies, M44 and M67, against the soluble CD30-Fc protein (Smith *et al.*, 1993; Gruss *et al.*, 1994c). These antibodies recognise cell surface CD30 and when immobilised induce proliferation of either anti-CD3, PMA/ionomycin or PHA activated T-cells in a dose-dependent manner. These data suggest that the bivalent antibody mimics ligand-induced receptor cross-linking. The M44 and M67 antibodies have no activity in the T-cell co-stimulation assay when added in solution.

TNF and CD40L have been shown to induce up-regulation of adhesion and activation molecules, particularly ICAM-1 (Armitage *et al.*, 1993). We have demonstrated that recombinant CD30L and CD30 monoclonal antibodies (M44 and M67) are also capable of enhancing ICAM-1 surface expression on PBT cells under co-stimulation conditions (Gruss *et al.*, 1995). The enhanced ICAM-1 expression is also associated with a co-stimulatory activity for PBT cells. CD30L promotes co-stimulatory proliferation (direct pathway), but also induces other molecules involved in the mitogenic cascade of T-cell activation (indirect pathway).

We were interested to test recombinant CD30L for activity on B-cells and monocytes/macrophages. Primary murine B-cells can be induced to proliferate and secrete polyclonal Ig by CD30L co-stimulation with IL-4 and IL-5 (Grabstein *et al.*, manuscript in preparation). This effect is so far not reproducible for human B-cells stimulated in a variety of ways. Further studies revealed that cytokine or mitogen stimulated human tonsil B-cells do not express detectable CD30 at the protein or mRNA level (Gruss *et al.*, 1995) and this may explain the different response to CD30L by human and murine B-cells. In our hands, CD30L does not play a significant biological role in human B-cell proliferation, maturation and/or differentiation, although CD30 is expressed by certain virus-transformed B-cell lines. Monocytes/macrophages, either untreated or stimulated with a panel of known monocytic activators, are negative for CD30 expression by Northern blot analysis and flow cytometry (Gruss et al., 1994b). It remains to be seen, however, if particular disease conditions may be able to induce CD30 expression on human monocytes or B-cells.

In summary, T-cells express both CD30 and CD30L and the CD30-CD30L interaction may be another critical pathway in the regulation of T-cell activation and functional activities. CD30L-expressing granulocytes or activated monocytes may be able to interact with the CD30 expressing PBT cells, mainly CD4$^+$, CD45RO$^+$ memory cells (Poppema, 1989; Ellis et al., 1993). IL-2 is further able to enhance mitogen or antigen induced CD30 expression on PBT cells (Ellis et al., 1993). More data are needed to elucidate the detailed role of CD30L in the immune system.

BIOLOGICAL ACTIVITIES OF RECOMBINANT CD30L
ON CD30 POSITIVE LYMPHOMA CELLS

In addition to H-RS cells of HD, CD30 is expressed to variable levels on different NHLs, such as LCAL, cutaneous T-cell lymphomas, nodular small cleaved-cell lymphomas, lymphocytic lymphomas, peripheral T-cell lymphomas, Lennert's lymphomas, immunoblastic lymphomas, T-cell leukaemia/lymphomas (ATLL) adult T-cell leukaemia (T-ALL), and centroblastic/centrocytic follicular lymphomas (Gruss et al., 1994c). The association of the CD30 antigen with lymphoid malignancies has proven to be a useful marker for the identification of malignant cells within lymphoid tissues, particularly lymph nodes. CD30 has also been reported, however, to be expressed by a proportion of embryonal carcinomas, non-embryonal carcinomas, malignant melanomas, mesenchymal tumours and some myeloid cell lines (Gruss et al., 1995).

For identification of biological targets for CD30L, we studied the expression of CD30 on a panel of Epstein-Barr virus (EBV) immortalised B-cell lines, EBV$^+$ or EBV$^-$ NHL cell lines, and HD-derived cell lines (Gruss et al., 1994c). Six of six EBV-immortalised normal lymphoblastoid B-cell lines studied expressed CD30. Three of five Burkitt NHL cell lines (BL-APB, Raji, Jijoye) were CD30 positive. All eight LCAL cell lines and six EBV$^-$ HD-derived cell lines strongly expressed CD30. The HD-derived cell lines HDLM-2 and L-540 showed the highest CD30 expression levels. Further, the EBV$^-$ adult T-ALL cell line KE-37 was positive for CD30. It is of interest that the EBV$^-$ cell lines, such as HDLM-2, L-540, Karpas 299, KE-37 expressed the highest surface levels of the CD30 antigen.

Recombinant human CD30L or anti-CD30 monoclonal antibodies were added to cultured CD30$^+$ EBV-immortalised B-cell lines and NHL cell lines, and examined for effects on cell growth. None of the cell lines responded with increased ^{3}H-thymidine uptake or cell number, regardless of the culture conditions. Furthermore, cytokine secretion (e.g., IL-6, IL-8, TNF) from these cell lines was not affected by the addition of CD30L or anti CD30 monoclonal antibodies. In addition, we examined the effects of CD30L on Ig isotype secretion by the CD30$^+$ non-tumour and tumour B-cell lines. The lymphoblastoid B-cell lines CB33 and B-MNK constitutively secreted IgM and IgA, respectively. The other B lymphoblastoid cell lines did not spontaneously secrete any Ig subtypes and no Ig secretion was induced by CD30L with or without co-stimulation with IL-2, IL-4 or IL-10. In contrast, the addition of CD30L or the anti-CD30 monoclonal antibodies M44 and M67 enhanced IgM and IgA secretion by CB33 and B-MNK B-cell lines, respectively. Co-stimulation with IL-2, IL-4, or IL-10 did not further enhance CD30L-induced Ig secretion. Two of the three CD30$^+$ Burkitt lymphoma cell lines constitutively secreted IgM, but there was no enhancement or induction by addition of the CD30L either in the presence or absence of IL-2, IL-4, or IL-10. In summary, the Ig secretion of some normal EBV-immortalised lymphoblastoid CD30$^+$ B-cell lines could be enhanced by the addition of CD30L or M44 and M67 anti-CD30 monoclonal antibodies and this was independent of co-stimulatory effects of IL-2, IL-4 or IL-10. In contrast, CD30$^+$ Burkitt lymphoma cell lines did not alter Ig secretion after addition of CD30L with or without IL-2, IL-4, or IL-10 present. The biological relevance of CD30 expression on Burkitt lymphoma cells therefore remains unclear.

CD30L-expressing CV-1/EBNA cells and anti-CD30 monoclonal antibodies (M44, M67, Ki-1, Ber-H2) were tested for their ability to enhance proliferation of the HD-derived cell lines HDLM-2, L-540 (both "T cell-like"), KM-H2, and L-428 (both "B cell-like") (Drexler, 1993; Gruss *et al.*, 1994c). Neither of the "B cell-like" HD-derived cell lines responded to either CD30L or CD30 monoclonal antibodies by showing altered growth rates after up to 72 hours stimulation. In contrast, proliferation of the "T cell-like" HD-derived cell lines was enhanced after addition of CD30L or the CD30 monoclonal antibodies M44 and M67. Both murine and human CD30L, expressed on CV-1/EBNA cells, enhanced ^{3}H-thymidine uptake by HDLM-2 and L-540 H-RS cells compared with cells cultured with medium or CV-1/EBNA cells transfected with the vector alone. Also, immobilised CD30 monoclonal antibodies M44 and M67 enhanced proliferation of HDLM-2 and L-540, but this effect was not seen with the CD30 antibodies Ki-1 and Ber-H2. The enhanced proliferation of HDLM-2 and L-540 cells with CD30L or M44 and M67 anti-CD30 treatment was time- and dose-dependent. CD30-mediated maximal proliferative effects were seen for a culture period of 48 to 72 hours and could be blocked by the addition of excess soluble CD30-Fc protein. Taken together, murine and human CD30L were mitogenic to the "T cell-like" cultured H-RS cells (HDLM-2, L-540), similar to the CD30L response seen for PBT cells. Both of our new CD30 monoclonal antibodies (M44 and M67) mimicked the CD30L-induced biological activities. In contrast, the CD30 monoclonal antibodies Ki-1 and Ber-H2 had no "agonistic" or "antagonistic" biological effects. The difference in CD30L responses of "T cell-like" and "B cell-like" HD-derived cell lines might be explained by different biological involvement and functional roles of the CD30-CD30L interaction for "T cell-like" and "B cell-like" primary HD cases, but also by the possibility that some HD-derived cell lines have lost the ability to respond to CD30L *in vitro*.

To exclude a CD30L-dependent autocrine growth loop for the HD-derived cell lines, cultured H-RS cells were analysed for CD30L mRNA and surface protein expression. None of the six HD-derived cell lines examined showed CD30L mRNA or protein expression, constitutively or after stimulation with PMA or cytokines (IL-1, IL-2, IL-4, IL-6, IL-9, TNF). Thus, the HD-derived cell lines do not appear to use the CD30-CD30L interaction in an autocrine fashion.

Recently, several groups have shown that primary and cultured H-RS cells produce and secrete a panel of different cytokines, which correlates with typical clinical and pathologic features of HD (Gruss *et al.*, 1994a). Moreover, HD is considered to be a tumour of cytokine-producing cells (Gruss *et al.*, 1994a). In addition, we have recently reported that primary and cultured H-RS cells express high levels of cell surface CD40, another member of the TNF receptor superfamily (Gruss *et al.*, 1994d). Recombinant soluble (leucine zipper trimer) or membrane-associated CD40L induced IL-8 secretion and enhanced IL-6, TNF, and LT-α release from cultured H-RS cells. CD40L had no mitogenic activity for cultured H-RS cells, but enhanced the expression of the surface molecules ICAM-1 and B7-1, both of which are overexpressed on primary H-RS cells (Delabie *et al.*, 1993; Gruss *et al.*, 1993a; Munro *et al.*, 1994; Pizzolo *et al.*, 1993). Further, CD40L induced a 40-60% reduction of CD30 antigen expression. CD40-CD40L interaction might be an element in the deregulated cytokine network and cell contact-dependent activation cascade characteristic of HD. We were interested to investigate whether recombinant CD30L is involved in similar biological pathways. CD30L was able to enhance IL-6, TNF and LT-α secretion from cultured H-RS cells, but did not induce IL-8 production. Furthermore, recombinant CD30L enhanced surface expression of ICAM-1 and B7 family members in a similar fashion to that seen for CD40L. It therefore seems likely that both CD30L and CD40L share similar biological activities in the pathogenesis of HD and represent a redundant network of signals involved in an unbalanced or deregulated cytokine cascade, typical for HD.

The adult T-ALL cell line KE-37 showed a strong constitutive surface and mRNA expression of the CD30 antigen. The addition of CV-1/EBNA cells expressing CD30L or the immobilised anti-CD30 monoclonal antibodies M44 and M67 also enhanced proliferation of KE-37 cells (Gruss *et al.*, 1994c). Again, Ki-1 and Ber-H2 antibodies did not alter proliferation. In summary, CD30L delivers a mitogenic activity to some "T cell-like" lymphoma/leukaemia cells, such as HD and adult T-ALL.

Some LCALs are characterised by the presence of strong CD30 surface expression (Stein *et al.*, 1985). We were interested to examine the biological activities of CD30L on this NHL subgroup using cell lines, for example Karpas 299, as a model. Karpas 299 cells incubated with CD30L, M44 or M67 anti-CD30 monoclonal antibodies showed a reduced proliferation in comparison to cells cultured with CV-1/EBNA cells transfected with the vector alone, isotype-matched control antibodies or medium. The reduction of proliferation was time- and dose-dependent, measurable after 24 hours in culture and was maximal 72 hours after initiation of the cultures. CD30L and both anti-CD30 monoclonal antibodies induced 15-23% specific ^{51}Cr release from labelled Karpas 299 cells as a measurement of cytolytic cell death. The CD30-mediated effect on Karpas 299 could be reversed by the addition of excess soluble CD30-Fc. Karpas 299 cells stimulated with control reagents showed exponential growth without significant reduction in viability. In contrast, Karpas 299 cells incubated with CD30L or M44 and M67 anti-CD30 monoclonal antibodies showed up to a 30-40% reduction of cell viability and arrested cell growth. These data were confirmed using multiparameter flow cytometric Hoechst staining and DNA cell cycle analysis. The mechanism for the anti-proliferative effect of CD30L on the LCAL Karpas 299 cells is presently unclear, but occurs rather late in comparison to the induction of programmed cell death through the FAS antigen and is not associated with apoptotic DNA fragmentation. CD30L seems to have a cytolytic and cytostatic biological component causing the anti-proliferative effect. Anti-proliferative effects of CD30L were found for seven of eight permanent CD30⁺ LCAL cell lines (Gruss *et al.*, 1994c). Further molecular and biological investigations of the significance of the CD30L-mediated anti-proliferative effect on LCAL cells are needed to determine the involvement in the clinical course and pathological presentation of this form of CD30$^+$ NHL.

In summary, the CD30-CD30L interaction may play a role in the pathophysiology of CD30$^+$ lymphomas, particular HD, LCAL and adult T-ALL, and may also play an important role in normal immune responses.

EXPRESSION OF CD30L IN PRIMARY CD30 POSITIVE LYMPHOMA CASES

None of the HD-derived cell lines examined expressed CD30L mRNA or surface protein (Gruss *et al.*, 1994c). CD30L is expressed on activated T-cells and monocytes/macrophages, but also constitutively on granulocytes (Smith *et al.*, 1993; Gruss *et al.*, 1994b). All three cell types are usually found in the reactive cell compartment surrounding H-RS cells in HD-involved tissue and might support the proliferation of primary H-RS cells, modulating cytokine expression and secretion, or cell-cell contact dependent interaction. It was of interest to determine if the *in vitro* situation is true for primary HD cases. After generation of a monoclonal antibody against the membrane-bound form of CD30L a preliminary study was performed. In a series of HD and LCAL cases the primary H-RS cells in HD and the neoplastic anaplastic lymphoma cells in LCAL stained CD30 positive, but were negative for CD30L expression. However, bystander cells surrounding H-RS cells showed significantly enhanced expression of CD30L in comparison to normal lymph nodes. HD cases of the mixed cellularity subform seem to have stronger expression of CD30L expression on T-cells and histiocytes, than nodular sclerosing subforms of HD cases. At the present time, *in vitro* data and expression in primary HD cases support a role for CD30L in the pathology of HD. CD30L seems to act as a paracrine factor on proliferation, cytokine secretion/production and expression of cell contact-dependent membrane antigens on H-RS cells.

HYPOTHESIS OF CD30L INVOLVEMENT IN THE PATHOLOGY OF CD30 POSITIVE LYMPHOMAS

HD and some LCAL NHLs are characterised by an abnormal expression of the cytokine receptor CD30, a member of the TNF/NGF receptor superfamily. The cloning of the cognate for CD30, named CD30L, allowed us to study biological properties of the overexpression of CD30 on these cells. As summarised in Fig. 3, CD30L has proliferative/mitogenic activity for H-RS cells, but anti-proliferative activity for CD30$^+$ anaplastic lymphoma cells. The TNFRs p60 and p80 have been associated with distinct biological functions of TNF (Smith *et al.*, 1994). Other members of the TNF/NGF receptor family, such as FAS, 4-1BB and as shown here CD30, seem to use only one receptor to mediate both increased and reduced proliferation on different cell types (Alderson *et al.*, 1993 and 1994; Suda *et al.*, 1993).

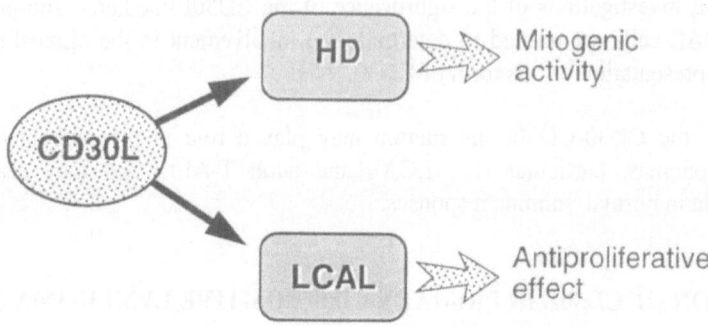

Fig. 3:   CD30L mediates different growth signals for CD30+ Hodgkin's disease-derived cell lines or large
cell anaplastic lymphoma cell lines. Some Hodgkin's disease-derived cell lines show increased
proliferation after CD30L stimulation. In contrast, CD30L has anti-proliferative effects on large
cell anaplastic lymphoma cell lines.

For HD, primary and cultured H-RS cells, the presumed neoplastic component, produce a
variety of cytokines (IL-1, IL-5, IL-6, IL-8, IL-9, TNF, LT-α, CD27L/CD70, M-CSF, GM-
CSF, TGF-β, LIF), the known biological activities of which correlate with the typical
clinical and pathological features of HD (Gruss *et al.*, 1994a). CD30L, and also CD40L, is
able to modulate secretion of cytokines (IL-6, TNF, LT-α) frequently found in HD cases.
IL-6 and TNF were recently found to be elevated in the serum of HD patients (Gause *et al.*,
1991; Gruss *et al.*, 1993b). A number of clinical and pathological features of HD are
consistent with characteristics of a tumour of cytokine producing cells, including
overexpression of cytokine receptors such as CD30, CD40, CD25 and CD71 on H-RS cells
(Gruss *et al.*, 1994a). Constitutional "B" symptoms with fever, night sweats and weight
loss, elevations of acute phase proteins, presence of mild thrombocytosis, decreased
immune response, increased alkaline phosphatase serum levels or certain histopathological
presentations (e.g., sclerosis, polykaryon formation, plasmacytosis, T-cell accumulation,
activation and rosetting, eosinophilia) can be related to an abnormal or unbalanced secretion
of cytokines. For instance, IL-6 is the major hepatocyte-stimulating factor, inducing release
of acute phase proteins, and has thrombopoietin activity and induces B-cell differentiation.
TNF can cause fever, weight loss and night sweats and might be a major mediator of the
development of constitutional "B" symptoms. TNF also causes elevation of fibrinogen
serum levels, also frequently seen in HD. Sclerosis can be associated with TGF-β and LIF
production, and the presence of tissue and blood eosinophilia with IL-5, GM-CSF and IL-3
secretion. As shown in the model depicted in Fig. 4, CD30L could function as an element
of the deregulated cytokine network involved in HD. CD30L is being produced by normal
reactive bystander cells surrounding H-RS cells and acting on the CD30[+] H-RS cells.
CD30L can modify cytokine secretion of H-RS cells, enhance proliferation and activation
of H-RS cells by up-regulation of cell contact-dependent antigens. The interaction between
H-RS cells and T-cells in HD-involved areas involves a complex network of interactive
signals with membrane-associated and cytokine-mediated events. The CD30
overexpression of most H-RS cells appears to be an important clinical, biological and
pathological marker for HD.

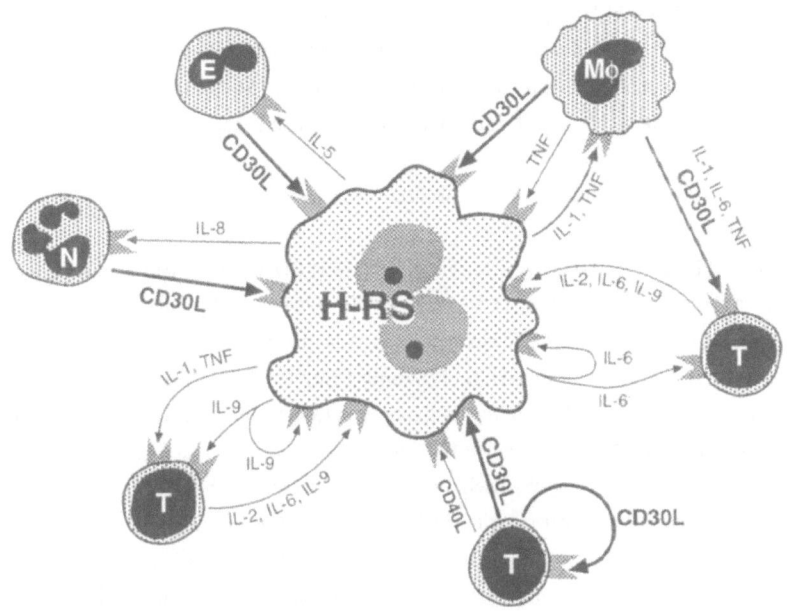

Fig. 4: Model for involvement of CD30/CD30L interaction in membrane-associated or cytokine-mediated effects between Hodgkin and Reed-Sternberg cells and surrounding bystander cells. Hodgkin and Reed-Sternberg cells are CD30L negative but surrounding bystander cells, such as activated T-cells and monocytes/macrophages or granulocytes, express CD30L which can act on CD30+ T-cells or Hodgkin and Reed-Sternberg cells. Signalling through CD30 induces the up-regulation of various cytokines and membrane-associated activation/adhesion antigens.

REFERENCES

Alderson MR, Armitage RJ, Maraskovsky E, Tough TW, Roux E, Schooley K, Ramsdell F, Lynch DH (1993) FAS transduces activation signals in normal human T lymphocytes. *J Exp Med* **178**:2231

Alderson MR, Smith CA, Tough TW, Davis-Smith T, Armitage RJ, Falk B, Roux E, Baker E, Sutherland GR, Din WS, Goodwin RG (1994) Molecular and biological characterisation of human 4-1BB and its ligand. *Eur J Immunol* **24**:2219

Andreesen R, Brugger W, Löhr GW, Bross KJ (1989) Human macrophages can express the Hodgkin's cell-associated antigen Ki-1 (CD30). *Am J Pathol* **134**:187

Armitage RJ, Maliszewski CR, Alderson MR. Grabstein KH. Spriggs MK, Fanslow WC (1993) CD40L. A multi-functional ligand. *Semin Immunol* **5**:401

Delabie J, Ceuppens JL, Vandenberghe P. de Boer M. Coorevits L. de Wolf-Peeters C (1993) The B7/BB-1 antigen is expressed by Reed-Sternberg cells of Hodgkin's disease and contributes to the stimulating capacity of Hodgkin's disease-derived cell lines. *Blood* **82**:2845

Drexler HG (1992) Recent results on the biology of Hodgkin and Reed-Sternberg cells. I. Biopsy material. *Leuk Lymphoma* **8**:283

Drexler HG (1993) Recent results on the biology of Hodgkin and Reed-Sternberg cells. II. Continuous cell lines. *Leuk Lymphoma* **9**:1

Drexler HG, Jones DB, Diehl V, Minowada J (1989) Is the Hodgkin cell a T- or B-lymphocyte? - Recent evidence from geno- and immunophenotypic analysis and in-vitro cell lines. *Hematol Oncol* **7**:95

Dürkop H, Latza U, Hummel M, Eitelbach F, Seed B, Stein H (1992) Molecular cloning and expression of a new member of the nerve growth factor receptor family that is characteristic for Hodgkin's disease. *Cell* **68**:421

Eck M, Sprang S (1989) The structure of tumour necrosis factor-gamma at 2.6 Å resolution. *J Biol Chem* **264**:17595

Ellis TM, Simms PE, Slivnick DJ, Jäck H-M, Fisher RI (1993) CD30 is a signal-transducing molecule that defines a subset of human activated CD45RO+ T cells. *J Immunol* **151**:2380

Farrah T, Smith CA (1992) An emerging cytokine family. *Nature* **358**:26

Froese P, Lemke H, Gerdes J, Havsteen B, Schwarting BR, Hansen H, Stein H (1987) Biochemical characterisation and biosynthesis of the Ki-1 antigen in Hodgkin-derived and virus-transformed human B and T lymphoid cell lines. *J Immunol* **139**:2081

Gause A, Scholz R, Klein S, Jung W, Diehl V, Tesch H, Pfreundschuh M (1991) Increased levels of circulating interleukin-6 in patients with Hodgkin's disease. *Haematol Oncol* **9**:307

Gruss H-J, Brach MA, Drexler HG, Bonnifer R, Mertelsmann RH, Herrmann F (1992) Expression of cytokine genes, cytokine receptor genes, and transcription factors in cultured Hodgkin and Reed-Sternberg cells. *Cancer Res* **52**:3353

Gruss H-J, Dölken G, Brach MA, Mertelsmann R, Herrmann F (1993a) Serum levels of circulating ICAM-1 are increased in Hodgkin's disease. *Leukemia* **7**:1245

Gruss H-J, Dölken G, Brach MA, Mertelsmann R, Herrmann F (1993b) The significance of 60 kDa soluble TNF receptors in patients with Hodgkin's disease. *Leukemia* **7**:1339

Gruss H-J, Herrmann F, Drexler HG (1994a) Hodgkin's disease, a tumor of cytokine producing cells. *Crit Rev Oncogen* **5**:473

Gruss H-J, DaSilva N, Hu ZB, Uphoff CC, Goodwin RG, Drexler HG (1994b) Expression and regulation of CD30 ligand and CD30 in human leukemia-lymphoma cell lines. *Leukemia* **8**:2083

Gruss H-J, Boiani N, Williams DE, Armitage RJ, Smith CA, Goodwin RG (1994c) Pleiotropic effects of the CD30 ligand on CD30-expressing cells and lymphoma cell lines. *Blood* **83**:2045

Gruss H-J, Hirschstein D, Wright B, Ulrich D, Caligiuri MA, Strockbine L, Armitage RJ, Dower SK (1994d) Expression and function of CD40 on Hodgkin and Reed-Sternberg cells and the possible relevance for Hodgkin's disease. *Blood* **84**:2305

Gruss H-J, Dower SK (1995) Tumor necrosis factor ligand superfamily: involvement in the pathology of malignant lymphomas. *Blood* **85**:3378

Herbst H, Stein H, Niedobitek G (1993) Epstein-Barr virus and CD30+ malignant lymphomas. *Crit Rev Oncogen* **4**:91

Jones EY, Stuart DI, Walker NP (1990) The three-dimensional structure of tumour necrosis factor. *Prog Clin Biol Res* **349**:321

Kaplan HS (1980) *Hodgkin's disease.* (Boston: Harvard University Press)

Munro JM, Freedman AS, Aster JC, Gribben JG, Lee NC, Rhynhart KK, Banchereau J, Nadler LM (1994) *In vivo* expression of the B7 co-stimulatory molecule by subsets of antigen-presenting cells and malignant cells of Hodgkin's disease. *Blood* **83**:793

Nadali G, Vinante F, Ambrosetti A, Todeschini G, Veneri D, Zanotti R, Meneghini V, Ricetti MM, Benedetti F, Vassanelli A, Perona G, Chilosi M, Menestrina F, Fiacchini M, Stein H, Pizzolo G (1994) Serum levels of soluble CD30 are elevated in the majority of untreated patients with Hodgkin's disease and correlate with clinical features and prognosis. *J Clin Oncol* **12**:793

Pizzolo G, Vinante F, Nadali G, Ricetti MM, Morosato L, Marrocchella R, Vincenzi C, Semenzato G, Chilosi M (1993) ICAM-1 tissue overexpression associated with increased serum levels of its soluble form in Hodgkin's disease. *Br J Haematol* **84**:161

Poppema S (1989) The nature of the lymphocytes surrounding Reed-Sternberg cells in nodular lymphocyte predominance and in other types of Hodgkin's disease. *Am J Pathol* **135**:351

Schwab U, Stein H, Gerdes J, Lemke H, Kirchner H, Schaadt M, Diehl V (1982) Production of a monoclonal antibody specific for Hodgkin and Sternberg Reed cells of Hodgkin's disease and a subset of normal lymphoid cells. *Nature* **299**:65

Schwarting R, Gerdes J, Dürkop H, Falini B, Pileri S, Stein H (1989) BER-H2: a new anti-Ki-1 (CD30) monoclonal antibody directed at a formol-resistant epitope. *Blood* **74**:65

Smith CA, Gruss H-J, Davis T, Anderson D, Farrah T, Baker E, Sutherland GR, Brannan CI, Copeland NG, Jenkins NA, Grabstein KH, Gliniak B, McAlister IB, Fanslow W, Alderson M, Falk B, Gimpel S, Gillis S, Din WS, Goodwin RG, Armitage RJ (1993) CD30 antigen, a marker for Hodgkin's lymphoma, is a receptor whose ligand defines an emerging family of cytokines with homology to TNF. *Cell* **73**:1349

Smith CA, Farrah T, Goodwin RG (1994) The TNF receptor superfamily of cellular and viral proteins: activation, co-stimulation and death. *Cell* **76**:959

Stein H, Gerdes J, Schwab U, Lemke H, Mason DY, Ziegler A, Schienle W, Diehl V (1982) Identification of Hodgkin and Sternberg-Reed cells as a unique cell type derived from a newly-detected small-cell population. *Int J Cancer* ***30***:445

Stein H, Mason DY, Gerdes J, O'Connor N, Wainscoat J, Pallesen G, Gatter K, Falini B, Delsol G, Lemke H, Schwarting R, Lennert K (1985) The expression of the Hodgkin's disease associated antigen Ki-1 in reactive and neoplastic lymphoid tissue: evidence that Reed-Sternberg cells and histiocytic malignancies are derived from activated lymphoid cells. *Blood* **66**:848

Suda T, Takahashi T, Golstein P, Nagata S (1993) Molecular cloning and expression of the FAS ligand, a novel member of the tumour necrosis factor family. *Cell* **75**:1169

SINGLE CELL PCR AT THE DNA AND RNA LEVEL: A NOVEL APPROACH TO THE STUDY OF HODGKIN'S DISEASE

Lorenz Trümper, Heiner Daus, Judith Roth, Uschi Loftin, Angela Gause and Michael Pfreundschuh

Department of Internal Medicine I,
University of Saarland at 66421 Homburg/Saar, Germany

Elucidation of the cellular origin of the Hodgkin and Reed-Sternberg cells (H-RS) cells, the tumour cells of Hodgkin's disease (HD), is made difficult by the fact that these cells represent only a small minority of the cellular infiltrate in tissues affected by HD. Neither the study of cell lines nor the study of DNA or RNA extracted from HD-lymph nodes has been able to conclusively solve some of the most interesting questions, e.g., the questions of origin and clonality of the H-RS cells, the role of viruses in the pathogenesis of HD and the role of soluble and cell-bound signals in the maintenance of the malignant phenotype.

Single cell-based studies represent a promising, albeit technically difficult, alternative to the previously mentioned studies. We have established single cell polymerase chain reaction (PCR)-based assays to examine gene expression and DNA conformation in single H-RS cells isolated from single cell suspensions or cytocentrifuge slides. This manuscript describes the different techniques established for the amplification of DNA and RNA sequences from single cells. These studies have enabled us to answer some of the questions surrounding the enigmatic nature of the H-RS cells and will make future studies aimed at the definition of H-RS-specific genes feasible.

INTRODUCTION

The diagnosis of HD is based on the characteristic histological finding of typical H-RS cells which are surrounded by a mixed lymphohistiocytic infiltrate. H-RS cells are believed to represent the malignant cell population in HD due to their ability to disseminate, their multinuclearity and the results of cytogenetic studies showing aneuploidy and clonality (Drexler, 1992). In contrast to non-Hodgkin's lymphomas (NHLs), which have been shown to be clonal proliferations of either T or B lymphocytes by immunohistology and molecular genetics, there is an ongoing debate on the cellular origin of H-RS cells in HD.

Since H-RS cells typically appear in very low numbers in affected nodes, functional studies and molecular genetic analyses have yielded conflicting results. Based on morphological and immunological studies, nearly all haematopoietic cells have been proposed as the normal counterpart of these cells (reviewed in Drexler et al., 1989). Results of immunological and molecular biological studies of HD-derived cell lines have also not been conclusive. Due to the presence of surface markers of B and T lymphocytes, a lymphoid origin of H-RS cells has been suggested (reviewed in Drexler et al., 1989). In addition, molecular genetic studies have attempted to resolve controversies concerning the clonality, origin and lineage of H-RS cells.

As the attempts of many research groups to obtain pure populations of sufficient numbers of H-RS cells have proven unsuccessful, only studies at the single cell level can be expected to solve some of the many unresolved questions surrounding Dorothy "Reed's riddle".

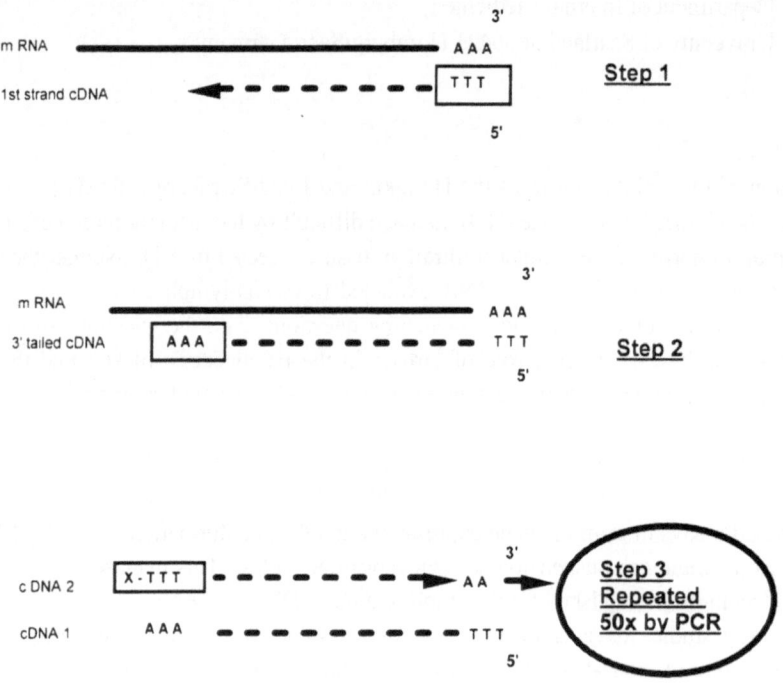

Fig. 1: Schematic representation of poly A PCR. 1st strand cDNA is synthesised in step 1, followed by poly A tailing and amplification with a poly T primer in steps 2 and 3. For technical details, refer to Brady et al. (1990).

In situ hybridisation was the first molecular tool used to study single H-RS cells (Weiss et al., 1988). It combines morphological analysis with information about the presence or expression of certain genes, and can also be used to assess chromosomal breakpoints (fluorescent in situ hybridisation). PCR (Saiki *et al.*, 1988) has revolutionised molecular biology and allows the analysis of genetic material at the single cell level (Schriever *et al.*, 1991). DNA and RNA sequences can be analysed by sequencing, and therefore genetic alterations in single tumour cells are now amenable to analysis. In addition, novel genes can be identified using degenerate primers or applying library screening techniques. The latter is only possible if a global PCR technique (Brady *et al.*, 1990) is used that allows the simultaneous amplification of all genes present at a given moment. We have employed a technique described by Brady *et al.* (1990) (Fig. 1) for the analysis of freshly isolated H-RS cells (Trümper *et al.*, 1993). This technique allows the generation of microgram quantities of cDNA from small samples (Cumano *et al.*, 1992). Samples can be renewed indefinitely by PCR, and can be cloned into vectors for sequence analysis and techniques based on library subtraction.

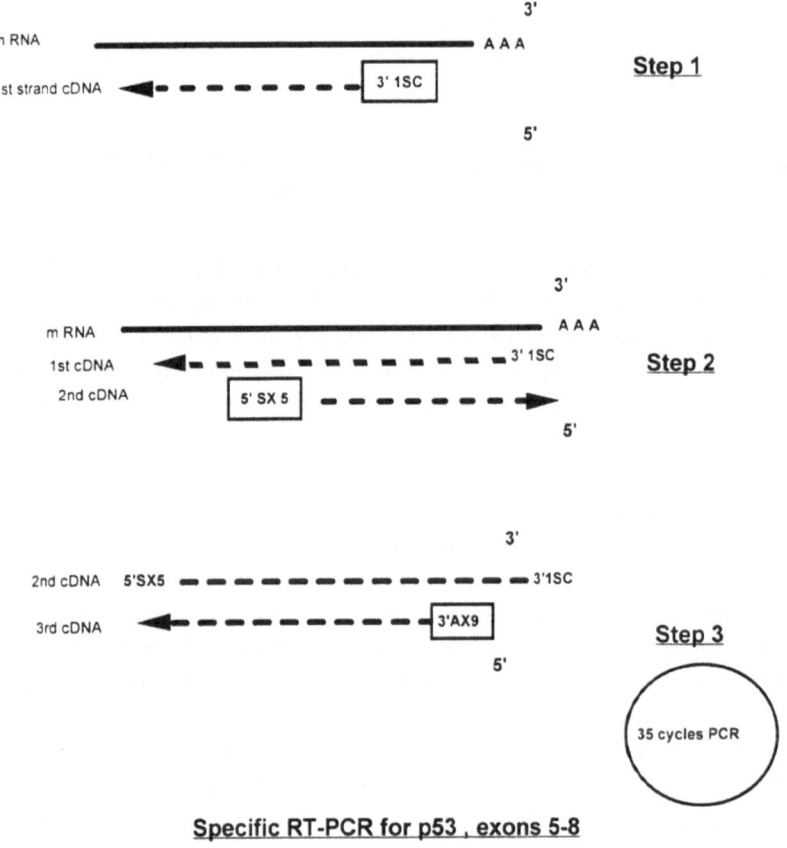

Specific RT-PCR for p53 , exons 5-8

Fig. 2:   Schematic representation of the p53 RT-PCR assay. For details, see Materials and Methods and Trümper *et al.*, (1993).

Amplification of DNA from single cells can be useful to identify specific chromosomal changes that are detectable by PCR using primers flanking the regions of interest. This can include the detection of characteristic breakpoints such as the t(14;18) breakpoint that is commonly found in follicular lymphoma, or the somatic rearrangements of the immunoglobulin (Ig) and T-cell receptor (TCR) genes that are characteristic of committed lymphoid cells. In addition, the presences of viral genomes and their sequences - such as the

Epstein-Barr virus (EBV) - can be examined by DNA PCR from single cells. Point mutations in certain oncogenes may also be examined, such as mutations in codons 12, 13 or 61 of the ras family genes (Bos, 1990) or mutations within exons 5 to 9 of the p53 gene (Vogelstein and Kinzler, 1992). p21 ras proteins are involved in cellular signalling and have been shown to become constitutively activated when mutations at positions that are involved in GTPase activity occur. Mutations of the N-ras genes are common in some haematopoietic malignancies (Bos, 1990) but have not been examined in uncultured H-RS cells as yet. The p53 protein - the so-called "guardian of the genome" (Lane, 1992) - acts as a DNA-binding protein and is important for the control of cellular proliferation. Mutations within structurally important regions may abolish its "tumour suppressive" function and have been described in a wide variety of human malignancies (Vogelstein and Kinzler, 1992). Detection of p53 protein in H-RS cells (Gupta et al., 1992) has been interpreted as evidence for the presence of mutations as non-mutated p53 protein is rarely detectable by immunohistochemistry due to its short half-life. We have developed a reverse-transcription (RT) PCR assay to examine the presence of p53 mutations in single cells (Trümper et al., 1993) (Fig. 2).

The role of the EBV in the aetiology of HD is an ongoing debate: EBV DNA can be detected in HD, however the proportion of positive cases ranges between 20-80% (Uhara et al., 1990; Herbst et al., 1990, 1992; Klein, 1992), depending on the technique used. Using the DNA in-situ hybridisation technique, EBV DNA sequences were detected in H-RS cells at a comparatively low rate (20-40%; Weiss et al., 1989). In an attempt to combine high sensitivity with specificity for H-RS cells, we amplified the BamHI-W region of EBV DNA by PCR from single H-RS cells. The BamHI-W region was chosen because it is known to be present in the EBV genome in several copies and should therefore be detectable by single cell PCR.

Somatic recombinations of genes coding for Ig and TCR molecules occur early in B and T-cell ontogeny and are a prerequisite for functionally active molecules in mature lymphocytes. Evidence for clonal rearrangements of Ig or TCR genes was found in less than 15% of Hodgkin's lymph nodes by Southern blot analysis (Griesser et al., 1987; Brinker et al., 1989; Roth et al., 1988; reviewed in Drexler, 1992). The vast majority of samples showed germline genes or polyclonal rearrangements. The lack of detectable rearrangements was attributed to the low sensitivity of Southern blot analysis, which is not able to detect clonal populations that make up less than 2 to 5% of a tissue sample. On the other hand, the rearrangements that were found in some HD-affected nodes may be derived either from H-RS cells, which sometimes make up more than 5% of the cellular infiltrate, or from oligoclonal proliferations of the lymphoid infiltrate and not the H-RS cells themselves.

We have started to employ single cell PCR for the detection of rearrangements of the Ig heavy chain (IgH) and the TCR gamma chain (TCRγ) genes in isolated H-RS cells (Roth et al., 1994; Daus et al., 1994). The PCR strategy is based on the fact that a PCR product can only be generated in cells with rearranged genes using PCR primers corresponding to the V-region and the J-region of the IgH gene. No product will be produced in non-rearranged, germline DNA, since the primer binding sites are too far apart to yield a PCR product.

MATERIALS AND METHODS

Patients

Patient data have been described elsewhere (Trümper *et al.*, 1993; Roth *et al.*, 1994; Daus *et al.*, 1995).

Preparation of Hodgkin's disease lymph nodes

Viable single cell suspensions were prepared from parts of fresh, HD-involved lymph nodes which were obtained at routine biopsy by mincing in RPMI 1640 medium with a sterile scalpel. The cells were suspended in phosphate buffered saline (PBS), and viable H-RS cells were isolated, along with the fraction of mononuclear cells, by density (Ficoll Hypaque) gradient centrifugation for 30 minutes. Cell suspensions were frozen in a solution containing 10% dimethylsulphoxide (DMSO), 10% fetal bovine serum and RPMI 1640 and stored in liquid nitrogen. Cells were rapidly thawed in RPMI 1640 at 37^0C and washed three times to remove DMSO. Alternatively, the cell suspension was cytocentrifuged on to glass slides after fixation (20 minutes) in a solution of paraformaldehyde (3% w/v in PBS). For DNA analysis, cells were stained for the presence of the CD30 antigen with the alkaline-phosphatase-anti-alkaline-phosphatase assay (APAAP) using the monoclonal antibody HRS-4. The H-RS cells were identified using morphological criteria (size, prominent nucleoli, multiple nuclei in case of Reed-Sternberg cells) as well as positive staining for the CD30 antigen. For RNA analysis, either "positive" or "negative" selection was used. Individual H-RS cells were identified as cells which were negative for CD3, CD14 and CD20 surface staining and exhibited the typical morphology of H-RS cells or their mononuclear variants: large mono-, bi- and multi-nucleated cells, with conspicuous nucleoli were clearly evident using phase contrast microscopy. Alternatively, cells were identified by staining positively for either CD15 or CD30 with a CD30 monoclonal antibody, HRS-3, that had been shown to be non-stimulatory by calcium ion flux analysis (W Jung, unpublished observations). The expression of each of the surface antigens (CD3, CD14, CD20) was detected using indirect immunofluorescence with FITC-conjugated secondary (goat anti-mouse) antibodies. Cells were placed in PBS in Petri dishes and single cells were drawn into a glass micropipette of 5-10 µm diameter with the help of a micromanipulator (Eppendorf, Hamburg, Germany). Cells were drawn into the pipette by applying suction through connected plastic tubing. After withdrawing the pipette containing a single cell (Fig. 3) and a small amount (1-2 µL) of PBS from the Petri dish, the cell and the PBS solution were ejected into a 500 µL Eppendorf tube containing 4 µL of lysis buffer. Control experiments, with the same procedure but without suction of cells, were performed as media controls. For DNA analysis, single H-RS cells were picked with glass capillaries from stained cytocentrifuge slides using a micromanipulator, transferred into reaction tubes and subsequently submitted to PCR (Roth *et al.*, 1993).

Poly A-cDNA preparation and amplification

RNA preparation, cDNA synthesis and cDNA amplification were performed as detailed elsewhere (Brady *et al.*, 1990). Briefly, single H-RS cells or Jurkat cells were dropped into 4 µL of lysis buffer, dNTPs, (dT)$_{24}$ primer and RNAse inhibitors. Reverse transcriptase was added to the lysis buffer containing the lysed cells. After incubation at 37^0C for 15 minutes, the reverse transcriptase was heat-inactivated at 65^0C. Tailing buffer containing dATP and terminal transferase, was added and incubation was continued at 37^0C for 30 minutes.

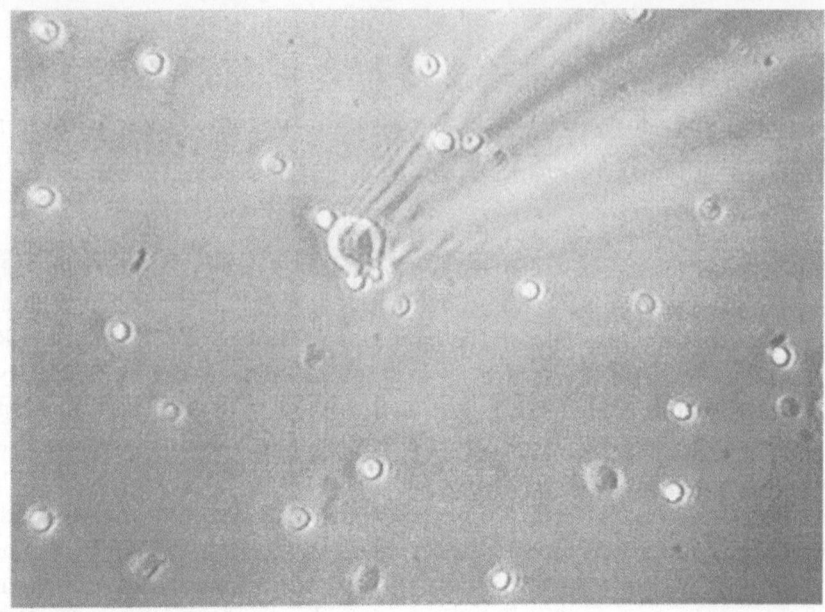

Fig. 3: A single Hodgkin cell with attached lymphocytes is shown while it is being manipulated into the glass capillary of the micromanipulator. Single tumour cells are transferred into a Petri dish and separated from attached lymphocytes before being transferred into an Eppendorf tube for PCR amplification

PCR buffer containing dNTPs, $(dT)_{24}X$ Primer and Taq polymerase was added to the tailed cDNA and amplification was carried through 25 cycles (1 minute 94^0C / 2 minutes 44^0C / 6 minutes 72^0C + 10 s extension/cycle), followed by another 25 cycles (1 minute 94^0C/ 2 minutes 44^0C/ 2 minutes 72^0C). Control samples for PCR reactions included samples lacking cDNA templates as well as culture medium without cells.

One tenth of the primary PCR product was loaded onto a 1.7% Tris-Acetate agarose gel and electrophoresed at 60 V for 90 minutes. DNA was transferred by alkaline Southern transfer to Hybond N^+ membranes and hybridised with different radioactively-labelled cDNA probes (Trümper et al., 1993).

RT-PCR from single H-RS cells
p53 RT-PCR was performed as described elsewhere (Trümper et al., 1993) (Fig. 2). Briefly, single cells were dropped into lysis buffer containing a first strand synthesis primer designated 3'1SC. After cDNA synthesis, amplification was performed using primers specific for sequences at the beginning of exon 5 and the end of exon 9 of the p53 gene. Products of approximately 600 bp size were obtained for cDNA, corresponding to a 1600 bp stretch of genomic DNA. Samples were analysed by gel electrophoresis and hybridisation with an internal oligonucleotide. Sequencing was performed after cloning the PCR products into the PCR 2000 (Invitrogen) vector using the Sequenase 2.0 kit (USB).

DNA single cell PCR
DNA analysis was performed as described elsewhere (Roth et al., 1994; Daus et al., 1994; Trümper et al., 1995). In general, PCR reactions were carried out in 50 μL volumes containing 0.25μM of the 5' and 3' primers, 200μM of each dNTP, 15mM Tris-HCl, pH 8.3, 50mM KCl, 1.5mM $MgCl_2$, 0.01% gelatin and 1U Taq polymerase (Boehringer Mannheim).

The samples were subjected to 40 temperature cycles using a Trioblock thermal cycler (Biometra). In a second round of PCR, a 1:25 dilution of the sample was submitted to an additional 40 PCR cycles under identical conditions. All experiments were performed using a reaction tube without DNA template as a negative control and a tube containing single cells of a control cell line or single cells of an NHL as a positive control. The reaction products were separated on agarose gels (2%), transferred to nylon membrane and hybridised with ^{32}P end-labelled oligonucleotide probes corresponding to internal sequences of the gene of interest. A sequence of the ß-actin gene was amplified from single H-RS cells in parallel as a positive control.

Primers for the detection of IgH gene rearrangements (Küppers *et al.*, 1993), TCRγ rearrangements (Daus *et al.*, 1994), N-ras sequences (Bos, 1990; Trümper *et al.*, 1994) and EBV-specific sequences (Roth *et al.*, 1994) have been described elsewhere.

RESULTS AND DISCUSSION

Poly A PCR
The amplification efficiency of poly A PCR depends on the purity of the primers chosen, the magnesium content of the buffer and other variables that have to be established for each new batch of primers. Ambient conditions - such as the addition of methylcellulose to cell suspensions - may drastically decrease the efficiency of the amplification. In general, poly A PCR yields reproducible results with 60-80% of amplification reactions positive for housekeeping genes. The PCR yields a "smear" of approximately 300 to 800 bp in length on agarose gels (Fig. 4) which will hybridise to cDNA probes containing the 3' end of the genes of interest (Fig. 4). In our experience, cDNA probes should be 500 - 1000 bp long and should lack repetitive sequences such as the ones found around Alu restriction sites. Hybridisation conditions have to be established for each new probe as the stringency of hybridisation and washing, as well as exposure times, may vary greatly from probe to probe. Apart from "conventional" Southern blots, dot blots can be used thereby eliminating the need for gel electrophoresis (Fig. 5). However, we sometimes find interpretation of results difficult as it may not be possible to distinguish non-specific hybridisation, e.g., to primer doublets, from *bona fide* expression signals.

Single cell poly A PCR will yield semi-quantitative results. The signal strength will vary greatly, even between cells of the same origin as demonstrated in Fig. 6. Nine cells from the Jurkat T-cell line were picked from the same Petri dish within 10 minutes and subjected to poly A PCR as described above. The cDNA was hybridised with probes for the ß and γ-Actin genes as well as the T-cell-specific genes p56-lck and TCRß chain (probe YT35). It is evident that there is significant fluctuation of gene expression at the single cell level. This type of fluctuation is not seen when PCR reactions with diluted RNA as a template are carried out in parallel and is therefore not due to differences of poly A PCR efficiency (provided the experiments are carried out in parallel under exactly the same conditions). These results suggest that the fluctuations seen represent a normal variation in cell to cell RNA content - possibly related to cell cycle effects - or that they are due to randomly introduced differences in the handling of the cells. We conclude from these observations that both the average expression level of all H-RS cells isolated from a patient's lymph node, as well as the individual expression patterns, have to be examined when looking at poly A PCR data. The former provide averaged data similar in quality and their limitations to a Northern blot analysis, whereas the individual cell expression patterns provide a unique means of determining the co-expression of genes in a single cell at a given moment.

Table 1: Genes universally expressed in Hodgkin and Reed-Sternberg cells. Average hybridisation data from Reed-Sternberg cells from six cases of Hodgkin's disease. For details, refer to Trümper *et al.*, 1993.

Case	HD1	HD2	HD3	HD5	HD6	HD7
Pathology	HDLP	HDNS	HDNS	HDNS	HDNS	HDMC
Time of study	1st relapse	Presentation	Presentation	Presentation	Presentation	2nd relapse
ß Actin	++++	++++	++++	++++	++++	++++
C 7761	nd	nd	+++	++++	+++	++++
GAPDH	++++	++++		nd	nd	nd
c-myc	++++	++++	+++	++++	++++	++++
c-fes/fps	+++	+++	+++	+++	+++	+++
CD4	+++	++	++	+++	++++	-
fyn	++	++	+++	+++	++++	++++
TNF ß	++++	+++	+++	+++	++	+++
IL 2 R ß	+++	+++	+++	+++	+++	++

Abbreviations: -, none of the Hodgkin and Reed-Sternberg cells from a case is positive for the respective gene; + , up to 25 % are positive; ++, up to 50% are positive; +++, up to 75% are positive; ++++, up to 100% are positive; nd, not done. Reproduced with kind permission of the publisher from Trümper *et al.*, 1993.

Tables 1 and 2 summarise the results obtained from the first series of experiments utilising poly A PCR. Single H-RS cells were obtained from 6 lymph nodes, and at least 7 single cells per patient were hybridised with a panel of 30 different cDNA probes (Trümper *et al.*, 1993). Only cells that were positive for at least two different housekeeping gene probes were included in the analysis. An example of such an analysis can be seen in Fig. 7. There is some variation in gene expression for the human CD4 gene. This gene codes for a cell surface receptor commonly associated with the helper subset of T-cells but also present on immature hematopoietic cells as well as non-lymphoid hematopoietic malignancies.

This analysis enabled us for the first time to provide a unique gene expression profile for single tumour cells. The results of this initial analysis show a significant case to case heterogeneity for these six cases of HD. Interestingly, there is a subgroup of genes that can be found in all cases examined. This subset of "Hodgkin's defining" genes characterises them as activated hematopoietic cells, especially, since genes such as c-fps/fes, vav, IL-2R ß chain and fyn are found in nearly all H-RS cells. High levels of c-myc expression are testament to the fact that H-RS cells are indeed actively cycling cells. However, conclusions as to which lineage H-RS cells are derived from are not yet possible. Examination of further lineage-specific genes as well as cytokine and cytokine receptor genes will provide further insight into the pathogenesis of HD.

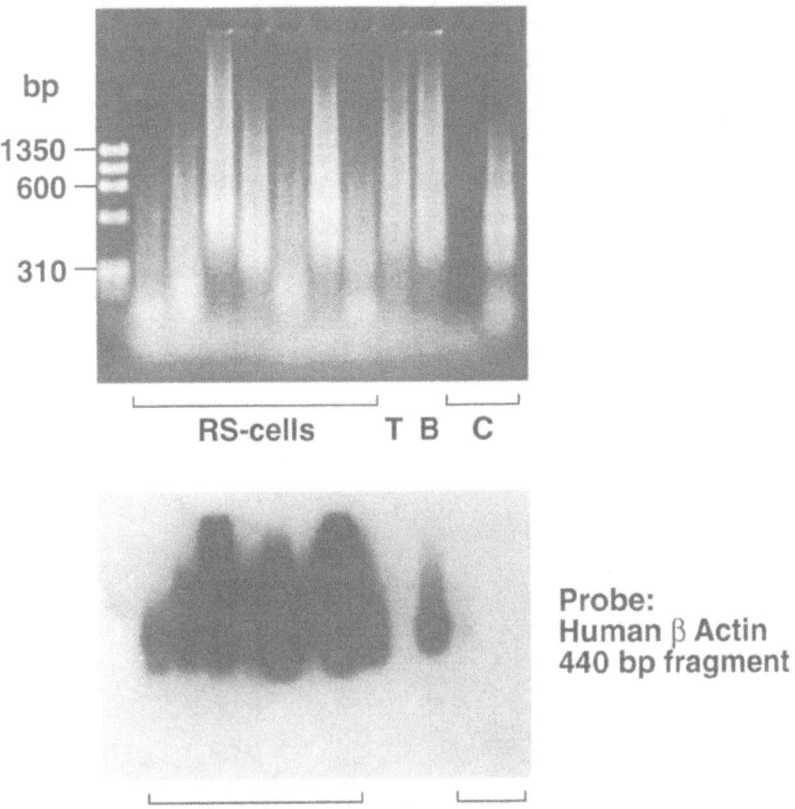

Fig. 4:   Agarose gel and Southern blot hybridisation of single Hodgkin and Reed-Sternberg cells after poly A PCR amplification. T and B-cells from the same lymph node suspension were also picked and processed by PCR. All cells except for one B-cell gave positive signals after hybridisation with a ß-Actin probe.

Table 2:   Genes expressed in some but not all cases that define molecular subgroups of Hodgkin's disease. Average hybridisation data from Hodgkin and Reed-Sternberg cells from six cases of Hodgkin's disease. For details, refer to Trümper *et al.*, 1993.

Case	HD1	HD2	HD3	HD5	HD6	HD7
Pathology	HDLP	HDNS	HDNS	HDNS	HDNS	HDMC
Time of study	1st relapse	Presentation	Presentation	Presentation	Presentation	2nd relapse
ß Actin	++++	++++	++++	++++	++++	++++
C 7761	nd	nd	+++	++++	+++	++++
GAPDH	++++	++++	nd	nd	nd	nd
lck	-	+++	+	+	+	-
hck	-	-	-	-	++++	-
TCR ß	+++	nd	++	+	+++	-
CD2	-	++++	-	-	++	-
IL-9	-	+++	-	+++	-	-

Abbreviations: -, none of the Hodgkin and Reed-Sternberg cells from a case is positive for the respective gene; +, up to 25 % are positive; ++, up to 50% are positive; +++, up to 75% are positive; ++++, up to 100% are positive; nd, not done. Reproduced with kind permission of the publisher from Trümper *et al.*, 1993.

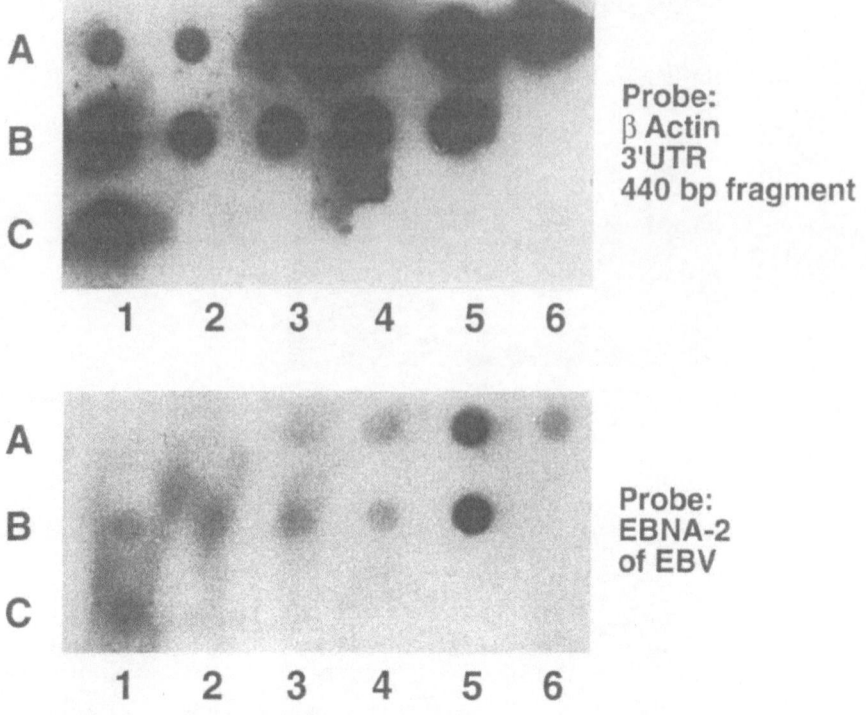

Fig. 5:   Single cell cDNA libraries. Hybridisation of dot blots of poly A PCR products generated from different cell lines. Actin is shown in the upper panel. The filter was stripped and rehybridised with a probe coding for the EBNA-2 gene of EBV. Only samples A5 and B5, derived from the Burkitt's lines Daudi and Raji respectively, hybridise with EBNA-2. Low background signals are due to insufficient stripping.

p53 RT-PCR assay

As shown in Fig. 8, the majority of single H-RS cells (all picked from one case of HD for the experiment shown here) do express p53 mRNA comparable to the levels observed in single cells of an AML line (AML M4) which is known to express high levels of p53 mRNA (S Benchimol, personal communication). In addition, a higher band was observed that corresponds to the 1600 bp genomic DNA band. So far, clones obtained from cells of three different cases of HD have been examined. In two cases, mutations at codon 246 (Slingerland *et al.*, 1991) were seen in 5/7 and 3/6 H-RS cells respectively, whereas a mutation at codon 312 was seen in 4/5 cells from the third case. Mutations were confirmed by sequencing at least two clones per cell in both directions.

These experiments show that p53 mutations seem to be a common event in HD. Interestingly, not all putative H-RS cells from a given case seem to carry mutations whereas the remaining cells carry identical mutations. Two conclusions can be drawn from these observations: first, that H-RS cells seem to be clonal, and second, that there is clonal heterogeneity, possibly with clonal evolution from wild type p53 to clones with mutated p53 as has been observed for other human tumours (Vogelstein and Kinzler, 1992).

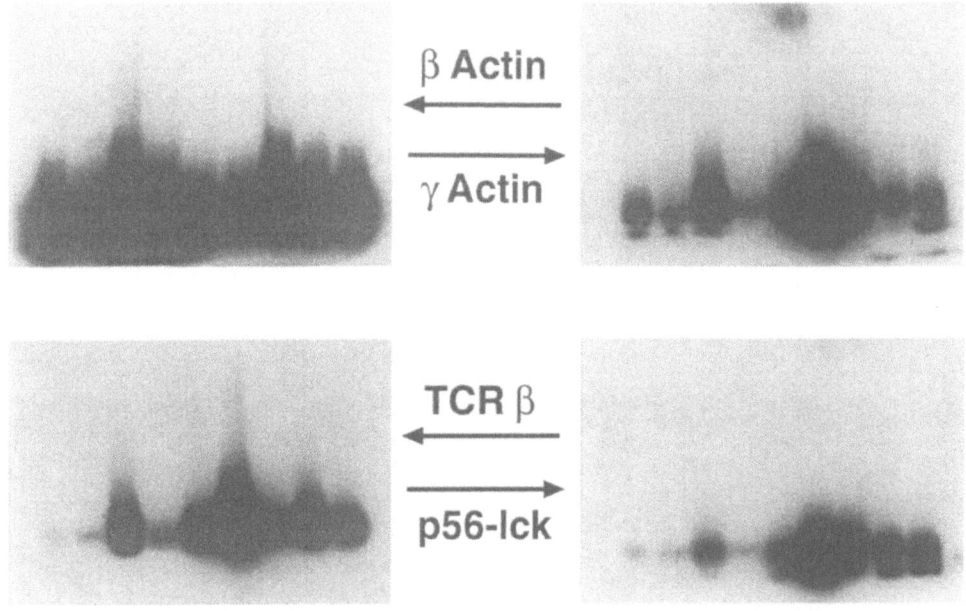

Fig. 6:   Gene expression in single Jurkat cells. Fluctuation of gene expression of four different genes is seen
with 9 randomly picked cells of the T-cell line Jurkat that were processed by poly A PCR and
hybridised with two different actin genes as well as the cell lineage specific genes T-cell receptor ß
(clone YT35) and p56-lck. Even though cells were picked at the same time from the same Petri dish,
levels of gene expression vary greatly from cell to cell. This may be due to cell cycle-specific effects
as well as variations in PCR efficiency.

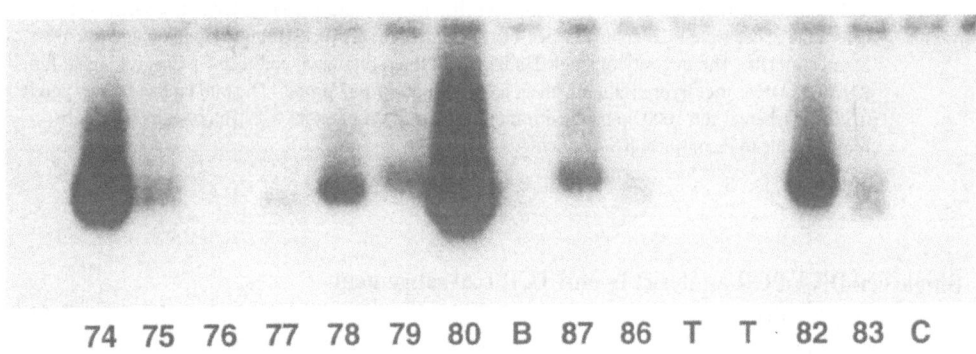

Fig. 7:   Hybridisation of single Hodgkin and Reed-Sternberg cells with a probe for human CD4. The majority
of Hodgkin and Reed-Sternberg cells show expression of this gene whereas neither of the two T-cells
nor the control lane show a hybridisation signal. Probe: human CD4, 1 kb Bam HI 3' UTR fragment;
the numbers correspond to the different Hodgkin and Reed-Sternberg cells as given in Trümper et al.
(1993); T, T-cell; C, control.

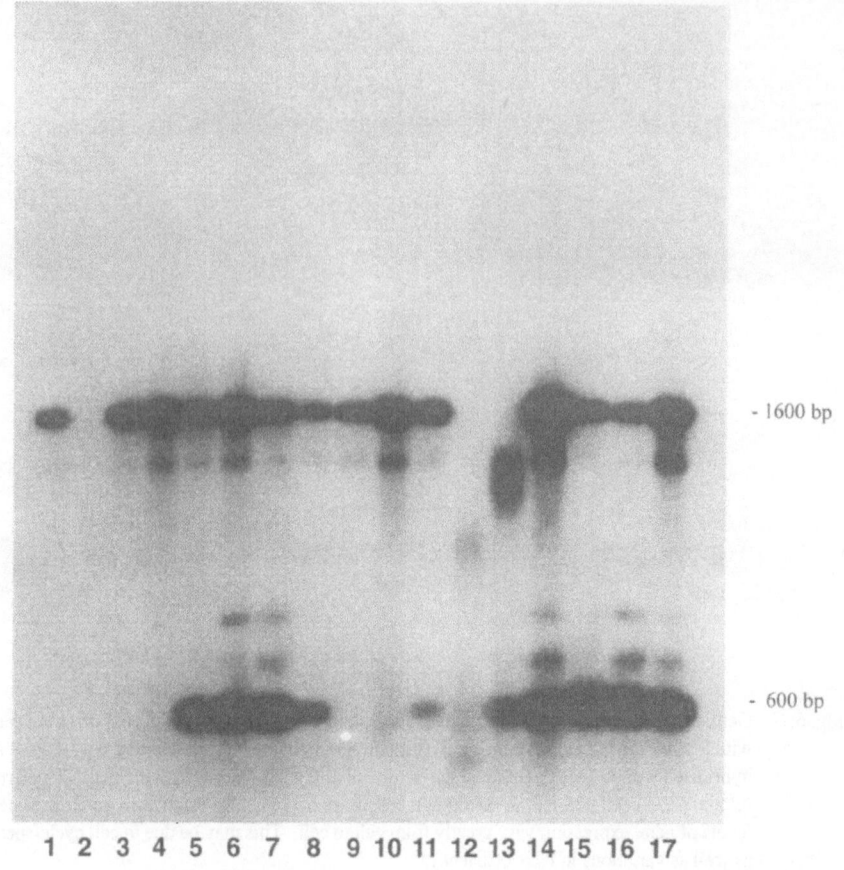

1 2 3 4 5 6 7 8 9 10 11 12 13 14 15 16 17

Fig. 8:   p53 expression in single Hodgkin and Reed-Sternberg cells. cDNA and genomic DNA - RT-PCR for
          p53 from single Hodgkin and Reed-Sternberg cells. Lanes 1 and 3, control DNA from AML-M4
          cells; lane 2, negative PCR control; lanes 4 to 28, consecutively picked single Hodgkin and Reed-
          Sternberg cells. The majority of single Hodgkin and Reed-Sternberg cells show a positive signal for
          RNA and DNA after hybridisation with an internal codon 7 p53 probe. The 600 bp band corresponds
          to cDNA whereas the 1600 bp band corresponds to the DNA of exons 5-8. Intervening bands are
          probably due to incomplete RNA splicing.

Single cell DNA PCR analysis: Ig and TCR rearrangements

As a control for the detection of Ig and TCR rearrangements, rearranged DNA from 2×10^4
peripheral blood lymphocytes was amplified by PCR. Six different "forward-primers" were
constructed corresponding to consensus sequences of the six known families of the IgH
variable region (Küppers et al., 1993) together with a mix of two "reverse primers"
corresponding to consensus sequences of the different J regions. Similarly, a set of four
primers binding to all known TCR Vγ families and two J primers were constructed.
Rearranged DNA of all families could be amplified from peripheral blood and each yielded a
PCR product of ~350 bp in length (Roth et al., 1994; Daus et al., 1994). Non-rearranged
DNA was not amplified, since the primer binding sites were too far from each other to yield a
PCR product.

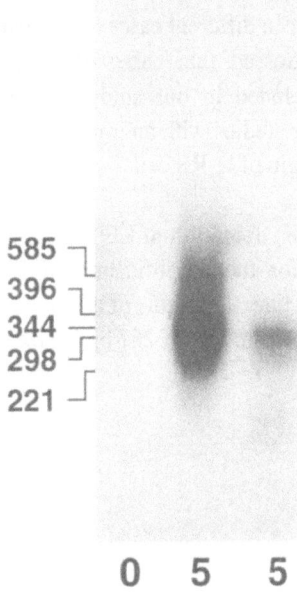

585 ⌐
396 ⌐
344 ≡
298 ⌐
221 ⌐

0   5   5

Fig. 9:   A V6DJ rearrangement can be detected in single cells from a non- Hodgkin's lymphoma case. Framework I DNA-PCR from cells of a non-Hodgkin's lymphoma case. Bands were visualised after hybridisation to an internal oligonucleotide probe corresponding to framework 3 sequences of the IgH gene.

Similarly, single cells were isolated from biopsy material (lymph nodes) of a patient with a B-cell NHL. Fig. 9 shows the product of a V6DJ rearrangement derived from a single NHL cell. Similarly, single cells from T-ALL and a series of B-cell NHLs were examined, demonstrating the sensitivity of the technique for the detection of Ig and TCR rearrangements in single tumour cells. Subsequently, single H-RS cells were subjected to Ig and TCR PCR. Whereas ß-actin was detectable in H-RS cells of all 17 cases examined, neither IgH nor TCRγ rearrangements were detectable in 16 (IgH) or 13 (TCRγ) cases. Three H-RS cells were examined per reaction to provide a "safety margin", and each experiment was repeated at least twice.

These initial studies indicate that H-RS cells may not possess complete Ig or TCR rearrangements, as this type of PCR analysis (Framework 1 Ig PCR) can only detect complete VDJ rearrangements. It is well known that several haematological malignancies, including acute myeloid malignancies, harbour incomplete or "illegitimate" Ig or TCR rearrangements that can be detected by Southern blot analysis but not by this type of PCR analysis. It is, however, possible that rearrangements are not detected due to technical reasons: the sensitivity of the PCR may be too low (although single cell control experiments were included with each set of experiments, and rearrangements in B and T-cells could be successfully amplified from the same lymph node suspensions), or the primers employed by us may not bind to the rearranged DNA found in H-RS cells due to mutations or deletions. It was recently reported that clonal IgH gene rearrangements were detected in approximately 30% of H-RS cells picked form histological sections of three cases of HD (Küppers et al., 1994). A different type of framework 1 Ig-PCR analysis was used in this study (Küppers et al., 1993).

Therefore, heterogeneity between different cases of HD with regard to Ig gene rearrangements may exist. It has to be stressed that cases of lymphocyte predominant HD (nodular paragranuloma) were not included in our study, as opposed to the study performed by Küppers *et al.* (1994). Further studies with an improved Ig-PCR are being performed to draw firm conclusions about the origin of H-RS cells.

Single cell DNA PCR analysis: detection of EBV DNA and Ras mutations

The single-cell PCR assay for the amplification of EBV-specific DNA sequences was established using the Raji cell line. Dilutions of a Raji cell suspension, as well as single Raji cells, were submitted to PCR. A product of 267 bp in size was obtained in each case.

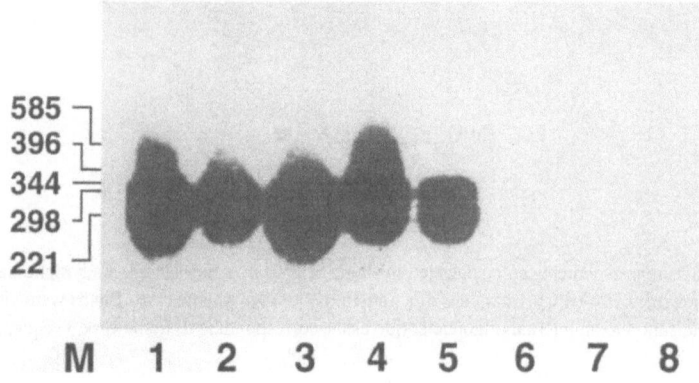

Fig. 10: Amplification of EBV-specific sequences from single Hodgkin and Reed-Sternberg cells. Each lane represents a representative amplification of cells from a different case. Single cells that had been stained with monoclonal antibodies against the CD30 antigen were picked from cytocentrifuge slides and subjected to PCR after micromanipulation into Eppendorf tubes. Reproduced with the publisher's permission from Roth *et al.*, 1994.

Four out of six patients examined were positive for EBV sequences, while no EBV product was seen in two patients (Fig. 10) (Roth *et al.*, 1994). A minimum of ten cells was examined from each patient. In all cases results were reproducible: the H-RS cells of a given patient yielded identical results, i.e. all cells were either positive or negative for EBV DNA. No EBV DNA was detected in lymphocytes from the same lymph node suspensions. Single-cell PCR allows the detection of EBV-specific DNA sequences with a sensitivity probably superior to *in situ* hybridisation and the origin of the amplified EBV DNA can be unequivocally attributed to the H-RS cells. Our results concur with other studies demonstrating the presence of EBV DNA in H-RS cells of all cells in a given case. The proportion of EBV-positive cases in different studies depends - among other variables - on the methods used (Uhara *et al.*, 1990).

Since we can unequivocally localise the EBV DNA to H-RS cells, we believe that we will be able to reliably determine the proportion of EBV-positive and negative cases in our series and will be able to correlate this with other molecular results. The results from the EBV study also demonstrate the reliability and reproducibility of single cell PCR and its applicability to the study of HD.

Ras mutations were detected with an assay originally designed for the detection of mutated cells within a population of wild type cells (Trümper *et al.*, 1994). N-ras was chosen as the gene of interest since most haematopoietic malignancies, notably acute myeloid leukemias and myelodysplasias, have been shown to harbour mutations in the N-ras genes. Single H-RS cells from 12 cases were examined for the presence of N-ras mutations at codons 12/13 and 61 by amplification, single strand conformation polymorphism assay on the PHAST minigel system (Pharmacia) and subsequent direct cycle sequencing using the Sequitherm cycle sequencing kit (Biozym Diagnostik). N-ras mutations were not seen in H-RS cells from these cases. We therefore conclude that N-ras mutations do not play a role in the pathogenesis of HD.

SUMMARY

We have shown that DNA and mRNA from single H-RS cells can be successfully and reproducibly amplified by global and specific PCR techniques. The results obtained by our initial studies confirm the fact that H-RS cells are activated haematopoietic cells. Corresponding to the well-known clinical heterogeneity of the disease there is a considerable case to case variation in the expression of certain genes. However, there are also genes that are present in all H-RS cells and can therefore be thought of as "Hodgkin's defining" genes. Among these are cytokines, such as TNF ß that may be responsible for some of the clinical symptoms of the disease, and oncogenes such as the c-fps/fes gene. Studies at the DNA level have confirmed the fact that approximately 50% of HD cases are positive for the EBV genome, and that the genome can be found in virtually all H-RS cells from a "positive" case. Mutations of the p53 tumour suppressor gene seem to play a crucial role in the pathogenesis of the disease, whereas ras gene mutations may be rare. The lack of rearrangements of either the IgH or TCRγ genes is an interesting and provocative finding. However, further studies confirming this finding are needed.

In summary, single cell PCR has opened new avenues for research into the enigmatic biology of HD that will almost certainly lead to new and exciting results.

ACKNOWLEDGEMENTS

I would like to thank Tak W. Mak, Ontario Cancer Institute, Toronto, Canada for his advice, help and valuable support in getting single cell PCR on HD started and in his belief that it would work a long time before it actually did! Ged Brady (Manchester, England), Sam Benchimol (Toronto, Canada) and Ralf Küppers (Cologne, Germany) provided advice and technical support. The technical help of Dawn Gray and Frederike von Bonin is gratefully acknowledged. Jeff Cossman (Washington DC), Randy Gascoyne (Vancouver, BC) and Peter Möller (Heidelberg, Germany) provided HD lymph nodes for the preparation of single cell suspensions. This laboratory is supported by grants from the Deutsche Krebshilfe, Bonn, Germany and the Deutsche Forschungsgemeinschaft, Bonn, Germany to LT, HD and MP.

REFERENCES

Bos J (1990) Ras gene mutations and human cancer. In *Molecular biology of human cancer*. Cossman J, ed (Elsevier) p273

Brady G, Barbara M, Iscove N (1990) Representative *in vitro* cDNA amplification from individual hemopoietic cells and colonies. *Methods Molec Cell Biol* **2**:17

Brinker MGL, Poppema S, Buys CHCM, Timens W, Osinga J, Visser L (1987) Clonal immunoglobulin gene rearrangements in tissues involved by Hodgkin's disease. *Blood* **70**:186

Cumano A, Paige CJ, Iscove NN, Brady G (1992) Bipotential precursors of B cells and macrophages in murine fetal liver. *Nature* **356**:612

Daus H, Trümper L, Roth J, Von Bonin F, Möller P, Ganse A, Pfreundschuh M (1995) Hodgkin and Reed-Sternberg cells do not carry T-cell receptor γ gene rearrangements: evidence from single-cell polymerase chain reaction examination. *Blood* **85**:1590

Drexler, HG (1992) Recent results on the biology of Hodgkin and Reed-Sternberg cells. I. Biopsy material. *Leuk Lymphoma* **8**:283

Drexler HG, Jones DB, Diehl V, Minowada J (1989) Is the Hodgkin cell a T or B lymphocyte. Recent evidence from geno- and immunophenotypic analysis and *in vitro* cell lines. *Hematol Oncol* **7**:95

Griesser H, Feller AC, Mak TW, Lennert K (1987) Clonal rearrangements of T-cell receptor and immunoglobulin genes and immunophenotypic antigen expression in different subclasses of Hodgkin's disease. *Int J Cancer* **40**:157

Gupta RK, Norton AJ, Thompson IW, Lister TA, Bodmer JG (1992) p53 expression in Reed-Sternberg cells of Hodgkin's disease. *Br J Cancer* **66**:649

Herbst H, Niedobitek G, Kneba M, Hummel M, Finn T, Anagnostopoulos I, Bergholz M, Krueger G, Stein H (1990) High incidence of Epstein-Barr virus genomes in Hodgkin's disease. *Am J Pathol* **137**:13

Herbst H, Pallesen G, Weiss LM, Delsol G, Jarrett RF, Steinbrecher E, Stein H, Hamilton-Dutoit S, Brousset P (1992a) Hodgkin's disease and Epstein-Barr virus. *Ann Oncol* **3**:27

Klein G (1992) Epstein-Barr virus-carrying cells in Hodgkin's disease. *Blood* **80**:299

Küppers R, Zhao M, Rajewsky K, Hansman ML (1993) Detection of clonal B cell populations in paraffin embedded tissues by PCR. *Am J Pathol* **143**:230

Küppers R, Zhao M, Rajewsky K, Fischer R, Hansman ML (1994) Hodgkin disease: Hodgkin and Reed-Sternberg cells picked from histological sections show clonal immunoglobulin gene rearrangements and appear to be derived from B-cells at various stages of development. *Proc Natl Acad Sci USA* **91**:10962

Lane DP (1992) p53, guardian of the genome. *Nature* **358**:15

Roth J, Daus H, Trümper L, Gause A, Pfreundschuh M (1994) Detection of immunoglobulin heavy chain gene rearrangement at a single cell level in malignant lymphomas: no rearrangement is found in Hodgkin and Reed-Sternberg cells. *Int J Cancer* **57**:1

Roth J, Daus H, Gause A, Trümper L, Pfreundschuh M (1994) Detection of Epstein-Barr virus DNA in Hodgkin- and Reed-Sternberg cells by single cell PCR. *Leuk Lymphoma* **13**:137

Roth M, Schnitzer B, Bingham EL, Harrnden CE, Hyder DM, Ginsburg D (1988) Rearrangement of immunoglobulin and T cell receptor genes in Hodgkin's disease. *Am J Pathol* **131**:331

Schriever F, Freeman G, Nadler LM (1991) Follicular dendritic cells contain a unique gene repertoire demonstrated by single-cell polymerase chain reaction. *Blood* **77**:787

Saiki RK, Gelfand DH, Faloona F, Mullis KB, Horn GT, Erlich HA, Arnheim N (1988) Primer-directed enzymatic amplification of DNA with a thermostable DNA polymerase. *Science* **239**:487

Slingerland JM, Minden MD, Benchimol S (1992) Mutation of p53 gene in human acute myelogenous leukemia. *Blood* **77**:1500

Trümper LH, Brady G, Bagg A, Gray D, Loke SL, Griesser H, Wagman R, Braziel R, Gascoyne RD, Vicini S, Iscove NN, Cossman J, Mak TW (1993) Single cell analysis of Hodgkin and Reed-Sternberg cells: molecular heterogeneity of gene expreession and p53 mutations. *Blood* **81**:3097

Trümper LH, Bürger B, von Bonin F, Hintze A, von Blohn G, Pfreundschuh M, Daus HD (1994) Diagnosis of pancreatic adenocarcinoma by polymerase chain reaction from pancreatic secretions. *Br J Cancer* **70**:278

Trümper LH, Jacobs G, von Bonin F, Möller P, Pfreundschuh M, Daus HD (1995) N-ras gene mutations are a rare event in Hodgkin's disease: results from single cell PCR examinations (submitted)

Uhara H, Sata Y Mukai K, Akao I, Matsuno Y, Furuya S, Hoshikawa T, Shimosato Y, Saida T (1990) Detection of Epstein-Barr virus DNA in Reed-Sternberg-cells of Hodgkin's disease using the polymerase chain reaction and *in situ* hybridisation. *Jpn J Cancer* **81**:272

Vogelstein B and Kinzler KW (1992) p53 function and dysfunction. *Cell* **70**:553

Weiss LM, Movahed LA, Warnke RA, Sklar J (1989) Detection of Epstein-Barr virus genomes in Reed-Sternberg cells of Hodgkin's disease. *N Engl J Med* **320**:502

THE NATURE OF THE LYMPHOCYTES
IN HODGKIN'S DISEASE

Sibrand Poppema, Lydia Visser

Department of Laboratory Medicine,
Cross Cancer Institute and University of Alberta
11560 University Avenue
Edmonton, Alberta, T6G 1Z2, Canada

*Supported by grants of the National Cancer Institute of Canada and
the Alberta Cancer Board Research Initiative Program*

CLONALITY OF REED-STERNBERG CELLS

Hodgkin's disease (HD) is characterised by the presence of Reed-Sternberg (RS) cells in the proper environment. This environment includes predominantly small lymphocytes and a variable admixture of histiocytes, plasma cells and eosinophils. It should be kept in mind that the RS cells and their variants constitute less than one percent of the cell population in the vast majority of cases. Nevertheless, these cells are the clonal and abnormal population in HD as shown by a combination of conventional karyotyping and chromosome specific *in situ* hybridisation (Poppema *et al.*, 1992). As well, in a minority of cases it has been possible to demonstrate clonal rearrangements of the immunoglobulin genes, especially in RS cell-enriched cell fractions (Brinker *et al.*, 1987; Weiss *et al.*, 1987; Sundeen *et al.*, 1987). Although no consistent chromosomal abnormality has been identified in HD, it is of interest that chromosome 14q, including the immunoglobulin heavy chain gene containing q32 region, is the area most frequently involved in translocations in HD (Cabanillas *et al.*, 1988; Poppema *et al.*, 1993). However, the non-Hodgkin's lymphoma associated translocations t(8;14), t(11;14) and t(14;18) are exceedingly rare in HD. The translocation t(2;5) that is found in anaplastic large cell lymphomas (Mason *et al.*, 1990) is not found in HD by cytogenetic analysis whereas the use of reverse transcriptase - polymerase chain reaction (PCR) has led to conflicting results (Poppema *et al.*, 1993). The absence of t(14;18) translocations is of interest since, using PCR-based methods, it has been possible to demonstrate rearrangements involving the major breakpoint region of the bcl-2 gene and the immunoglobulin heavy chain gene in 30% of cases (Stetler-Stevenson *et al.*, 1990). This suggests that the bcl-2 rearrangements are probably present in small bystander B lymphocytes, a notion supported by the finding that a similar proportion of individuals with hyperplastic tonsils can be shown to have such bcl-2 rearrangements by a PCR based method (Limpens *et al.*, 1991). The only exception to this general rule is in patients that previously, or at the same time, have t(14;18) positive follicular lymphoma (Lebrun *et al.*, 1994), although it is extremely difficult to prove that the rearrangement is present in the RS cells and not in a small admixture of the follicular lymphoma cells. The multitude of chromosomal abnormalities found in HD suggests that none of these is the primary driving force of the disease and that other factors, such as viruses in combination with a particular form of immune response, should be considered. The chromosomal abnormalities as well as several oncogenes can be considered as co-factors that may provide some survival or proliferation advantage to the RS cells.

ROLE OF EPSTEIN-BARR VIRUS

The evidence for the involvement of Epstein-Barr virus (EBV) in the pathogenesis of HD is extensively reviewed elsewhere in the proceedings of this meeting and there is epidemiological evidence that other infectious agents may be involved in the EBV-negative cases, specifically in the adolescent and young adult age group with the nodular sclerosis subtype of HD. The most direct histologically documented link between mononucleosis and HD remains the report of a patient who over a period of 6 months showed the transition from mononucleosis with predominantly EBNA-positive B immunoblasts, via an intermediate stage with predominantly activated CD4-positive T-cells, to mixed cellularity HD with EBNA-positive RS cells (Poppema *et al.*, 1985). In comparison to primary infectious mononucleosis this case, and also cases of HD in general, lack a prominent CD8-positive response (Poppema *et al.*, 1982; Aisenberg and Wilkes, 1982; Martin and Warnke, 1984).

There are a number of potential explanations for this lack of CD8 cells. First, the patients might lack CD8 cells, however in the peripheral blood they generally have normal or even increased numbers of CD8+ T-cells. Secondly, they might lack EBV-specific cytotoxic T-cells. This is unlikely since they initially overcame the acute mononucleosis phase and also are able to control the latently infected small lymphocytes, in contrast to immunosuppressed transplantation patients or HIV-positive individuals. A third explanation might be the lack of appropriate antigen expression on the RS cells.

Several potential targets for the specific cytotoxic immune response such as EBNA-2 and EBNA-3A, -3B, -3C are not expressed on the RS cells. However, cases of EBV-positive HD do express LMP-1 and, based on the presence of mRNA, are likely to express LMP-2 on their surface (Deacon *et al.*, 1993). The immune response against LMP-2 in normal individuals is usually directed through HLA A2.1 (Murray *et al.*, 1992) and it has been hypothesised therefore that individuals who are HLA A2 negative might be predisposed to EBV-positive HD since they would also lack the capability of responding to LMP-2.

LACK OF CORRELATION BETWEEN EBV POSITIVITY
AND HLA A2 NEGATIVE PHENOTYPE

We have recently investigated this hypothesis by testing a series of 72 consecutive patients diagnosed with HD in Northern and Central Alberta during 1990 and 1991. EBV-positive RS cells, as demonstrated by EBER *in situ* hybridisation, were found in 26% of these cases with 86% of the mixed cellularity, 13% of the nodular sclerosis and 0% of the lymphocyte predominance cases being positive. All EBV-positive and negative cases were subsequently tested for HLA A2 with a monoclonal antibody reactive with the polymorphic HLA A2 determinant. The results indicated that the number of HLA A2 positive cases was approximately 50% in the EBV-positive and -negative cases, similar to the distribution in the general population. Therefore, no correlation between HLA A2 expression and presence or absence of EBV in the RS cells of HD was identified (Poppema *et al.*, 1994).

LACK OF HLA CLASS I EXPRESSION ON REED STERNBERG CELLS

Since it appeared in the immunohistological stains that in some of the cases RS cells were HLA A2 negative, whereas the lymphocytes and histiocytes were positive, we decided to study the expression of monomorphic determinants of HLA class I and II, employing three monoclonal anti-monomorphic HLA class I, an anti-ß2 microglobulin, an anti-HLA class II and an anti-invariant chain reagent. The results on tissue sections with several adjoining RS cells and on cytospins indicate that RS cells, with the exception of L&H type RS cells in the lymphocyte predominance subtype, generally are HLA class I negative and strongly HLA class II positive, independent of the EBV status of the case. This finding suggests that down-regulation of HLA class I expression is a major contributor to the lack of a CD8-positive cytotoxic immune response in HD (Poppema *et al.*, 1994).

FLOW CYTOMETRIC ANALYSIS OF T-CELLS
IN TISSUES AND BLOOD OF HODGKIN PATIENTS

The question remains what the nature is of the prominent lymphoid infiltrate/immune response that is invariably present in HD. It has been known for several years that the majority of these lymphocytes are CD4-positive and express some early activation markers such as CD38 (Poppema *et al.*, 1982). It has also been demonstrated that the majority of these cells are CD45R0-positive, consistent with memory T-cells (Poppema *et al.*, 1989). These lymphocytes are capable of γ-interferon and IL-2 production *in vitro*. We have recently extended these studies on a series of cases of nodular sclerosis and and mixed cellularity HD, comparing the tissue lymphocytes with the peripheral blood lymphocytes in individual patients. Flow cytometric analysis was performed with multiple combinations of antibodies including PerCP-labelled CD3, CD4 or CD8 as the T-cell defining reagents in combination with FITC and or phycoerythrin-labelled reagents for CD45 subsets and several activation markers. The results (Table 1) indicate that the T-cell population is enriched for a specific subset and depleted of other subsets. As previously found, CD4+ cells clearly predominate over CD8+ cells with CD4/CD8 ratios of 3.2 in mixed cellularity, 7 in nodular sclerosis subtype I, and 22 in nodular sclerosis subtype II, indicating that the more extensively tissue is involved, the higher the CD4/CD8 ratio becomes.

Table 1:   Lymphocyte subsets in Hodgkin's disease tissue and peripheral blood.

	NS I PBL (8 cases)		NS II PBL (4 cases)		MC PBL (4 cases)		control PBL
Age	30yr(24-39)		32yr(24-44)		62yr(37-86)		34yr(24-44)
CD3	75%	74%	77%	70%	67%	63%	73%
CD4	65%	51%	67%	37%	45%	47%	41%
CD8	9%	23%	3%	32%	14%	18%	32%
CD4/CD8	7.0	2.2	22.0	1.2	3.2	2.6	1.3
CD26	29%		8%	41%	30%	36%	65%
CD69	63%		86%	23%	54%	12%	10%

Abbreviations: NS I, nodular sclerosis Hodgkin's disease subtype I; NS II, nodular sclerosis Hodgkin's disease subtype II; MC, mixed cellularity Hodgkin's disease; PBL, peripheral blood lymphocytes; yr, years.

We also studied a number of so-called T-cell activation antigens. These included CD69, CD38, CD25, CD71, CD26 and HLA class II. CD69, the so called activation inducer molecule, a phosphorylated disulphide-linked 27/33 kD homodimeric protein is the earliest inducible cell surface glycoprotein known on T-cells. It is a type II membrane protein with a C-type animal lectin domain suggesting a role in signal transduction (Lopez-Cabrera et al., 1993). CD69 was found to be expressed on 54% of mixed cellularity, 63% of nodular sclerosis subtype I and 86% of nodular sclerosis subtype II cases, again indicating a higher percentage in more extensively involved tissues. CD69 is expressed on approximately 10% of peripheral blood T-cells and up to 23% of peripheral blood cells in nodular sclerosis subtype II patients (Fig. 1). In normal and reactive lymphoid tissues CD69 is expressed on a sub-population of thymus medullary T-cells, on germinal centre T-cells and on inflammatory T-cells. It has been hypothesised that CD69 plays a role in tissue positioning and localisation and/or migration towards sites of inflammation. Whether the T-cells in HD have been activated in the circulation and then migrate to the involved tissue or whether they are activated in situ is not clear at the moment.

CD38 is a 46 kD type II trans-membrane glycoprotein with a short N-terminal cytoplasmic tail. Although originally described as a marker for thymocytes, CD38 is expressed on different cell lineages. Resting and circulating T-cells are CD38- and activated T-cells are CD38+. Recently CD38 has been found to bear a resemblance to NAD+ glycohydrolase and ADP-ribosyl cyclase, enzymes involved in the synthesis of cyclic ADP-ribose, which acts as an endogenous secondary messenger regulating the mobilisation of intracellular calcium. The function of CD38 is not clear although it can be shown to play a role in T-cell activation and proliferation in synergy with CD3 and CD2. It is of interest that activation through CD38 is dependent on the presence of accessory cells and IL-2 (Malavasi *et al.*, 1994). In tissues involved by HD the percentage of CD38 positive cells is slightly lower than that of CD69, whereas in peripheral blood of these patients the percentage is approximately 10%. Other activation markers, like CD25 the IL-2 receptor, CD71 the transferrin receptor and HLA class II are also consistently expressed on the lymphocytes of tissues involved by HD.

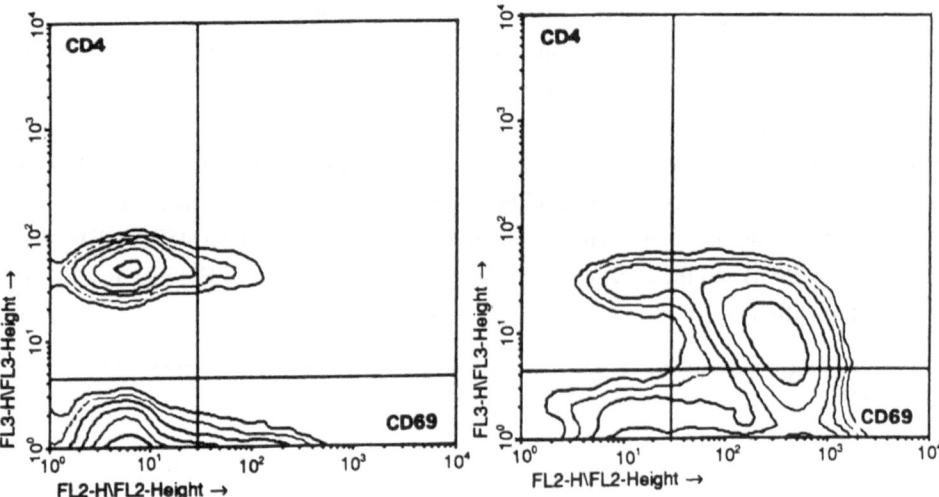

Fig. 1:   CD69 expression in peripheral blood CD4 T lymphocytes (left) and in CD4 lymphocytes from a lymph node involved by nodular sclerosis subtype of HD (right). The FACScan analysis indicates that there are approximately 10% CD69+CD4+ lymphocytes in the blood versus over 90% CD69+CD4+ T lymphocytes in the lymph node.

Another marker that is frequently described as an activation antigen is CD26. This is a 110 kD T-cell antigen with dipeptidyl peptidase IV enzyme activity that plays a role in T-cell co-stimulation (Morimoto and Schlossman, 1994). In normal peripheral blood 65% of the T-cells express CD26, including CD4 and CD8 subsets, with a small subset of bright staining cells, predominantly in the CD8 T-cell and NK cell populations. In HD the percentage of CD26 cells is reduced to approximately 40% in peripheral blood and to 30%, 29% and 8% in tissues involved by mixed cellularity, nodular sclerosis subtype I and nodular sclerosis subtype II respectively. This indicates that the majority of T-cells in the involved tissues are CD26 negative (Fig. 2). This is of interest since the CD4+, CD45R0+, CD26+ T-cell population can respond to recall antigens, induce B-cell immunoglobulin synthesis and activate MHC-restricted cytotoxic T-cells, whereas the CD4+, CD26- T-cells can only respond to mitogens and allo-antigens (Morimoto and Schlossman, 1994). The enrichment of this CD26 population may be another reason for the apparently ineffective nature of the immune response in HD.

Reagents reactive with restricted forms of leukocyte common antigen (CD45) have been found to allow the distinction of functionally different T-cell subsets. This is not surprising since CD45 is one of the most abundant leukocyte cell surface antigens and is exclusively expressed on haematopoietic cells. The cytoplasmic portion of CD45 has been shown to have protein tyrosine phosphatase activity and this allows the regulation of protein tyrosine phosphorylation. CD45 plays an important role in antigen-stimulated proliferation of T lymphocytes (Trowbridge and Thomas, 1994). Within the CD4+ T-cell population, subsets of CD45RA++R0-, CD45RA+R0+, and CD45RA-R0++ can be distinguished in normal peripheral blood (Visser *et al.*, 1993). The CD45RA++ cells comprise a virginal T-cell subset that in other studies has been shown to have a suppressor-inducer function. The CD45R0++ cells are memory T-cells and have helper-inducer activities. The CD45RA+R0+ cells are probably a transitional population and their functional activity has not been defined. In neonates and young individuals the CD45RA++ virginal T-cells predominate whereas with ageing the proportion of CD45R0++ cells increases (Lai *et al.*, 1994). Our results confirm that CD45R0++ cells are clearly predominating in Hodgkin lesions (Fig. 3). This is of some clinical relevance since it has been shown that CD45R0+CD4 cells are more sensitive to radiation therapy than CD45RA+CD4 cells (Uzawa *et al.*, 1994) and the excellent radiosensitivity of most cases of HD may result from this T-cell sensitivity.

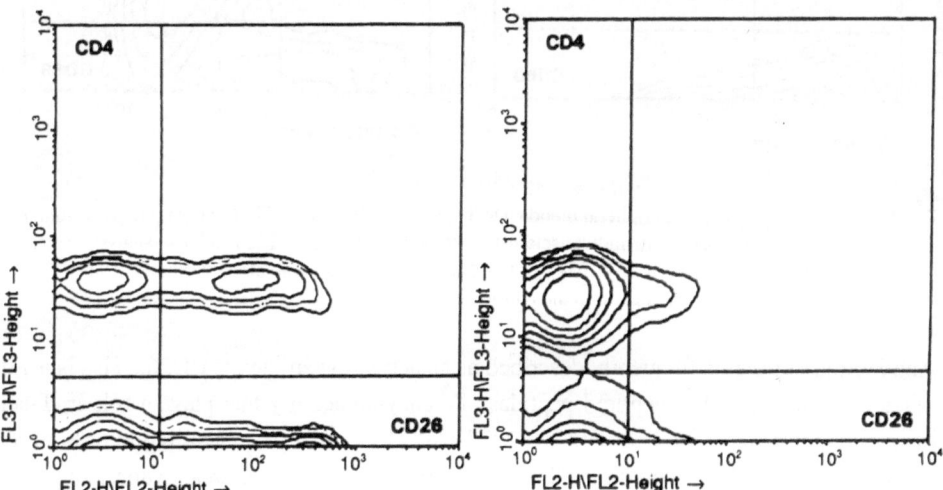

Fig. 2: CD26 expression in peripheral blood CD4 T lymphocytes (left) and in CD4 lymphocytes from a lymph node involved by nodular sclerosis Hodgkin's disease (right). The FACScan analysis indicates that there are approximately 65% CD26+CD4+ lymphocytes in the blood versus less than 10% CD26+CD4+ T lymphocytes in the lymph node.

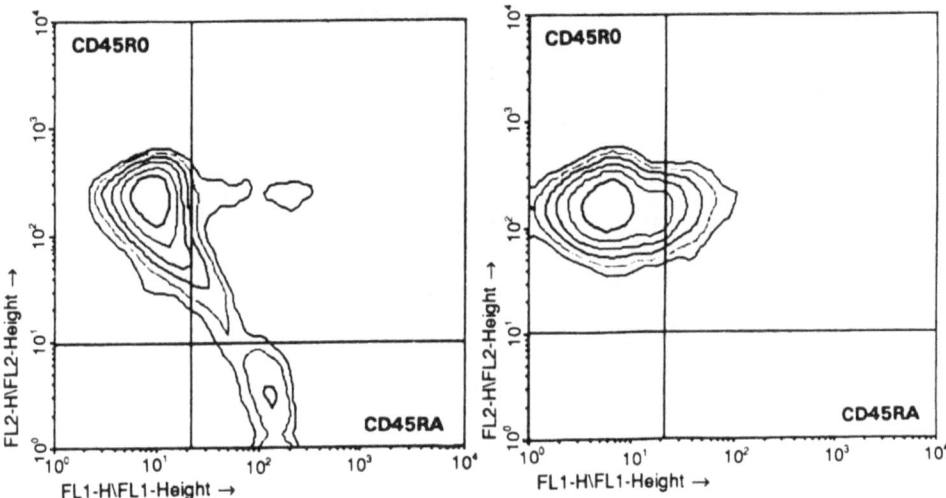

Fig. 3:   Comparison of CD45R0/CD45RA staining in peripheral blood CD4+ T lymphocytes (left) and CD4+ lymphocytes derived from a lymph node involved by nodular sclerosis Hodgkin's disease (right).   FACScan analysis demonstrates the enrichment of CD45R0+/CD45RA- CD4 + T lymphocytes in the lymph node.

A further distinction in CD45R0++ cells can be made based on the expression of CD45RB with RB[bright] and RB[dim] sub-populations. According to some studies in rodents this might allow a distinction between Th1 cells, producing γ interferon and IL-2, acting predominantly in delayed type hypersensitivity reactions and leading to predominantly cellular immune responses and Th2 cells, producing IL-4 and IL-10, and acting predominantly as helpers of B-cells, leading to predominantly humoral immune responses (Mason and Powrie, 1990). In addition to the Th1 and Th2 subsets other subsets, such as the Th0 subset that is a precursor of the other two have been described and the phenotype of such cells has not been established clearly. Some macrophage produced factors like IL-12 (Trinchieri, 1993) promote Th1 type responses whereas other factors such as IL-10 and its EBV-induced homologue (Mosmann, 1991) inhibit Th1 responses.   The functional significance of differential CD45RB staining in human lymphocytes is not clear at this time.   However, our results clearly demonstrate an enrichment of CD45RB[dim] CD4+CD45R0+ T-cells in HD involved tissues (Fig. 4).

Bcl-2 protein is a mitochondrial membrane protein that is encoded by the *bcl-2* gene on chromosome 18. Its function is associated with the prevention of apoptosis. Normally, Bcl-2 protein is expressed in virtually all small B and T lymphocytes, but not in cortical thymocytes and germinal centre B-cells, that are in cell stages that require high proliferation and the possibility to select for desired and against non-desired specificities. The small lymphocytes surrounding RS cells have a very low expression of Bcl-2 protein. This is perhaps related to the fact that CD45R0 cells have lower Bcl-2 protein levels than CD45RA cells. It has been shown that CD45R0+ cells in acute viral infections undergo apoptosis *in vitro* if not rescued by IL-2 or fibroblasts (Akbar *et al.*, 1993). The low expression of Bcl-2 in the T-cells surrounding RS cells suggest that they are in a state that requires contact with other cells or cytokine support to survive.

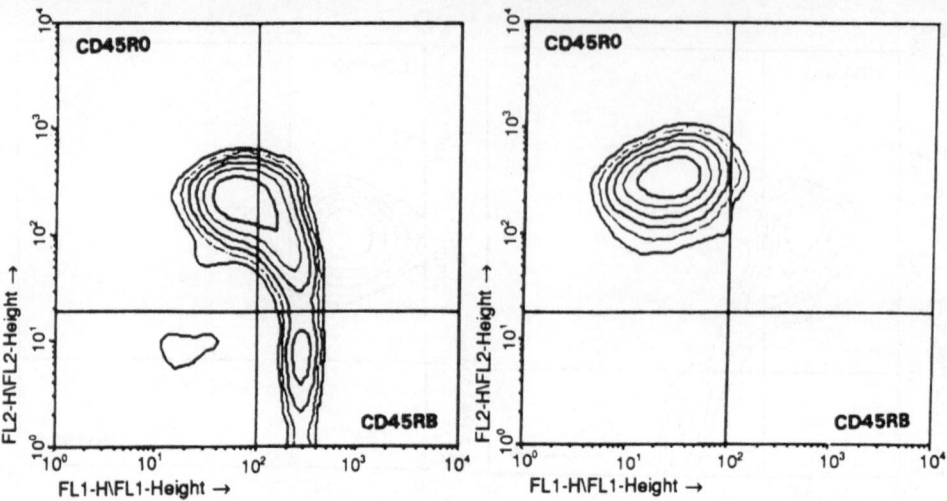

Fig. 4:   Comparison of CD45R0/CD45RB staining in peripheral blood CD4+ T lymphocytes (left) and CD4+ lymphocytes derived from a lymph node involved by nodular sclerosis Hodgkin's disease (right). FACScan analysis demonstrates the enrichment of CD45RBdim/CD45R0+ lymphocytes and the virtual absence of CD45RBbright/CD45R0+ lymphocytes in the lymph node.

NODULAR LYMPHOCYTE PREDOMINANCE
SUBTYPE OF HODGKIN'S DISEASE

The lymphocytes in the nodular lymphocyte predominance subtype of HD (NLPHD) or nodular paragranuloma have been shown to differ from those in the other subtypes. First of all, the majority of the lymphocytes are B-cells with a general phenotype similar to that of mantle cell B lymphocytes (Poppema et al., 1979). Their phenotype is IgM+, IgD+, CD23+, KiB3+, MT3-, CD5-, CD43- with a polyclonal staining pattern of immunoglobulin light chains. The only difference with mantle zone lymphocytes is the relatively strong staining of the lymphocytes in NLPHD nodules for CD23. The RS cells of nodular paragranuloma, the so called L&H cells, are immediately surrounded by a rosette of T-cells with a phenotype that is the same as that of T-cells found in reactive germinal centres, namely CD4+CD57+ (Poppema, 1987). CD57 was originally described as a marker for large granular lymphocytes and natural killer cells, but was subsequently found to be co-expressed on this population of CD4+ T-cells present in the light zones of germinal centres (Poppema et al., 1989). These are activated T-cells that also strongly express CD69. It has been demonstrated that they express mRNA for IL-4 (Butch et al., 1993). This is of interest since IL-4 is known to induce CD23 (Delespesse, 1992) as found on the B-cells in the nodules of NLPHD.

NLPHD has a precursor lesion consisting of follicular hyperplasia with progressively transformed germinal centres (PTGCs) (Poppema et al., 1979). These are large follicles without germinal centres which contain increased numbers of small B lymphocytes and T lymphocytes similar to those of the nodules of NLPHD. In fact, the only difference is in the absence of L&H type RS cells in this condition. The lymphocytes show no mitoses and predominantly are Ki-67 and PCNA-cyclin negative. Therefore, the accumulation of small B lymphocytes in PTGCs and NLPHD nodules must result from increased migration through high endothelial venules (HEVs), or from retainment, or from a combination of these mechanisms. The size of follicles is normally determined by a balance between entry and exit of the recirculating follicular B-cell population, the degree of transformation to follicular centre blasts and the amount of apoptosis. The small mantle zone lymphocytes have a lifespan of several weeks and new cells are continuously produced in the bone marrow.

The CD4+CD57+ T-cells are also increased in PTGCs and may in fact be an important factor since they appear to be already increased in the mantle zones of some apparently normal reactive follicles in these cases. We hypothesise that a dysregulation of the CD4+CD57+ T-cell population may be the underlying cause of the accumulation and retainment of small B-cells in these nodules. We have analysed cases of NLPHD and of PTGC for the presence of HEVs surrounding the nodules and the presence of B and T-cells in these venules. Our results indicate an increased frequency of HEVs immediately surrounding the nodules and the presence of increased numbers of small B-cells in these vessels.

Another cell population that may be of relevance is the dendritic reticulum cells that are present in the nodules of NLPHD and in PTGC. These DRC express C3d receptors (CD21) and C3b receptors (CD35), but unlike those in the light zones of reactive germinal centres, they do not stain for CD23 and for immune complexes. As a result, they resemble the DRC of mantle zones, rather than those of germinal centres.

CONCLUSION

In conclusion, there are at least two different types of HD with different types of lymphocyte responses: nodular lymphocyte predominance type and others, such as nodular sclerosis and mixed cellularity. The latter group includes EBV-positive cases that are mostly confined to cases with mixed cellularity morphology. NLPHD is a disease of follicles including the polyclonal small B-cells, the CD57-positive T-cells and dendritic reticulum cells in addition to the L&H type RS cells that can be considered atypical germinal centre blasts. These are likely to be clonal, although this has not been proven satisfactorily. Nodular sclerosis and mixed cellularity cases predominantly involve parafollicular and T-cell areas. The T lymphocytes surrounding the RS cells are a selected population that excludes CD8 cells. They are CD4 and CD45R0 positive memory helper cells and in addition express several early activation markers. However, they lack CD26 and are CD45RB^{dim}, suggesting that they are a functionally inactive population. This finding, combined with the lack of HLA class I expression on the RS cells and their variants may explain the lack of an efficient cytotoxic immune response versus the RS cells and variants, despite the presence of at least one of the latent membrane proteins in EBV-positive cases.

Further research is needed; this should include attempts to define other viruses potentially involved in the pathogenesis of EBV-negative cases, particularly the nodular sclerosis cases in adolescents and young adults. Also, it needs to be elucidated whether the CD4 cells recognise a specific, EBV-associated antigen, presented with HLA class II. Next, it would be useful to define functional activity of the CD26 negative T-cell population and the reasons for its selective influx into the tissues involved by HD. Results of such studies would allow us to define immunological approaches towards the treatment and prevention of HD.

REFERENCES

Aisenberg AC, Wilkes BM (1982) Lymph node T cells in Hodgkin's disease: analysis of suspensions with monoclonal antibody and rosetting techniques. *Blood* **59**:522

Akbar AN, Bowick N, Salmon M, Gombert W, Bofill M, Shamsadeen N, Pilling D, Pett S, Grundy JE, Janossy G (1993) The significance of low bcl-2 expression by CD45R0 T cells in normal individuals and patients with acute viral infections. The role of apoptosis in T cell memory. *J Exp Med* **178**:4

Brinker MG, Poppema S, Buys CH (1987) Clonal immunoglobulin gene rearrangements in tissues involved by Hodgkin's disease. *Blood* **70**:186

Butch AW, Chung G-H, Hoffmann J, Nahm MH (1993) Cytokine expression by germinal centre cells. *J Immunol* **150**:39

Cabanillas P, Pathak S, Trujillo J (1988) Cytogenetic features of Hodgkin's disease suggest possible origin from a lymphocyte. *Blood* **71**:1615

Deacon EM, Pallesen G, Niedobitek G, Crocker J, Brooks L, Rickinson AB (1993) Epstein-Barr virus and Hodgkin's disease: transcriptional analysis of virus latency in the malignant cells. *J Exp Med* **177**:339

Delespesse G, Sarfati M, Wu CY (1992) The low affinity receptor for IgE. *Immunol Rev* **125**:77

Khanna R, Burrows SR, Kurilla MG, Jacob CA, Misko IS, Sculley TB, Kieff E, Moss DJ (1992) Localisation of Epstein-Barr virus cytotoxic T cell epitopes using recombinant vaccinia: implications for vaccine development. *J Exp Med* **176**:169

Lai R, Visser L, Poppema S (1994) Postnatal changes of CD45 expression in peripheral blood T and B cells. *Br J Haematol* **87**:251

Limpens J, de Jong D, van Krieken JH (1991) Bcl-2/Jh rearrangements in benign lymphoid tissues with follicular hyperplasia. *Oncogene* **6**:2271

Lopez-Cabrera M, Santis AG, Fernandez-Ruiz E, Blacher R, Esch F, Sanchez-Mateos P, Sanchez-Madrid F (1993) Molecular cloning, expression, and chromosomal localisation of the human earliest lymphocyte activation antigen AIM/CD69, a new member of the C-type animal lectin superfamily of signal-transmitting receptors. *J Exp Med* **178**:537

Malavasi F, Funaro A, Roggero S, Horenstein A, Calosso L, Mehta K (1994) Human CD38: a glycoprotein in search of a function. *Immunol Today* **15**:95

Martin JM, Warnke RA (1984) A quantitative comparison of T-cell subsets in Hodgkin's disease and reactive hyperplasia. Frozen section immunohistochemistry. *Cancer* **53**:2450

Mason DY, Bastard C, Rimokh R (1990) CD30 positive large cell lymphomas (Ki-1 lymphomas) are associated with a chromosomal translocation involving 5q35. *Br J Haematol* **74**:161

Mason D, Powrie F (1990) Memory CD4+ T cells in man form two distinct sub-populations defined by their expression of isoforms of the leukocyte common antigen. *Immunology* **70**:427

Morimoto C, Schlossman SF (1994) CD26 a key co-stimulatory molecule on CD4 memory cells. *The Immunologist* **2**:4

Mosmann TR (1991) Role of a new cytokine, interleukin-10, in the cross-regulation of T helper cells. *Ann NY Acad Sci* **628**:337

Murray RJ, Wang D, Young L, Wang F, Rowe M, Kieff E, Rickinson AB (1988) Epstein-Barr virus-specific cytotoxic T-cell recognition of transfectants expressing the virus-coded latent membrane protein LMP. *J Virol* **62**:3747

Murray RJ, Kurilla MG, Brooks JM, Thomas WA, Rowe M, Kieff E, Rickinson AB (1992) Identification of target antigens for the human cytotoxic T cell response to Epstein-Barr virus (EBV): implications for the immune control of EBV-positive malignancies. *J Exp Med* **176**:157

Poppema S, Elema JD, Halie MR (1979) The localisation of Hodgkin's disease in lymph nodes. A study with immunohistological, enzyme histochemical and rosetting techniques on frozen sections. *Int J Cancer* **24**:532

Poppema S, Visser L, de Leij L (1983) Reactivity of presumed natural killer cell antibody Leu 7 with intrafollicular T lymphocytes. *Clin Exp Immunol* **54**:834

Poppema S, Kaiserling E, Lennert K (1979) Hodgkin's disease with lymphocytic predominance, nodular type (nodular paragranuloma) and progressively transformed germinal centres. A cytohistological study. *Histopathol* **3**:295

Poppema S, Visser L, de Jong B, Brinker M, Atmosoerodjo J, Timens W (1989) The typical Reed-Sternberg phenotype and Ig gene rearrangement of Hodgkin's disease derived cell line ZO indicating a B cell origin. *Rec Res Cancer Res* **117**:67

Poppema S (1989) The nature of the lymphocytes surrounding Reed-Sternberg cells in nodular lymphocyte predominance and in other types of Hodgkin's disease. *Am J Pathol* **135**:351

Poppema S, Kaleta J (1992) Biology of Hodgkin's disease, a model based on viral infection and delayed type hypersensitivity reaction. *Ann Oncol* **3**:5

Poppema S, Bhan AK, Reinherz EL, Posner MR, Schlossman SF (1982) *In situ* immunological characterisation of cellular constituents in lymph nodes and spleens involved by Hodgkin's disease. *Blood* **59**:226

Poppema S, Kaleta J, Hepperle B (1992) Chromosomal abnormalities in patients with Hodgkin's disease: evidence for frequent involvement of the 14q chromosomal region but infrequent bcl-2 gene rearrangement in Reed-Sternberg cells. *J Natl Cancer Inst* **84**:1789

Poppema S, van Imhoff G, Torensma R, Smit, J (1985) Lymphadenopathy morphologically consistent with Hodgkin's disease associated with Epstein-Barr virus infection. *Am J Clin Pathol* **84**:385

Poppema S, Visser L (1994) Epstein-Barr virus positivity in Hodgkin's disease does not correlate with HLA A2 negative phenotype. *Cancer* **73**:3059

Poppema S, Visser L (1994) Absence of HLA class I expression by Reed-Sternberg cells. *Am J Pathol* **145**:37

Romagnani S, Del Prete GF, Maggi E (1983) Displacement of T lymphocytes with the "helper/inducer " phenotype from peripheral blood to lymphoid organs in untreated patients with Hodgkin's disease. *Scand J Haematol* **31**:305

Stetler-Stevenson M, Crush-Stanton S, Cossman J (1990) Involvement of the bcl-2 gene in Hodgkin's disease. *J Natl Cancer Inst* **82**:855

Sundeen J, Lipford E, Uppenkamp M (1987) Rearranged antigen receptor genes in Hodgkin's disease. *Blood* **70**:96

Trinchieri G (1993) Interleukin-12 and its role in the generation of Th1 cells. *Immunol Today* **14**:335

Trowbridge IS, Thomas ML (1994) CD45: an emerging role as a protein tyrosine phosphatase required for lymphocyte activation and development. *Ann Rev Immunol* **12**:85

Uzawa A, Suzuki G, Nakata Y, Akashi M, Ohyama H, Akanuma A (1994) Radiosensitivity of CD45R0+ memory and CD45R0- naive T cells in culture. *Radiation Research* **137**:25

Visser L, Lai R, Poppema S (1993) Patterns of leukocyte common antigen expression in peripheral blood T cell populations. *Cell Immunol* **151**:218

Weiss LM, Strickler JG, Hu E (1986) Immunoglobulin gene rearrangements in Hodgkin's disease. *Hum Pathol* **17**:1009

PROPAGATION OF HODGKIN AND REED-STERNBERG CELLS

Ursula Kapp, Jürgen Wolf, Christof von Kalle and Volker Diehl
Med. Universitätsklinik I, Joseph-Stelzmann-Straße 9, 50924 Köln, Germany

SUMMARY

Tumour tissue derived from Hodgkin's disease (HD) has been propagated *in vitro* and *in vivo*. Until now, 14 *in vitro* cell lines have been established, which are likely to be derived from the tumour cells of HD. These cell lines have gained importance in the diagnosis and therapy of HD by helping to define the antigens CD30, CDw70 and Ki-67, by the establishment of an animal model and by the development of new immunotherapeutic strategies.

The heterogeneity of phenotype, genotype, karyotype, cytokine and oncogene expression of the cell lines reflects the heterogeneity of Hodgkin and Reed-Sternberg (H-RS) cells found in primary biopsy specimens. HD-derived cell lines have not fully explained the aetiology of HD. As a new experimental approach to clarify the origin of H-RS cells, we attempted to propagate HD-derived primary tumour tissue in SCID mice. Nine Epstein-Barr virus (EBV) positive B-cell tumours grew progressively from 3 of 15 HD biopsy specimens transplanted into SCID mice. The tumours were derived from EBV-positive bystander cells. In 5/6 tumours examined cytogenetically, chromosomal aberrations were detected. We conclude that these bystander cells might differ from normal lymphocytes and possibly play a role in the pathogenesis of HD.

INTRODUCTION

To date, little is known about the origin of Hodgkin and Reed-Sternberg (H-RS) cells, the putative malignant cell population in Hodgkin's disease (HD). Investigation of these cells has been hampered because of their scarcity (0.1-1%) among apparently normal bystander cells in lymphatic tissue affected by HD. Many attempts have been made to separate or enrich these cells by density gradient centrifugation, magnetic activated cell sorting (MACS), *in vitro* or *in vivo* culture and recently by picking H-RS cells from single cell suspensions or slides. Here we will focus on the *in vitro* and *in vivo* propagation of H-RS cells.

PROPAGATION OF HODGKIN'S DISEASE-DERIVED CELL LINES

Rare *in vitro* growth of Hodgkin and Reed-Sternberg cells

Only 14 cell lines have been established which are considered to be derived from H-RS cells (Table 1). The low efficiency of propagating permanent cell lines from HD-derived lymphoma tissue is not fully understood as yet.

Table 1: Hodgkin's disease-derived cell lines.

Cell line	Histological subtype	Cultured material	References
CO	nodular sclerosis	lymph node	Jones, 1985, 1989
DEV	nodular sclerosis	pleural effusion	Poppema, 1985, 1989
HD-70	nodular sclerosis	peripheral blood	Kanzaki, 1992
HDLM-1/2/3	nodular sclerosis	pleural effusion	Drexler, 1986, 1989
HD-MyZ	nodular sclerosis	pleural effusion	Bargou, 1993
HO	nodular sclerosis	lymph node	Jones, 1989
HuT-11	mixed cellularity	lymph node	Roberts, 1978
KM-H2	mixed cellularity	pleural effusion	Kamesaki, 1986
L-428	nodular sclerosis	pleural effusion	Schaadt, 1979, 1980
L-538/540	nodular sclerosis	peripheral blood bone marrow	Diehl, 1981, 1982
L-591	nodular sclerosis	pleural effusion	Diehl, 1982
SUP-HD1	nodular sclerosis	pleural effusion	Naumovski, 1989
SU/RH-HD-1	nodular sclerosis	spleen	Olsson, 1984, 1985, 1988
ZO	nodular sclerosis	pericardial fluid	Poppema, 1989

One explanation could be the dependence of H-RS cells upon cell to cell contact or growth factors produced by non-malignant bystander cells. Alternatively, at least in initial stages of the disease, cells might be targets for the host immune response rather than truely malignant cells.

Another problem is the difficulty in determining whether a new cell line is derived from the tumour cells in HD. Interestingly from the 14 established HD-derived cell lines only one (L591) is EBV-positive. In contrast, the EBV genome is found by EBER *in situ* hybridisation in the H-RS cells of about 45% of HD cases in industrialised countries (Hummel *et al.,* 1992). Thus, EBV-negative HD-derived cell lines represent cases which either were originally negative for EBV or lost EBV during the course of lymphoma development. There is no special marker to discriminate HD-derived cell lines from EBV immortalised lymphoblastoid cell lines (LCL), which often grow from *in vitro* cultures of primary HD specimens. Even the HD-associated CD30 antigen, which is regularly expressed on H-RS cells *in vivo*, is also expressed on LCLs. The only criterion, which might discriminate HD-derived cell lines from LCLs is aneuploidy after short term *in vitro* culture. It cannot be excluded that numerous cell lines derived from H-RS cells were not recognized to be HD-derived and disappeared into the dustbin, because they were diploid, had minor chromosomal abnormalities and were positive for EBV.

In vitro cultured tumour specimens

In vitro cell lines have been propagated from only two histological subtypes. Twelve of fourteen were established from nodular sclerosis and two of fourteen from mixed cellularity disease. The cell lines were derived from lymph nodes in only 3/14 cases and in 1 case from a spleen; all the other cell lines stem from pleural or pericardial effusions, peripheral blood or bone marrow, where the tumour cells are already in suspension and possibly less dependent on microenvironmental factors (see Table 1).

Phenotype, genotype and karyotype of Hodgkin's disease-derived cell lines

HD-derived cell lines regularly express several antigens such as the X-Hapten (CD15), the IL-2 receptor alpha chain (CD25), the transferrin receptor (CD71) and the activation antigen Ki-1 (CD30), which also can be demonstrated on activated or virally transformed lymphocytes and monocytes. The expression of cell lineage associated antigens is heterogeneous. The majority of the cell lines shows a B- or T-lymphoid phenotype, some express neither B- nor T-cell markers (Drexler, 1993) and the recently established cell line HD-MyZ expresses the myeloid/monocytic surface markers CD13 and CD68 (Bargou, 1993). The cell lines with B- or T-cell phenotype show heterogeneous immunoglobulin (Ig) or T-cell receptor (TCR) rearrangements (see Table 2).

Table 2:   Characteristics of Hodgkin's disease-derived cell lines.

Cell line	Phenotype	CD30	CD15	CDw70	Genotype
B-cell association					
DEV	CD20, CD22	+	+		$Ig_{H,L}$
HD-70	-	+	+		$Ig_{H,L}$
KM-H2	CD21	+	+	+	Ig_H
L-428	-	+	+	+	Ig_H
L-591	CD19, CD20	+		+	$Ig_{H,L}$
SUP-HD1	-	-	+		$Ig_{H,L}, TCR_{beta}$
ZO	-	+	+		$Ig_{H,L}$
T-cell association	CD3, CD5, CD7				
CO		+	+		$TCR_{beta, gamma}$
HDLM-2	CD2	+	+		TCR
HO	CD3, CD4, CD5, CD7	+	-	+	$TCR_{beta, gamma}$
L-540	CD2, CD4	+	+	+	TCR
Myeloid association					
HD-MyZ	CD13, CD68	-	-	+	-

These results are in line with the analysis of primary H-RS cells which also show a marked heterogeneity with regard to lineage-associated surface marker expression and Ig/TCR rearrangement (Drexler, 1992).

Numerous non-random structural and numeric cytogenetic aberrations are found in HD-derived cell lines, but no consistent cytogenetic markers have been identified. By comparison, cytogenetic aberrations are found in about 50% of cases of primary HD tissue (Schouten et al., 1989; Tilly et al., 1991). Similiar to the findings with the HD-derived cell lines, no consistent marker was found among these structural and numerical aberrations.

BENEFITS OF HODGKIN'S DISEASE-DERIVED CELL LINES

Generation of new monoclonal antibodies
New monoclonal antibodies have been generated against the cell line L428. These antibodies recognise the Hodgkin-associated antigen Ki-1 (CD30) (Schwab et al., 1982), the CDw70 antigen (Stein et al., 1983) and the proliferation associated antigen Ki-67 (Gerdes et al., 1983).

CD30 expression, in combination with morphology, is routinely used to diagnose HD. In addition, a new non-Hodgkin's lymphoma entity was defined by the expression of CD30 (Stein et al., 1985), the anaplastic large cell lymphoma (ALCL, Ki-1 lymphoma). Furthermore, the CD30 antigen is used as a target for immunotherapy in experimental strategies for the treatment of HD (see below).

ESTABLISHMENT OF AN ANIMAL MODEL

For preclinical testing of new therapeutic regimens as well as for studying the in vivo biology of HD cells, it was desirable to develop an in vivo model in a rodent system. In athymic, T-cell deficient nude mice HD-derived cell lines could not be reproducibly propagated. We, therefore, tested tumourigenicity of HD-derived cell lines in untreated and pre-treated severe combined immunodeficient (SCID) mice and in beige/nude/X-linked immunodeficient (BNX) mice (Von Kalle et al., 1992; Kapp et al., 1992).

SCID mice lack functional B- and T-lymphocytes due to an autosomal recessive defect of the VDJ recombination mechanism (Bosma et al., 1983; Schuler et al., 1989). They have been shown to be suitable recipients for the xenotransplantation of malignant haematopoetic cells (Dillmann et al., 1985; Charly et al., 1990; Kamel-Reid et al., 1989) and even of normal lymphoid (McCune et al., 1988) or haematopoetic cells (Lapidot et al., 1992; Kamel-Reid and Dick, 1988).

Seven HD-derived cell lines (L428, L540, L540cy, L591, DEV, HDLM-2 and KM-H2) were injected subcutaneously and intraperitoneally into SCID and BNX mice. Only two of the six lines (DEV, L540cy) led to subcutaneous tumour growth in BNX mice. Three additional HD-derived cell lines could be transplanted successfully into BNX mice only after growth supportive treatments such as fibrosarcoma co-transplantation, intraperitoneal mineral oil injection or the injection of anti-asialo-GM1-antibody inhibiting natural killer (NK) cells.

In contrast, in SCID mice tumours developed after inoculation of all the HD-derived cell lines tested. The best tumour take (100%) could be induced with the mutant subline L540cy, which was established from a nude mouse following treatment of L540 cells, which had been injected by the intramuscular route, with cyclophosphamide. The take rate of the other cell lines varied between 20% and 60% in untreated SCID mice with a latency period ranging between 7 and 20 days. There was no change of immunophenotype or karyotype of the HD-derived cell lines after mouse passage. From these experiments the SCID mouse was considered to be the favourable rodent system for xenotransplantation of HD-derived cells.

In a second series of experiments the HD cell lines L428, KM-H2, L540 and its subline L540cy were injected intravenously into SCID mice (Kapp *et al.*, 1994). We found disseminated tumour growth after transplantation of the cell line L540 in 40% of the inoculated animals and in 57% following inoculation of the subline L540cy (Table 3). With the exception of the spleen, the tumour dissemination resembled the dissemination of HD in man. In all mice with positive tumour growth, tumours developed in the lymph nodes (Table 4). Immunostaining for CD30 antigen, and DNA dot blots from the different mouse organs hybridised with COT-DNA specific for human DNA, showed that the tumours grown in the SCID mice were of human origin. The lymph nodes affected by tumour growth were mainly the cervical nodes (Table 5) (other organs were less frequently involved). Histologically the tumours resembled human anaplastic large cell lymphomas. Immunostaining against the CD30 antigen showed human tumour cells side by side with only a few mouse cells among them.

Table 3: Growth of the intravenously transplanted Hodgkin's disease-derived cell lines L540, L540cy, L428 and KM-H2.

Cell lines	L540	L540cy	L428	KM-H2
Number of injected mice	10	21	10	10
Number of mice with detectable tumours	4	12	1	1
Growth pattern	d.g.	d.g.	s.t.	s.t.
Take (%)	40	57	10	10

Abbreviations: s.t., single tumour; d.g., disseminated growth

Table 4: Organ distribution of the 'disseminating' Hodgkin's disease-derived cell lines L540 and L540cy in tumour bearing SCID mice.

Site of tumour growth	L540	L540cy
Lymph nodes	4/4	12/12
Liver	1/4	9/12
Lungs	3/4	8/12
Thymus	0/4	3/12
Bone marrow	1/4	3/12
Spleen	0/4	1/12

Table 5: Lymph node distribution of the 'disseminating' Hodgkin's disease-derived cell lines L540 and L540cy in tumour bearing SCID mice.

Lymph nodes	L540	L540cy
Cervical	3/4	10/12
Parathymic	0/4	7/12
Para-aortic	2/4	5/12
Portal	2/4	7/12
Renal	1/4	7/12
Iliac	1/4	11/12

To address the question whether specific adhesion molecules contribute to the growth of the HD-derived cell lines L540 and L540cy in the SCID mouse, lymph nodes were analysed for expression of selectins, ß2 integrins, ß1 integrins, the Ig super family and the supposed lymphocyte homing receptor CD44 (Gunthert et al., 1991) by flow cytometry. Significant differences in the expression of CD30, CD2, ICAM-1 and CD44 were found between disseminating and non-disseminating cell lines. L540 and L540cy expressed the CD30 antigen more strongly than L428 and KM-H2. CD2 expression could be demonstrated only on L540 and L540cy. The ICAM-1 molecule was more strongly expressed on L428 and KM-H2. Interestingly the putative homing receptor CD44 was found only on the non-disseminating cell lines L428 and KM-H2. Thus, none of the adhesion molecules tested seems to contribute to the disseminated growth pattern of the HD-derived cell lines L540 and L540cy.

NEW TREATMENT STRATEGIES FOR ADVANCED STAGES OF HODGKIN'S DISEASE

The probability of curing patients with HD is about 70% at the present time (Diehl, 1993). Nevertheless, 30-50% of patients suffering from HD at an advanced stage will still die from their disease (Longo et al., 1986). For patients cured from HD there is a risk of about 19% of getting secondary neoplasia 20 years after treatment (Henry-Amar, 1992).

Antigens regularly expressed on H-RS cells, like the CD30 antigen, can be used as a target for experimental immunotherapeutic reagents that are selectively toxic to the putative malignant cells. HD-derived cell lines were an important tool for the development of new non-mutagenic, immunological treatment strategies, because they were used to conduct the preclinical in vitro and in vivo experiments.

Approaches to create a new immunological therapy of HD are: the development of immunotoxins consisting of the ribosome destroying A chain of ricin or other toxins coupled to antibodies directed against H-RS cell-associated antigens such as CD30 or the IL-2 receptor alpha-chain (CD25) (Falini et al.,1992; Engert et al., 1991), the preparation of bispecific monoclonal antibodies, which bind to the CD30 antigen and antigens on the surface of T-cells (Renner et al., 1994) or NK-cells (Hombach et al., 1993) and finally the construction of an anti-idiotype vaccine against HD (Pohl et al., 1992). In this last approach the anti-CD30 monoclonal antibody HRS-4 was used to generate monoclonal anti-idiotypic antibodies, which were shown to carry the internal image of the CD30 antigen and perhaps can be used as vaccines against CD30 antigen-expressing lymphomas.

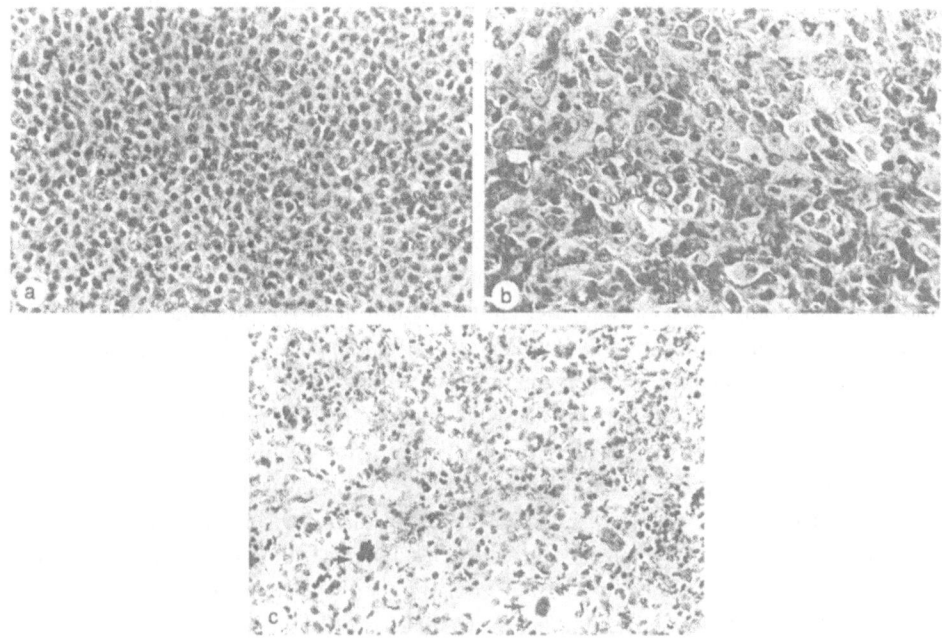

Fig. 1:   Histological patterns found in SCID mouse tumours grown after xenotransplantation of primary
Hodgkin's disease tumour tissue: a. Lymphoproliferative disease, haematoxylin & eosin stain,
original magnification x 180. b. Anaplastic large cell lymphoma, Giemsa stain, original
magnification x 450. c. Hodgkin-like lesion, frozen section, with the Reed-Sternberg-like giant
cells arrowed, original magnification x 180.

PROPAGATION OF HODGKIN'S DISEASE-DERIVED TUMOUR TISSUE *IN VIVO*

Propagation of HD-derived tumour tissue *in vitro* led to the establishment of 14 permanently growing cell lines. The alternative experimental approach is the propagation of Hodgkin's lymphoma specimens *in vivo*. As mentioned above, the SCID mouse was found to be the most favourable host for *in vivo* growth of HD-derived cell lines. Therefore we selected this mouse strain for the xenotransplantation of HD-derived primary tumour tissue (Kapp *et al.*, 1993). Altogether 15 tumour samples from 13 patients were transplanted into the mice. Ten of these specimens were transplanted into the subrenal capsule and 5 were intrahepatic. The tumour samples stemmed from 13 lymph nodes and 2 spleens histologically affected by HD. A total of 52 mice were inoculated and 10 developed human tumours. These were characterised by the following attributes: they all were of human origin and consisted of CD20 and CD30-positive B-cells. By Southern blotting, oligoclonal and monoclonal heavy chain (IgH) gene rearrangements were detected. Three histological patterns were described: 1. Lymphoproliferative disease, as seen in post-transplant lymphomas in humans, 2. the very bizzare picture of anaplastic large cell lymphomas and 3. a pattern termed Hodgkin-like lesion, because large CD30-positive cells resembling RS cells were surrounded by mouse macrophages (see Fig. 1). All the tumours were EBV-positive as shown by EBER *in situ* hybridisation and, using immunohistochemical methods, all were postive for LMP-1 and EBNA-2. The most important finding was that the majority of the tumours (5/6 in the first mouse passage) had numerical karyotypic aberrations such as trisomy of chromosome 3, 7, 9, 10, 11, 12, 13 and 20 and some structural aberrations, for example a translocation between the long arms of chromosomes 6 and 11. When these tumours were cultured *in vitro*, B-cell lines, exhibiting the same chromosomal abnormalities, grew out after *in vitro* recultivation (see Fig. 2).

Because the H-RS cells in 2 of the 3 specimens which led to tumour growth in the mice were EBV-negative, the tumours grown in the SCID mice were likely to stem from EBV-positive bystander cells. The HD-derived SCID mouse tumours resembled EBV-induced B-cell tumours from healthy donors as the latter are propagated in SCID mice and have been well characterised by other groups such as Purtilo *et al.,* (1991), Cannon *et al.,* (1990) and Rowe *et al.,* (1991). These tumours, which commonly are termed LPD, are also EBV-positive in 100% of cases. The tumours grown in SCID mice from EBV-positive peripheral blood lymphocytes exhibit the same pattern of EBV-expression as the tumours induced by transplantation of HD tumour tissue (EBNA-2+, LMP-2+), whereas in primary HD EBNA-2 is down-regulated (EBNA-2-, LMP-2+). Nevertheless, there are some differences between the well described B-cell lymphomas grown in SCID mice after injection of EBV and peripheral blood lymphocytes of healthy donors, and the tumours propagated after xenotransplantation of HD-derived tumour tissue. The B-cell lymphomas that developed in the mice from lymphocytes of normal donors were histopathologically diagnosed as lymphoproliferative disorder, whereas HD SCID mouse tumours partly exhibited the patterns of anaplastic large cell lymphomas and Hodgkin-like lesions.

Furthermore, the incidence of chromosomal aberrations detected after transplantation of HD tumour specimens is significantly higher in comparison to SCID mouse B-cell lymphomas derived from healthy donors. After the first mouse passage of the HD tumours 6 samples of the explanted tumours could be examined cytogenetically. In 5 of these, clonal karyotypic aberrations were found. From a total of 14 tumour samples, 11 exhibited clonal cytogenetic abnormalities in the course of 3 mouse passages (see Table 6). As described by Pisa *et al.* (1992), no cytogenetic aberrations were detected in EBV-positive B-cell tumours derived from healthy donors, which were propagated in SCID mice. Thangavelu *et al.* (1992) observed clonal abnormalities in only 4/27 human LPD tumours in SCID mice following the injection of peripheral blood leukocytes or LCL from 14 EBV-seropositive healthy donors and from an EBV-positive patient with CLL.

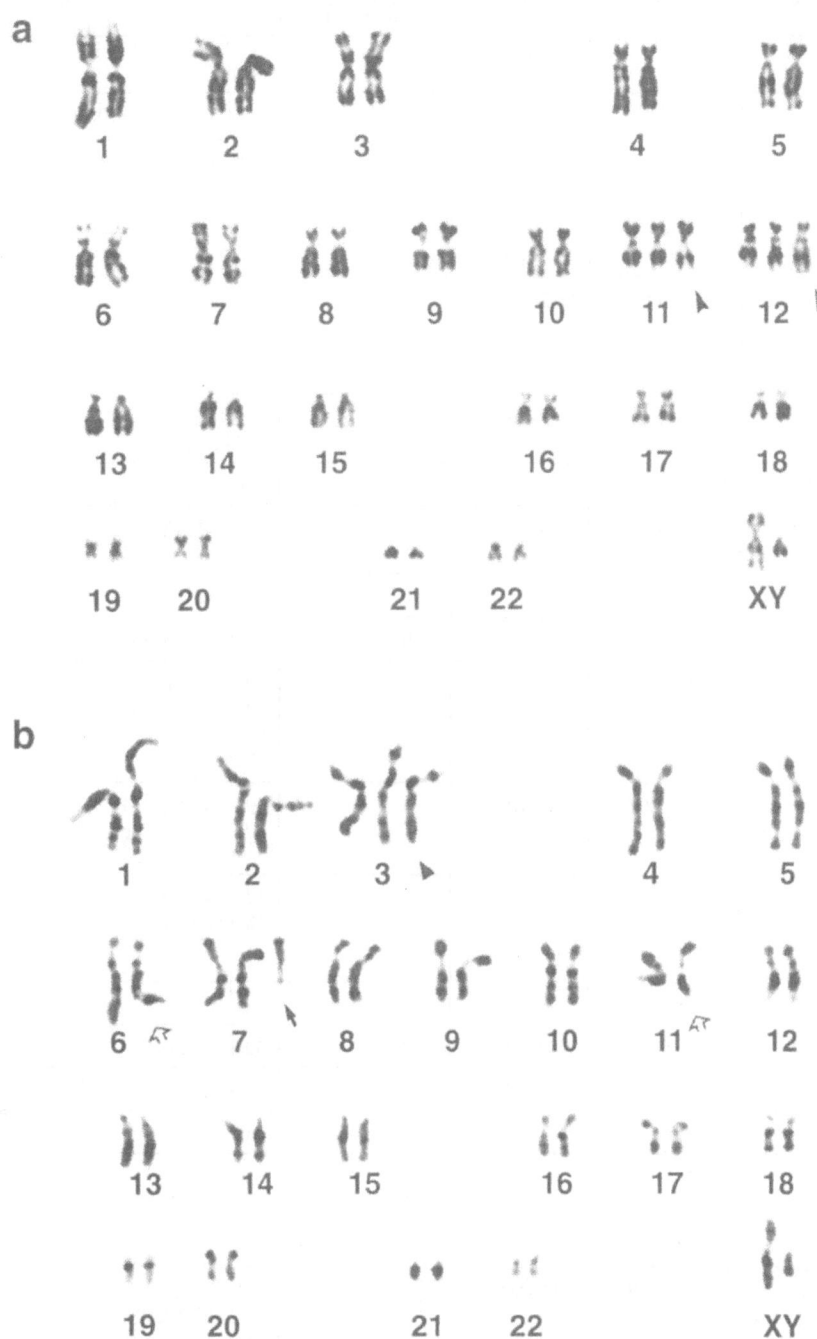

Fig. 2a: Karyotype of a human B-cell line established from a SCID mouse tumour in the first tumour passage: 48XY, + 11, + 12.

b: Karyotype of human tumour cells from a SCID mouse tumour in the 4th tumour passage: 48XY, + 3, + 7, de (7)(q11.2/21), t(6;11)(q21;q13).

Table 6:   Characteristics of the human tumours grown in SCID mice.

Tumour	Histology	Cell line	Time in culture (months)	Karyotype
DS1-MT2	ALCL	DS1-ML2	1	30/30 48,XY,+11,+12, sporadic structural anomalies
DS1-MT3	LPD, ALCL HDLL	DS1-ML3	5	9/34 52,XY,+9,+10,+11, +12,+13,+20 5/34 46,XY,14p+ 8/34 46,XY,14p++ 12/34 normal
DS1-MT5	LPD, ALCL	DS1-MT5	2	30/30 normal
DS2-MT1	LPD, ALCL	DS2-ML1	3	3/35 47,XY,+9 7/35 46,XY,3q+ 1/35 46,XY,14p++ 24/35 normal
DS2-MT2	LPD, ALCL, HDLL	DS2-ML2	2	29/34 48,XY,+3,+7 5/34 normal
DS3-MT1	LPD, ALCL	-	-	n.d.
DS3-MT2	LPD, ALCL, HDLL	DS3-ML2	2	30/30 48,XY,+3,+7
DS3-MT3	LPD, ALCL	-	0	31/31 48,XY,+3,+7
DS4-MT1	LPD	-	0	25/25 48,XY,+3,+7
DS4-MT2	LPD	-	0	1/25 48,XY,+3,+7 22/25 48,XY,+3,+7,del(7) (q11.2/21), t(6;11)(q21;q13) 2/25 normal
AS1-MT1	LPD, ALCL	AS1-ML1	4	29/33 46,XY,1p+ 4/33 normal, sporad. structural anomalies
AS1-MT2	LPD, ALCL	AS1-ML2	4	12/30 47,XY,+9 18/30 normal
AS1-MT3	ALCL	-	-	n.d.
AS2-MT1	ALCL	-	-	n.d.
AS2-MT2	LPD, ALCL	AS2-ML2	3	30/30 normal
AS2-MT3	LPD, ALCL	AS2-ML3	2	30/30 normal
AS2-MT4	ALCL	-	-	n.d.
AS3-MT1	LPD	-	-	n.d.
AS3-MT2	LPD, ALCL, HDLL	-	-	n.d.
OH1-MT1	LPD, ALCL	OH1-ML1	3	11/25 normal 14/25 92,XXYY, sporad. structural anomalies

Abbreviations: DS, AS, OH, initials of the patients; number following the initials, number of the mouse passage; MT, mouse tumour; ML, mouse cell line; number following MT/ML, number of mouse tumour/cell line; LPD, lymphoproliferative disease; ALCL, anaplastic large cell lymphoma; HDLL, Hodgkin-like lesion; n.d., not done.

CONCLUSIONS

HD-derived tumour tissue has been propagated *in vitro* and *in vivo*. Until now, only 14 permanent growing cell lines have been established by *in vitro* propagation despite numerous cultivation attempts over many years. These cell lines have been used as tools for improving diagnosis and therapy of HD. New antigens have been discovered using reagents raised against these cell lines. The CD30 antigen is used as a marker to define the H-RS cells, and together with the cellular morphology is used to reach a histological diagnosis of HD. *In vivo* growth of HD-derived cell lines offers a reproducible experimental system to study the biology of HD-derived cells *in vivo* and for preclinical testing of new therapeutic strategies. The HD-derived cell lines themselves are an important tool for the development of new therapeutic reagents such as immunotoxins, bispecific antibodies and the anti-idiotype vaccine.

HD-derived cell lines are heterogeneous in phenotype, genotype, karyotype and their expression of cytokines and oncogenes. This heterogeneity reflects the heterogeneity found in primary biopsy specimens. So far no marker has defined the origin of H-RS cells. Possibly a specific feature will be found if more HD-derived cells can be permanently grown or enriched by *in vitro* or *in vivo* culture.

We therefore tried to propagate HD-derived tumour tissue *in vivo*. Xenotransplantation of HD biopsy specimens into untreated SCID mice did not lead to the growth of Hodgkin's lymphoma. Considering the fact that SCID mice do not have functional lymphocytes, a histology typical of HD was not to be expected. Instead of H-RS cells, EBV-positive bystander cells were propagated in the mice. SCIDhu mice reconstituted with a human immune system, by transplantation of human bone marrow (Lapidot *et al.*,1992) or human fetal tissue (McCune *et al.*, 1988), could be better recipients for H-RS cells. They could provide the required growth factors, bystander cells and cell to cell contact. Future attempts will go in this direction.

The human LPD tumours we observed in SCID mice differ from tumours derived from peripheral lymphocytes of healthy donors, because the HD-derived SCID mouse tumours showed cytogenetic aberrations. We conclude that EBV-positive bystander cells in HD-derived tumour tissue might differ from normal EBV-positive B-cells. The increased incidence of structural and numerical aberrations observed in the SCID-tumours might characterise them as "not so innocent" bystander cells, possibly even clonogenic HD-precursor cells. Focusing the investigation of HD on H-RS cells might not tell us the whole story. H-RS cells may be just the peak of an iceberg.

REFERENCES

Bargou RC, Mapara MY, Zugck C, Daniel PT, Pawlita M, Döhner H, Dörken B (1993) Characterization of a novel Hodgkin cell line HD-MyZ, with myelomoncytic features mimicking Hodgkin's disease in severe combined immunodeficient mice. *J Exp Med* 177:1257

Bosma C, Custer RP, Bosma MJ (1983) A severe combined immunodeficient mutation in the mouse. *Nature* **301**:527

Cannon MJ, Pisa P, Fox RI, Cooper NR (1990) Epstein-Barr virus induces aggressive lymphoproliferative disorders of human B cell origin in SCID/hu chimeric mice. *J Clin Invest* **85**:1333

Charly MR, Tharp M, Locker J, Jau-Shyong Deng PD, Golsen JB, Mauro T, McCoy P, Abell E, Jegasothy B (1990) Establishment of a human cutaneous T-cell lymphoma in CB-17 SCID mice. *J Invest Dermatol* **94**:381

Diehl V, Kirchner HH, Schaadt M, Fonatsch C, Stein H, Gerdes J, Boie C (1981) Hodgkin's disease. Establishment and characterization of four *in vitro* cell lines. *J Cancer Res Clin Oncol* **101**:111

Diehl V, Kirchner HH, Burrichter H (1982) Characteristics of Hodgkin's disease derived cell lines. *Cancer Treat Rev* **66**:615

Diehl V (1993) Dose-escalation study for the treatment of Hodgkin's disease. *Ann Hematol* **66**:139

Dillman RO, Johnson DE, Shawler DL, Halpern SE, Leonard E, Hagan PL (1985) Athymic mouse model of a human T-cell tumour. *Cancer Res* **45**:5632

Drexler HG, Gaedicke G, Lok MS, Diehl V and Minowada J (1986) Hodgkin's disease derived cell lines HDLM-2 and L428: comparison of morphology, immunological and isoenzyme profiles. *Leuk Res* **10**:487

Drexler HG, Gignac SM, Hoffbrand AV, Leber BF, Norton J, Lok MS, Minowada J (1989) Characterisation of Hodgkin's disease derived cell line HDLM-2. In *New aspects in the diagnosis and treatment of Hodgkin's disease.* Diehl V, Pfreundschuh M, Loeffler M, eds (Berlin: Springer) p75

Drexler HG (1992) Recent results on the biology of Hodgkin and Reed-Sternberg cells. I. Biopsy material. *Leuk Lymphoma* **8**:283

Drexler HG (1993) Recent results on the biology of Hodgkin and Reed-Sternberg cells. II. Continuous cell lines. *Leuk Lymphoma* **9**:1

Engert A, Martin G, Amlot P, Wijdenes J, Diehl, V, Thorpe P (1991) Immunotoxins constructed with anti CD25 monoclonal antibodies and deglycosylated ricin A-chain have potent anti-tumour effects against human Hodgkin cells *in vitro* and solid Hodgkin tumours in mice. *Int J Cancer* **49**:450

Falini B, Bolognesi A, Flenghi L, Tazzari PL, Broe MK, Stein H, Dürkop H, Aversa F, Corneli P, Pizzolo G, Barbabietola G, Sabattini E, Pileri S, Martelli MF, Stirpe F (1992) Response of refractory Hodgkin's disease to monoclonal anti-CD30 immunotoxin. *Lancet* **339**:1195

Gerdes J, Schwab U, Lemke H, Stein H (1983) Production of a mouse monoclonal antibody reactive with a human nuclear antigen associated with cell proliferation. *J Clin Path* **36**:167

Gunthert U, Hofmann M, Rudy W, Reber S, Zoller M, Haussmann I, Matzku S, Wenzel A, Ponta H, Herrlich P (1991) A new variant of glycoprotein CD44 confers metastatic potential to rat carcinoma cells. *Cell* **65**:13

Henry-Amar M (1992) Second cancer after the treatment for Hodgkin's disease. A report from the international database on Hodgkin's disease. *Ann Oncol* **3(Suppl 1)**:117

Hombach A, Jung W, Pohl C, Renner C, Sahin U, Schmits R, Wolf J, Kapp U, Diehl V, Pfreundschuh M (1993) A CD16/CD30 bispecific monoclonal antibody induces lysis of Hodgkin's cells by unstimulated natural killer cells *in vitro* and *in vivo*. *Int J Cancer* **55**:830

Hummel M, Anagnostopoulos I, Dallenbach F, Korbjuhn P, Dimmler C, Stein H (1992) EBV infection patterns in Hodgkin's disease and normal lympjoid tissue: expression and cellular localization of EBV gene products. *Br J Hematol* **82**:689

Jones DB, Scott CS, Wright DH, Stein H, Beverly PCL, Payne SV, Crawford DH (1985) Phenotypic analysis of an established cell line derived from a patient with Hodgkin's disease (HD). *Hematol Oncol* **3**:133

Jones DB, Furley AJW, Gerdes J, Greaves MF, Stein H, Wright DH (1989) Phenotypic and genotypic analysis of two cell lines derived from Hodgkin's disease tissue biopsies. In *New aspects in the diagnosis and treatment of Hodgkin's disease.* Diehl V, Pfreundschuh M, Loeffler M, eds (Berlin: Springer) p62

Kamel-Reid S, Dick JE (1988) Engraftment of immune-deficient mice with human hematopoietic stem cells. *Science* **242**:1706

Kamel-Reid S, Letarte M, Sirard C, Noedens M, Grunberger T, Fulop G, Freedman G, Philips RA, Dick JE (1989) A model of acute lymphoblastic leukemia in immune-deficient SCID mice. *Science* **246**:1597

Kamesaki H, Fukuhara S, Tatsumi E, Uchino H, Yamaha H, Miwa H, Shirakawa S, Hatanaka M, Honjo T (1986) Cytochemical, immunologic, chromosomal, and molecular genetic analysis of a novel cell line derived from Hodgkin's disease. *Blood* **68**:285

Kanzaki T, Kugonishi I, Eguchi T, Yano S, Sonobe H, Ohyashiki JH, Ohyashiki K, Toyama K, Ohtsuki Y, Miyoshi I (1992) Establishment of a new Hodgkin's cell line (HD-70) of B-cell origin. *Cancer* **69**:1034

Kapp U, Dux A, Schell-Frederick E, Banik N, Hummel M, Mücke S, Fonatsch C, Bullerdiek J, Gottstein C, Engert A, Diehl V, Wolf J (1994) Disseminated growth of Hodgkin's-derived cell lines L540 and L540cy in immune deficient SCID mice. *Ann Oncol* **5 (Suppl 1)**:121

Kapp U, Wolf J, Hummel M, Pawlita M, von Kalle C, Dallenbach F, Schwonzen M, Krueger GRF, Müller-Lantzsch N, Fonatsch C, Stein H, Diehl V (1993) Hodgkin's lymphoma-derived tissue serially transplanted into severe combined immunodeficient mice. *Blood* **82**:1247

Kapp U, Wolf J, Von Kalle C, Tawadros S, Röttgen A, Engert A, Fonatsch C, Stein H, Diehl V (1992) Preliminary report: growth of Hodgkin's lymphoma derived cells in immunocompromised mice. *Ann Oncol* **3 (Suppl 4)**:21

Kapp U, Wolf J, Von Kalle, Stein H, Fonatsch C, Schell-Frederick E, Diehl V (1992) Recent efforts to establish an *in vivo* model as a new experimental tool in the study of Hodgkin's disease. *Eur J Cancer* **28A**:1408

Lapidot T, Pflumio F, Doedens M, Murdoch B, Williams DE, Dick JE (1992) Cytokine stimulation of multilineage hematopoiesis from immature human cells engrafted in SCID mice. *Science* **255**:1137

Longo IL, Young RC, Wesley, M (1986) Twenty years of MOPP therapy for Hodgkin's disease. *J Clin Oncol* **4**:1295

McCune JM, Namikawa R, Kaneshima H, Shultz LD, Liebermann M, Weissman IL (1988) The SCID-hu-mouse: murine model for the analysis of human hematolymphoid differentiation and function. *Science* **241**:1632

Naumovski L, Utz PJ, Bergstrom SK, Morgan R, Molina A, Toole JJ, Glader BE, McFall P, Weiss LM, Warnke R, Smith SD (1989) SUP-HD1: a new Hodgkin's disease derived cell line with lymphoid features produces interferon-gamma. *Blood* **74**:2733

Olsson L, Behnke O, Pleibel N, D'Amore F, Werdelin O, Fry K, Kaplan HS (1984) Establishment and characterization of a cloned giant cell line from a patient with Hodgkin's disease. *J Natl Cancer Inst* **73**:809

Olsson L, Behnke O (1985) Phenotypic attributes of the malignant cell population in Hodgkin's disease indicate a monocyte/macrophage origin. *Cancer Surveys* **4**:421

Olsson L, Behnke O (1988) Emergence of a retrovirus in a cloned cell line established from a lesion of Hodgkin's disease. *Hematol Oncol* **6**:213

Pisa P, Cannon MJ, Pisa EK, Cooper NR, Fox RI (1992) Epstein-Barr virus induced lymphoproliferative tumours in severe combined immunodeficient mice are oligoclonal. *Blood* **79**:173

Pohl C, Renner C, Schwonzen M, Sieber M, Lorenz P, Pfreundschuh M, Diehl V (1992) Anti -idiotype vaccine against Hodgkin's lymphoma: induction of B- and T-cell immunity across species barriers against CD30 antigen by murine monoclonal internal image antibodies. *Int J Cancer* **50**:958

Poppema S, Visser L, de Jong B, Brinker M, Atmosoerodjo J, Timens W (1989) The typical Reed-Sternberg phenotype and Ig rearrangement of Hodgkin's disease derived cell line ZO indicate a B-cell origin. In *New aspects in the diagnosis and treatment of Hodgkin's disease*. Diehl V, Pfreundschuh M, Loeffler M, eds (Berlin: Springer) p67

Poppema S, de Jong B, Atmosoerodjo J, Idenburg V, Visser L, de Ley L (1985) Morphologic, immunologic, enzymehistochemical and chromosomal analysis of a cell line derived from Hodgkin's disease. Evidence for a B-cell origin of Sternberg-Reed cells. *Cancer* 55:683

Purtilo DT, Falk K, Pirruccello SJ, Nakamine H, Kleveland K, Davis JR, Okano M, Taguchi Y, Sanger WG, Beisel KW (1991) SCID mouse model of Epstein-Barr virus induced lymphomagenesis of immunodeficient humans. *Int J Cancer* 47:510

Roberts AN, Smith KL, Dowell BL, Hubbard AK (1978) Cultural, morphological, cell membrane, enzymatic, and neoplastic properties of cell lines derived from Hodgkin's disease lymph node. *Cancer Res* 38:3033

Renner C, Jung W, Sahin U, Denfeld R, Pohl C, Trümper L, Hartmann F, Diehl V, van Lier R, Pfreundschuh M (1994) Cure of xenografted human tumours by bispecific monoclonal antibodies and human T cells. *Science* 264:833

Rowe M, Young LS, Crocker J, Stokes H, Henderson S, Rickinson AB (1991) Epstein-Barr virus (EBV)-associated lympho-proliferative disease in the SCID mouse model: implications for the pathogenesis of EBV-positive lymphomas in man. *J Exp Med* 173:147

Schaadt M, Fonatsch C, Kirchner HH, Diehl V (1979) Establishment of a malignant, Epstein-Barr-virus (EBV)-negative cell line from the pleural effusion of a patient with Hodgkin's disease. *Blut* 38:185

Schaadt M, Diehl V, Stein H, Fonatsch C, Kirchner HH (1980) Two neoplastic cell lines with unique features derived from Hodgkin's disease. *Int J Cancer* 26:723

Schouten HC, Sanger WG, Duggan M, Weisenburger DD, MacLennan KA, Armitage JO (1989) Chromosomal abnormalities in Hodgkin's disease. *Blood* 73:2149

Schuler W, Bosma MJ (1989) Nature of the SCID defect: a defective VDJ recombinase system. In *The SCID Mouse characterisation and potential uses*. Bosma MJ, Phillips RA, Schuler W eds (Heidelberg: Springer) p55

Schwab U, Stein H, Gerdes J, Lemke H, Kirchner H, Schaadt M, Diehl V (1982) Production of monoclonal antibody specific for Hodgkin and Sternberg-Reed cells on Hodgkin's lymphoma and a subset of normal lymphoid cells. *Nature* 299:65

Stein H, Gerdes J, Schwab U, Lembke H, Diehl V, Mason DY, Bartels H, Ziegler A (1983) Evidence for the detection of the normal counterpart of Hodgkin and Reed-Sternberg cells. *Hematol Oncol* 1:21

Stein H, Mason DY, Gerdes J, O'Connor N, Wainscoat J, Pallesen G, Gatter K, Falini B, Delsol G, Lemke H, Schwarting R, Lennert K (1985) The expression of the Hodgkin's disease associated antigen Ki-1 in reactive and neoplastic lymphoid tissue: evidence that Reed-Sternberg cells and histiocytic malignancies are derived from activated lymphoid cells. *Blood* 66:848

Thangavelu M, Snyder L, Anastasi J, Le Beau MM, Kirven M, Picchio G, Mosier DE, Rowley JD (1992) Cytogenetic characterization of B-cell lymphomas from severe combined immunodeficiency disease mice given injections of lymphocytes from Epstein-Barr virus-positive donors. *Cancer Res* 52:4678

Tilly H, Bastard C, Delastre T, Duval C, Bizet M, Lenormand B, Daucé JP, Monconduit M, Piguet H (1991) Cytogenetic studies in untreated Hodgkin's disease. *Blood* 77:1298

Von Kalle C, Wolf J, Becker A, Sckaer A, Munck M, Engert A, Kapp U, Fonatsch C, Komitowski D, Feaux de Lacroix W (1992) Growth of Hodgkin cell lines in severely combined immunodeficient mice. *Int J Cancer* 52:887

A SCID MOUSE MODEL OF HODGKIN'S DISEASE ? TRANSPLANTATION OF HODGKIN'S AND NON-HODGKIN'S LYMPHOMAS INTO SEVERE COMBINED IMMUNODEFICIENT MICE

Andrew S Krajewski[1], Jacqueline Lowrey[1], Sarah E M Howie[1], Alice Gallagher[2], Ruth F Jarrett[2]

[1]Department of Pathology, University of Edinburgh, UK
[2]LRF Virus Centre, University of Glasgow, UK

This work is supported by a grant from the Scottish Hospitals Endowment Research Trust (Grant no.1041)

SUMMARY

Severe combined immunodeficient (SCID) mice were transplanted with biopsy material from 18 cases of Hodgkin's disease (HD) and 23 biopsies from non-Hodgkin's lymphoma. Five of the HD (3 lymphocyte predominant, 1 nodular sclerosing and 1 mixed cellularity) and 7 of the non-Hodgkin's lymphomas (3 centroblastic, 1 immunoblastic, 1 follicular lymphoma and 2 biopsies from a case of large cell anaplastic lymphoma) produced tumours in mice. All the SCID HD-derived tumours had a B-cell phenotype and expressed the Epstein-Barr virus (EBV) gene products, EBER RNA, EBNA-2 and LMP-1. Morphologically these tumours resembled polymorphous immunoblastic lymphomas. The EBV-positive tumours all expressed CD23 and CD43. All HD-derived SCID tumours showed clonal immunoglobulin heavy chain rearrangements. One tumour derived from a follicular lymphoma showed similar morphology and phenotype and an identical pattern of EBV gene expression to the HD-derived tumours. Three centroblastic and one immunoblastic lymphoma derived tumours were EBV-negative. These showed a near identical B-cell phenotype to original biopsies and showed identical immunoglobulin gene rearrangements to the original tumours. Tumours derived from the large cell anaplastic lymphoma showed two types of lymphoproliferation in SCID mice; first immunoblastic tumours expressing EBER RNA, EBNA-2 and LMP-1, with immunoglobulin gene rearrangements differing from those of the original tumour; secondly large cell anaplastic tumours showing EBER RNA with little or no LMP-1 or EBNA-2 expression, and immunoglobulin gene rearrangements identical to those of the original tumour.

INTRODUCTION

Transplantation of lymphomas into immunodeficient animals has the potential for providing a useful model for the study of the biology of human lymphoid neoplasia. CB-17 SCID mice (Bosma *et al.*, 1983) have been used to study Epstein-Barr virus-induced human B-cell lymphoproliferations, since between 30-80% develop human B-cell lymphoproliferative disorder following transplantation of lymphocytes from Epstein-Barr virus (EBV) seropositive donors (Mosier, 1991; Nakamine *et al.*, 1991; Rowe *et al.,* 1991), and have also been used to support *in vivo* growth of human lymphoma cell lines (Chang *et al.*, 1992) and lymphomas from immunosuppressed post-transplantation patients (Waller *et al.*, 1993). Recently Kapp *et al.* (1993) have shown that SCID mice support the growth of transplanted HD-derived cell lines and have shown in three cases of Hodgkin's disease (HD) that transplantation of biopsy material into SCID mice gives rise to EBV-positive human B-cell lymphoproliferations. EBV-negative lymphomas derived from HD have not been described.

The aim of the study reported here was to determine whether SCID mice would support growth of transplanted human lymphomas apart from EBV driven lymphoproliferations. In particular, we wished to produce an animal model to study the immunology and molecular biology of HD. In common with Kapp *et al.* (1993), we believed that transplantation of HD tissue into immunodeficient SCID mice would allow selective proliferation of tumour cells (H-RS cells) free from a host response thus facilitating the study of their histogenesis.

In this study we show that transplantation of both HD and non-Hodgkin's lymphoma into SCID mice results in growth of high grade human lymphomas. In cases of HD these tumours have characteristics of clonal or oligoclonal EBV lymphoproliferations which are probably derived from normal EBV-infected bystander cells. In contrast, most cases of non-Hodgkin's lymphoma transplanted into SCID mice gave tumours which were EBV-negative B-cell lymphomas with immunophenotypic and genotypic identity to the original tumours. Tumours derived from a case of large cell anaplastic lymphoma were EBV-positive tumours some of which were derived from bystander cells and others which were genotypically identical to the original tumour.

MATERIALS AND METHODS

HD and non-Hodgkin's lymphoma lymph nodes and one spleen were obtained as routine biopsies from the Royal Infirmary and the Western General Hospital, Edinburgh. SCID mice were injected both intraperitoneally and subcutaneously with 250 µL of tumour cell suspension (containing 0.7-7 x 10^7 cells). Mice were sacrificed when there was detectable tumour or poor condition (4-20 weeks). Tumour blocks were taken for frozen section immunophenotyping, and into 10% buffered formalin for histological assessment, paraffin section immunophenotyping and *in-situ* hybridisation. A block of frozen tumour was used for DNA extraction in order to carry out molecular analyses. Mouse serum was stored and used for detection of secreted human immunoglobulin (Ig) using an ELISA method.

Paraffin and frozen section immunophenotyping was carried out using a panel of T-cell, B-cell, monocyte and EBV markers, LMP-1 and EBNA-2 (Dako, UK). Immunostaining was carried out using an ABC method with diaminobenzidine (DAB) as the substrate for peroxidase. Frozen section immunophenotyping was carried out on 3 μm acetone-fixed cryostat sections using an indirect peroxidase method with DAB as the substrate.

In situ hybridisation was used to demonstrate EBER RNAs (Dako, UK). Three micrometer paraffin sections were mounted on TESPA coated slides and dried overnight at 37°C. Sections were then hybridised overnight using a 1/20 dilution of fluorescein-conjugated oligonucleotide probe complementary to the two nuclear EBER RNAs. Hybridisation was visualised using an anti-FITC, alkaline phosphatase-conjugated antibody and BCIP/NBT (5-bromo-4-chloro-3-indoyl phosphate/nitro blue tetrazolium) as a substrate for alkaline phosphatase (Dako, UK).

DNA samples from frozen biopsies or tissue cell suspensions were examined for the presence of immunoglobulin heavy chain (IgH) gene rearrangements using Southern blotting and PCR (Gledhill *et al.*, 1990).

Southern blot analysis was also used to determine whether EBV genomes were present. The EBV BamHI-W probe which detects the major internal repeat of the viral genome was used to determine the EBV status. In some cases the clonality of EBV-positive samples was determined by examining the viral terminal repeat sequences using the EBV EcoRI-D fragment, or a subclone derived from the 5' end of this fragment, as probe (Armstrong *et al.*, 1992).

Ploidy analysis by DNA flow cytometry was carried out on both paraffin-embedded and frozen tissue (Hedley *et al.*, 1983; Vindelov *et al.*, 1982).

RESULTS

Cell suspensions from a total of 44 fresh biopsy specimens were injected into SCID mice. Human lymphomas developed, after a latent period of between 11-20 weeks, from 5/18 cases of HD and 7/23 cases of non-Hodgkin's lymphoma (Table 1). No tumours developed in animals injected with cells from reactive lymph nodes (3 cases).

Only one animal developed a subcutaneous tumour. All other cases showed a large peritoneal tumour mass invading liver and pancreas and bowel wall. Mesenteric nodes were commonly involved with less frequent involvement of para-aortic and mediastinal nodes. Pulmonary, thymic and cervical node involvement was occasionally seen. There was no splenic, or gastrointestinal mucosal involvement except in tumours derived from the large cell anaplastic lymphoma.

Microscopic and phenotypic findings in transplanted non-Hodgkin's lymphomas

Four cases of high grade, large cell, non-Hodgkin's lymphoma produced tumours in SCID mice (Table 2). These showed near identical morphology and immunophenotype to the original tumours. The lymphomas in the mice also showed identical patterns of Ig expression and IgH gene rearrangement to the original tumour biopsies. In these cases no Ig secretion was detected in mouse serum, excepting the one case of immunoblastic lymphoma which showed small amounts of secreted human IgG. All of these tumours were tetraploid or aneuploid. None of these tumours contained EBV.

Table 1:   Tumours developing in SCID mice transplanted with Hodgkin's and non-Hodgkin's lymphomas

Number of cases	Histological subtype	Number of tumours developing in SCID mice
18 HD	3 LP, 1 NS, 1 MC	6 EBV$^+$ lymphoproliferations
23 NHL	3 CB, 1 IB	4 EBV$^-$ high grade B-NHLs
	1 follicular CBCc	1 EBV$^+$ lymphoproliferation
	1 EBV$^+$ large cell anaplastic lymphoma	
	Biopsy 1	4 EBV$^+$ lymphoproliferations
	Biopsy 2	1 EBV$^+$ lymphoproliferation
		4 EBV$^+$ large cell anaplastic lymphomas
3 reactive nodes		0 cases of lymphoma

HD, Hodgkin's disease; NHL, non-Hodgkin's lymphoma; LP, lymphocyte predominant HD; NS, nodular sclerosis HD; MC, mixed cellularity HD; CB centroblastic; IB, immunoblastic; CBCc, centroblastic centrocytic.

One case of follicular lymphoma also produced a high grade lymphoma in a SCID mouse. Histologically this was a B-cell immunoblastic lymphoma which showed a different pattern of Ig expression and IgH gene rearrangement to that of the original biopsy and showed secretion of human Ig into mouse serum. This tumour was diploid and showed expression of EBV gene products (Table 2).

Two biopsies from a case of large cell anaplastic lymphoma were transplanted into SCID mice. All mice injected with tumour cells developed lymphomas. All four tumours derived from the first biopsy and one of five tumours derived from the second biopsy were CD20-positive B-cell immunoblastic lymphomas expressing LMP-1, EBNA-2 and EBER. These tumours were diploid and all showed different patterns of Ig expression with different IgH gene rearrangements to the original biopsy. In contrast four of the tumours derived from the second biopsy were CD20-negative large cell lymphomas containing multinucleated cells, similar to those in the original biopsy. In two of these tumours the only EBV product detectable was EBER (as in the original tumour) with one tumour containing a small number of LMP-1 positive cells and the other tumour showing foci of LMP-1 and EBNA-2 positive cells. These tumours were tetraploid and showed identical IgA expression to the original tumour and, in the three cases examined, identical IgH gene rearrangements and EBV terminal repeat fragments.

Table 2:   Details of non-Hodgkin's lymphoma cases producing tumours in SCID mice.

Diagnosis	Original Biopsy			Transplanted tumour				
	Ig expression	IgH-rearrangements	EBV status	Latent period	Ig expression	Ig secretion	IgH-PCR	EBV status
Centroblastic	Mκ	R*	neg	20 weeks	Mκ	neg	R	neg
Centroblastic	Mκ	R	neg	14 weeks	Mκ	neg	R	neg
Centroblastic	Gλ	R	neg	21 weeks	Gλ	neg	R	neg
Immuno-blastic	Gκ	neg	neg	16 weeks	Gκ	G	neg (R-SB)	neg
Follicular lymphoma	Gκ	neg	neg	20 weeks	Gλ(MAκ)	Gλ(MAκ)	R1	EBER/LMP/EBNA2
Large cell Anaplastic								
Biopsy 1	A	neg	EBER only	9 weeks	Gλ(M) Mλ Gκ Mκλ(GA)	Gλ(M) Mλ Gκ not done	R1 R2 R3 R4	EBER/LMP/EBNA2 EBER/LMP/EBNA2 EBER/LMP/EBNA2 EBER/LMP/EBNA2
Biopsy 2	A	neg (R-SB)	EBER only (clonal)	11+ weeks	A A A MA	neg neg neg neg M	neg(R-SB) neg neg(R-SB) neg(R-SB) R5	EBER only EBER only EBER/LMP EBER/LMP/EBNA2 EBER/LMP/EBNA2

*R = rearrangement in original tumour detected by PCR; R1-5 = differing rearrangements. R-SB = rearranged on Southern Blot analysis. Immunoglobulins in parentheses indicate a minor population.

Table 3:   Details of Hodgkin's disease cases producing tumours in SCID mice.

Diagnosis	Original Biopsy			Transplanted tumour				
	RS cell Phenotype	IgH-rearrangements	EBV status	Latent period	Ig expression	Ig secretion	IgH-PCR	EBV status
Lymphocyte predominant	CD20- CD15+	neg	neg	14 weeks	Aλ(κ)	Aλ	R1	EBER/LMP/EBNA2
Lymphocyte predominant	CD20+	neg	neg	11 weeks	Gκ(MA)	GκM(A)	R2	
Lymphocyte predominant	CD20+	neg	neg	12 weeks 12 weeks	MGλ MGκλ	MGλ MGκλ	R3 R4	EBER/LMP/EBNA2 EBER/LMP/EBNA2
Nodular sclerosis	CD15+ CD30+	neg	neg	15 weeks	Gλ(Mκ)	GMλ(κ)	R5	EBER/LMP/EBNA2
Mixed cellularity	CD15+ CD30+	neg	EBER/LMP (clonal)	13 weeks	Mκ(Gλ)	Mκ(λ)	R6	EBER/LMP/EBNA2 (oligoclonal)

*R = rearrangement in original tumour detected by PCR; R1-6 = differing rearrangements. Immunoglobulins in parentheses indicate a minor population.

Microscopic and phenotypic findings in transplanted Hodgkin's disease

Tumours developed in SCID mice following injection of tumour cells from 5 cases of HD (Table 3). All of the tumours were high grade, large cell, B-cell immunoblastic lymphomas that showed variable pleomorphism and usually foci of plasmacytoid differentiation. In two of the cases (derived from two cases of lymphocyte predominant HD) occasional multinucleated cells resembling Reed-Sternberg cells were found. As well as expressing B-cell antigens (CD19, CD20 and CD22) tumour cells in all cases expressed CD23 and CD43 with weak CD30 expression. All cases showed Ig expression, usually with marked predominance of a clone expressing a single heavy and light chain which was also present as secreted Ig in mouse serum. All of the SCID tumours showed clonal IgH gene rearrangements while none of the original tumours contained detectable rearrangements. All of the mouse tumours showed expression of LMP-1, EBNA-2 and EBER. In the original tumours only one case of mixed cellularity HD showed LMP-1 and EBER expression by H-RS cells. In this case the SCID mouse tumour contained oligoclonal EBV genomes which differed from the clonal genome that was present in the original biopsy. All of these tumours were diploid.

DISCUSSION

This study shows that following transplantation of HD and non-Hodgkin's lymphoma into SCID mice three types of lymphoproliferative disorder can be defined:

1. EBV-negative lymphomas derived from high grade non-Hodgkin's lymphomas. These tumours are aneuploid and show phenotypic and genotypic identity to the original biopsy specimens.

2. EBV-positive lymphomas derived from a case of large cell anaplastic lymphoma. These tumours also show abnormal ploidy and show phenotypic and genotypic identity to the original tumour. In these lymphomas an unusual pattern of EBV gene expression with strong EBER expression but minimal or no LMP-1 and EBNA-2 was seen. This pattern of EBV gene expression is similar to type I or type II latency as found in HD and in other cases of large cell anaplastic lymphoma (Rowe et al., 1992; Hamilton-Dutoit et al., 1993).

3. EBV-induced lymphoproliferations derived from HD biopsies, one case of follicular lymphoma and from the large cell anaplastic lymphoma. These tumours are diploid, show clonal or oligoclonal IgH gene rearrangements and show expression of EBER, LMP-1 and EBNA-2. This pattern of EBV gene expression corresponds to type III latency and is similar to that seen in lymphoblastoid cell lines and cases of lymphoma that develop in SCID mice following injection of normal EBV-infected human lymphocytes (Rowe et al., 1992). Thus these tumours are almost certainly derived from normal B-cells in tumour biopsies and not from tumour cells.

The only previously reported study of transplantation of HD into SCID mice by Kapp et al. (1993) described 3 cases out of 13 in which B-cell lymphoma developed in SCID mice following transplantation of tumour cell fragments. All of these SCID tumours were EBV-positive. Kapp et al. (1993) defined three histological categories: lymphoproliferative disease containing medium and large immunoblasts with some plasmacytoid cells, anaplastic large cell lymphomas and HD-like lesions containing H-RS-like cells surrounded by murine macrophages. The lymphoproliferative disease described by this group corresponds closely to the polymorphous immunoblastic proliferations that we have seen, but we found no tumours with features of anaplastic large cell lymphomas and HD-like lesions. However, areas of increased cellular pleomorphism could be found in most cases and in two cases we found occasional binucleated or multinucleated Reed-Sternberg-like cells within the tumours.

Extreme caution is required before accepting that the HD-induced EBV-positive SCID lymphomas described in this paper, and by Kapp et al., are derived from the neoplastic H-RS. It is well established that normal blood or tonsillar B-cells from individuals infected with EBV will, in most cases, produce B-cell lymphoproliferation following inoculation into SCID mice (Canon et al., 1990; Mosier, 1991; Rowe et al., 1991; Nakamine et al., 1991; Nakamine et al., 1993). These cases appear to have a similar mature B-cell phenotype, express CD23 and show similar EBV gene product expression (EBER, EBNA-2 and LMP-1) to those seen in our cases. It therefore remains a strong probability that the EBV-positive lymphoproliferation seen following transplantation of HD or non-Hodgkin's lymphoma may simply be secondary to outgrowth and clonal proliferation of normal B-cells containing EBV, rather than due to proliferation of neoplastic cells. Evidence supporting this was seen in the normal diploid DNA content of these tumours. Kapp et al. (1993) reported that their HD-derived tumours showed a high incidence of chromosomal abnormality, however of the karyotypes they reported some were diploid and others showed a near diploid chromosome number and it is possible that we may have failed to detect minor chromosomal abnormalities by flow cytometry. In one case that we have examined cytogenetically a normal diploid 46XY karyotype was present.

Further evidence that the EBV-positive tumours are derived from bystander rather than malignant cells was provided by the one case of EBV-positive mixed cellularity HD, which had a clonal EBV genome, detectable by Southern blotting, in the original biopsy but multiple EBV genomes of differing size in the SCID lymphoma. None of the cases of HD leading to growth of EBV-positive SCID lymphomas showed detectable IgH gene rearrangements in the original tumours and comparisons of Ig clonality between the original biopsy and SCID lymphomas were therefore not available to further resolve this issue.

In conclusion, we believe that HD transplanted into SCID mice has not yet been proven to be a satisfactory model for the study of HD as there is at present no evidence that HD-derived SCID tumours have any relationship to neoplastic cells in the original tumour. Further studies are required, using cases of HD with clonotypic markers, such as clonal EBV, Ig gene rearrangements or cytogenetic markers, in original tumours that can be compared to those in SCID tumours. It is further possible that the growth of HD requires a functioning immune system with cell-cell or cytokine interactions that are absent in the SCID mouse.

REFERENCES

Armstrong AA, Weiss LM, Gallagher A, Jones DB, Krajewski AS, Brown G, Jack AS, Wilkins BS, Onions DE, Jarrett RF (1992) Criteria for the definition of Epstein-Barr virus association in Hodgkin's disease. *Leukémia* **6**:869

Bosma GC, Custer RP, Bosma MJ (1983) A severe combined immunodeficiency mutation in the mouse. *Nature* **301**:527

Cannon MJ, Pisa P, Fox RI, Cooper NR (1990) Epstein-Barr virus induces aggressive lymphoproliferative disorders of human B cell origin in SCID/hu chimeric mice. *J Clin Invest* **85**:1333

Chang H, Leeder S, Cook VA, Patterson B, Dosch M, Minden MD, Messner HA (1992) Growth of human lymphoma cells in SCID mice. *Leuk Lymphoma* **8**:129

Gledhill S, Krajewski AS, Dewar AE, Onions DE, Jarrett RF (1990) Analysis of T-cell receptor and immunoglobulin gene rearrangements in the diagnosis of Hodgkin's and non-Hodgkin's lymphoma. *J Pathol* **161**:245

Hamilton-Dutoit SJ, Rea D, Raphael M, Sandvej K, Delecluse HJ, Gisselbrecht C, Marelle L, van Krieken JHJM, Pallesen G (1993) Epstein-Barr virus latent gene expression and tumour cell phenotype in acquired immunodeficiency syndrome related non-Hodgkin's lymphoma. Correlation of phenotype with three distinct patterns of viral latency. *Am J Pathol* **143**:1072

Hedley DW, Friedlander ML, Taylor IW, Rugg CA, Musgrove EA (1983) Method for analysis of cellular DNA content of paraffin-embedded pathological material using flow cytometry. *J Histochem Cytochem* **31**:1333

Kapp U, Wolf J, Hummel M, Pawlita MV, Kalle C, Dallenbach F, Schwonzen M, Krueger GRF, Muller-Lantzsch N, Fonatsch C, Stein H, Diehl V (1993) Hodgkin's lymphoma-derived tissue serially transplanted into severe combined immunodeficient mice. *Blood* **4**:1247

Mosier DE (1991) Adoptive transfer of human lymphoid cells to severely immunodeficient mice: models for normal human immune function, autoimmunity, lymphomagenesis, and AIDS. *Adv Immunol* **50**:303

Nakamine H, Masih AS, Okano M, Taguchi Y, Piruccello SJ, Davis JR (1993) Characterisation of clonality of Epstein-Barr virus-induced human B-cell lymphoproliferative disease in mice with severe combined immunodeficiency. *Am J Pathol* **142**:139

Nakamine H, Okano M, Taguchi Y, Pirruccello SJ, Davis JR, Mahloch ML, Beisel KW, Kleveland K, Sanger WG, Purtilo DT (1991) Hematopathologic features of Epstein-Barr virus-induced human B-lymphoproliferation in mice with severe combined immuno deficiency. A model of lymphoproliferative diseases in immunocompromised patients. *Lab Invest* **65**:389

Rowe M, Young LS, Crocker J, Stokes H, Henderson S, Rickinson AB (1991) Epstein Barr Virus (EBV)-associated lymphoproliferative disease in the SCID mouse model: implications for the pathogenesis of EBV-positive lymphomas in man. *J Exp Med* **173**:147

Rowe M, Lear AL, Croom-Carter D, Davies AH, Rickinson AB (1992) Three pathways of Epstein-Barr virus gene activation from EBNA1-positive latency in B lymphocytes. *J Virol* **66**:122

Vindelov LL, Christensen IBJ, Nissen NI (1982) A detergent-trypsin method for the preparation of nuclei for flow cytometric DNA analysis. *Cytometry* **3**:323

Waller EK, Ziemianska M, Bangs CD, Cleary M, Weissman I, Kamel OW (1993) Characterisation of post-transplant lymphomas that express T-cell associated markers: immunophenotypes, molecular genetics, cytogenetics, and heterotransplantation in severe combined immunodeficient mice. *Blood* **82**:247

REFERENCES

THE PATHOGENESIS OF HODGKIN'S DISEASE: ONCOGENE, TUMOUR SUPPRESSOR GENE, AND EPSTEIN-BARR VIRAL STUDIES

Lawrence M Weiss

Division of Pathology, City of Hope National Medical Center, Duarte, California, USA

Supported in part by NCI Grant CA 50341

ROLE OF ONCOGENES AND TUMOUR SUPPRESSOR GENES IN THE PATHOGENESIS OF HODGKIN'S DISEASE

The role of oncogenes and tumour suppressor genes in Hodgkin's disease (HD) has been difficult to study for several reasons. First, the percentage of neoplastic cells in involved tissues is generally quite low, usually less than 0.5%. Therefore, any methodology that homogenises the tissues (such as the conventional DNA extraction procedures) must have a high degree of sensitivity to analyse these cells. In most cases, one can only gain information concerning the status of the host cells. Secondly, very few *bona fide* HD cell lines are available for study. Even these are of questionable significance, since cell lines are derived from unusual cases and may not reflect the ordinary case of HD and because the genetic composition of the cell lines may change *in vitro*, especially over time.

BCL-2 PROTO-ONCOGENE

Interest in the status of the *bcl-2* proto-oncogene was kindled by data from several groups suggesting that the t(14;18)(q32,q21) translocation may be detected in some cases of HD (Weiss *et al.*, 1987; Shibata *et al.*, 1990; Stetler-Stevenson *et al.*, 1990; Said *et al.*, 1991; Louie *et al.*, 1991; Masih *et al.*, 1991; Whitney *et al.*, 1992; Weiss and Chang, 1992; Athan *et al.*, 1992; Segal *et al.*, 1992; Poppema *et al.*, 1992; Gupta *et al.*, 1992; Reid *et al.*, 1993;

LeBrun *et al.*, 1994). In this translocation, the *bcl-2* proto-oncogene on chromosome 18 becomes juxtaposed to the joining region of the immunoglobulin heavy chain (IgH) gene, leading to deregulation of Bcl-2 protein (Weiss *et al.*, 1987). Bcl-2 protein has a relatively unique method of action among proto-oncogenes (Weiss and Chang, 1993). Rather than increasing the rate of cell proliferation, it acts by decreasing the rate of cell death by preventing programmed cell death (apoptosis). Normally, Bcl-2 plays a physiological role in preventing long-lived post-mitotic and stem cells from undergoing apoptosis. In development, it may be important in inducible cell survival as a mechanism of morphogenesis. In neoplasia, Bcl-2 may act by providing an abnormal pool of long-lived cells at risk for secondary genetic events.

A summary of several published studies reporting on the incidence of the t(14;18) in HD is presented in Table 1. Initially, we studied 10 cases of HD by Southern blotting and were unable to find translocations in the major or minor breakpoint region, or at the 5' region of the *bcl-2* gene (Weiss *et al.*, 1987). Although cases with large numbers of Reed-Sternberg (RS) cells were studied, there may not have been sufficient sensitivity to detect genetic changes present in the RS cells since the sensitivity of Southern blotting is in the order of 1-2%.

Table 1:   Summary of studies of t(14;18) in classical Hodgkin's disease by Southern blot hybridisation and polymerase chain reaction.

Author	Results	Technique
Weiss *et al.*, 1987	0/8 (0%)	mbr and mcr SB
Shibata *et al.*, 1990	0/6 (0%)	mbr PCR
Stetler-Stevenson *et al.*, 1990	17/53 (32%)	mbr PCR
Said *et al.*, 1991	0/19 (0%)	mbr SB and PCR
Louie *et al.*, 1991	0/25 (0%)	mbr and mcr PCR
Masih *et al.*, 1991 (abstr.)	12/44 (27%)	mbr PCR
Whitney *et al.*, 1992 (abstr.)	0/44 (0%)	mbr PCR
Athan *et al.*, 1992	0/34 (0%)	mbr and mcr SB and PCR
Segal *et al.*, 1992	0/3 (0%)	mbr PCR
Poppema *et al.*, 1992	11/28 (39%)	mbr PCR
Gupta *et al.*, 1992	4/21 (19%)	mbr PCR
Reid *et al.*, 1993	6/47 (13%)	mbr PCR
LeBrun *et al.*, 1994	2/32 (6%)	mbr and mcr PCR

Abbreviations: mbr, major breakpoint region; mcr, minor cluster region; SB, Southern blot hybridisation; PCR, polymerase chain reaction.

The polymerase chain reaction (PCR) potentially represents a much more sensitive method for the detection of any specific translocation, because it theoretically can detect a product from a single cell. Table 1 illustrates the wide variation in the results of PCR studies that have examined the status of the *bcl-2* breakpoint in HD, with detection of the translocation in 0% to 39% of cases (Shibata *et al.*, 1990; Stetler-Stevenson *et al.*, 1990; Said *et al.*, 1991; Louie *et al.*, 1991; Masih *et al.*, 1991; Whitney *et al.*, 1992; Athan *et al.*, 1992; Segal *et al.*, 1992; Poppema *et al.*, 1992; Gupta *et al.*, 1992; Reid *et al.*, 1993; LeBrun *et al.*,

1994). In the study of Stetler-Stevenson and colleagues, nearly one-third of cases of HD were shown to contain bcl-2/J_H sequences using a PCR analysis designed to detect breakpoints in the major breakpoint region (mbr), a region of the bcl-2 gene in which 50-70% of the t(14;18) translocations are known to occur (Settler-Stevenson et $al.$, 1990). In contrast, over 70 control specimens, including T cell lymphoma, benign lymphadenopathies and normal peripheral blood lymphocytes, showed no evidence of the translocation product. These investigators minimised the possibility of false positive results due to contamination by performing studies on multiple samples of the same case and by ensuring that the amplified translocation products varied in molecular weight from case to case.

In a subsequent study, the Stetler-Stevenson group did not find a correlation between the identification of t(14;18) by PCR and expression of Bcl-2 protein, using a monoclonal antibody directed against an epitope of Bcl-2 (Bhagat et $al.$, 1993). Thirteen cases of HD were studied; seven cases had a detectable t(14;18) and six cases did not. Bcl-2 protein was detected immunohistochemically in the RS cells in eight cases, including three with the translocation and five in which the translocation was not detected. These data suggested that the translocation, if present, either does not reside in the RS cells or does not lead to overexpression of Bcl-2 as it does in follicular lymphoma.

Louie and colleagues used a PCR methodology similar in sensitivity to the Stetler-Stevenson study to look for both bcl-2 major and minor cluster region breakpoints (in which, together, over 85% of translocations occur). They were unable to detect translocation products in 26 cases of HD (Louie et $al.$, 1991). Moreover, they employed immunohistochemical studies using a highly specific anti-Bcl-2 antiserum and were unable to detect overexpression of the bcl-2 gene in the RS cells of any case. In the study of Poppema et $al.$, 11 of 28 cases (39%) of HD were found to have detectable t(14;18) by PCR. However, they were able to detect a t(14;18) by cytogenetics in only one case (Poppema et $al.$, 1992).

LeBrun et $al.$ studied 32 cases of HD by PCR using primers to detect breakpoints in the major and minor cluster regions (LeBrun et $al.$, 1994). PCR evidence of breakpoints occurring within the mbr was found in two cases, whereas none of the cases showed breakpoints within the minor cluster region. Interestingly, both of the cases with detectable mbr breakpoint had a history of prior follicular lymphoma, as did one other case without evidence of a translocation. Tissue from the initial follicular lymphoma was available from one of the two patients with a detectable breakpoint; PCR revealed an identical translocation product to the one found in the HD tissue, implying the presence of the same clone of cells. Dilution studies determined that fewer than 1% of the cells in the corresponding HD tissue carried the translocation, consistent with the estimated abundance of RS cells. Alternatively, there could have been an occult population of follicular lymphoma cells, below the threshold of detection by morphological and immunophenotypic assessment.

LeBrun also performed immunohistochemical studies on the 32 cases using a monoclonal antibody to Bcl-2 that is probably more sensitive than the anti-Bcl-2 antiserum used by

Louie *et al.* (Louie *et al.*, 1991; LeBrun *et al.*, 1994). Cytoplasmic Bcl-2 staining was found in 20/32 cases (63%), with strong staining found in seven cases (22%). Interestingly, all three cases with a history of follicular lymphoma, including both cases with a detectable translocation product, showed strong cytoplasmic staining for Bcl-2. A retrospective immunohistochemical analysis of additional cases in which HD was associated with prior or concurrent follicular lymphoma showed strong staining of Hodgkin's cells for Bcl-2 in all seven cases for which the staining was interpretable. Although this evidence is indirect, the data suggest that a small subset of HD may derive from transformation of follicular lymphoma. Prior series studying the relationship of follicular lymphoma and HD have shown an association of less than 1%, so this mechanism can only account for a small minority of cases of HD. Similar transformation of chronic lymphocytic leukaemia/small lymphocytic lymphoma and mycosis fungoides to HD have been described, and may also contribute to the pathogenesis of a small percentage of cases of HD.

One might conclude from the above studies that *bcl-2* rearrangements may be occasionally detected in HD. However, even if *bcl-2* rearrangements are truly present in tissues involved by HD, it does not necessarily follow that they are derived from the RS cells. Although Stetler-Stevenson and colleagues were unable to detect the translocation product in a variety of negative control tissues, others have reported its detection by PCR in a majority of benign lymph nodes and tonsils, raising the possibility that rare non-neoplastic B, or even T, lymphocytes may carry the translocation (Limpens *et al.*, 1991; Aster *et al.*, 1992). In rare cases associated with prior or concurrent follicular lymphoma, the t(14;18) may reside within the Hodgkin's cells and be associated with increased Bcl-2 protein expression.

Expression of Epstein-Barr virus (EBV) latent membrane protein-1 (LMP-1) in B-cells has been shown to protect lymphoid cells from apoptotic cell death; this action is mediated through the induction of the *bcl-2* proto-oncogene (Henderson *et al.*, 1991). Thus, another possible mechanism for the increased expression of Bcl-2 protein in some cases of HD could be via EBV infection of Hodgkin's cells, as LMP-1 protein is consistently expressed in the Hodgkin's cells of EBV-associated HD. The Stetler-Stevenson group study showed no correlation between Bcl-2 protein positivity and the detection of EBV RNA using EBER *in situ* hybridisation studies (Bhagat *et al.*, 1993). Similarly, in LeBrun's study, there was no good correlation between Bcl-2 protein positivity and the detection of EBV using either LMP-1 immunohistochemistry or EBER *in situ* hybridisation studies (LeBrun *et al.*, 1994). However, Khan and colleagues found a 30% incidence of both Bcl-2 protein positivity and EBV positivity, but only 4% of cases in which both were expressed (Khan *et al.*, 1993). The authors speculated that EBV infection and Bcl-2 expression may each have a role in the pathogenesis of HD, and that they represent different and exclusive events in a multistep pathway of oncogenesis. Work by Armstrong *et al.* and our recent data do not support these conclusions (Armstrong *et al.*, 1992). We found a 32% incidence of Bcl-2 expression and a 45% incidence of EBV infection; there was a 10% incidence of both Bcl-2 expression and EBV positivity, similar to what one might expect solely by chance (Weiss, unpublished observations). The different findings of Khan *et al.* (1993) may be explained by the fact that most of their EBV-positive cases were found in the category of mixed cellularity while most of the Bcl-2 positive cases were found in the category of nodular sclerosis. Regardless of whether or not an inverse correlation between Bcl-2 and EBV exists, it is clear that EBV LMP-1 expression in HD does not lead to induction of the Bcl-2 protein in Hodgkin's cells.

p53 TUMOUR SUPPRESSOR GENE

The normal p53 gene codes for a protein whose exact function is still unknown, but is associated with control of the cell cycle and the transcription of multiple genes. The p53 gene does not appear to be essential for cell viability; it induces cell cycle arrest in response to DNA damage. If DNA repair cannot be accomplished, p53 has an important role in inducing apoptosis. If cells lack functional p53, then progressive DNA mutations and rearrangements may occur, ultimately leading to the production of malignancies.

Mutant p53 protein typically has a much longer life than the wild type protein, leading to increased amounts of inactive p53 protein. The inactive p53 protein may then form complexes with the wild type protein, to render the wild type protein also functionally ineffective. Therefore, a mutation of one p53 allele may exert effects in a dominant fashion. Most mutations of the p53 gene are missense mutations, occurring in exons 5 through 9, that cause a change in the amino acid selected for the affected codon. Since it is impractical to sequence the p53 genes of every tumour, the most common way to assess p53 mutations is by performing PCR on exons 5 to 9 of the gene, followed by single strand conformation polymorphism (SSCP), and DNA sequencing of any abnormal fragments. Obviously, this technique may miss mutations occurring outside of exons 5 to 9. Another drawback to this technique is that it is time-consuming and impractical for all but a research setting. Since mutation of the p53 gene generally leads to accumulations of abnormal p53 proteins, immunohistochemistry to detect overexpression of p53 protein has been used as an easy screen for p53 mutations. Several monoclonal antibodies are available that react with either mutant or wild-type p53 protein, or both, in paraffin sections. The presence of strong staining of the nucleus is taken as indirect evidence of p53 mutation. However, some p53 mutations may lead to absence of protein product or nonsense protein that lacks immunoreactivity with p53 antibodies; these changes cannot be detected by immunohistochemistry. In addition, some non-neoplastic cells, particularly cells undergoing repair from DNA damage, may potentially have detectable levels of p53 protein by this assay.

Despite the shortcomings described above, several groups including ours have studied p53 protein expression in HD using immunohistochemistry as a screen for p53 mutation (Gupta et al., 1992; Niedobitek et al., 1993; Doglioni et al., 1991). All groups have found a high incidence of p53 protein expression within HD, ranging from 32% to 100% of cases. For example, Gupta and colleagues found a 72% incidence, with similar rates of p53 protein expression found for nodular sclerosis and mixed cellularity (Gupta et al., 1992). In our study, we found 100% incidence of p53 protein staining in RS cells, using the monoclonal antibody D0-7 in a staining protocol that included microwave-based antigen retrieval (Weiss, unpublished observations). There was often case-to-case variability, with some cases showing uniform and intense nuclear staining of Hodgkin's cells (Fig. 1), and other cases showing less intense staining of lesser numbers of Hodgkin's cells. Occasionally, there was variability of staining within single multinucleated RS cells, with several nuclei showing positive staining and one or more remaining unlabelled.

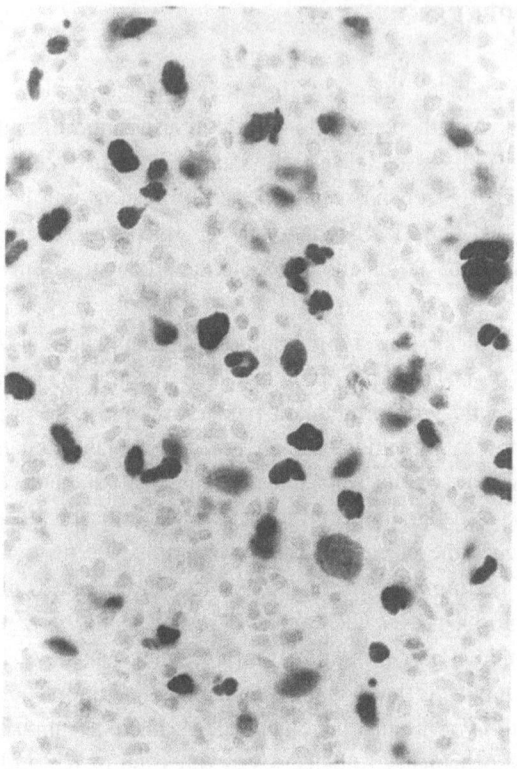

Fig. 1:   p53 protein expression in a case of Hodgkin's disease. Note strong nuclear staining of Hodgkin's cells.

As mentioned above, immunohistochemical studies of p53 protein expression are only indirect measures of the presence of p53 mutations and may be potentially misleading without confirmation by molecular methods. Gupta and colleagues employed a strategy involving extraction of nuclei followed by enrichment by flow cytometry to determine whether p53 overexpression in the Hodgkin's cells was due to mutations in this gene (Gupta et al., 1993). SSCP analysis revealed additional bands, and DNA sequence analysis confirmed the presence of point mutations in two of two cases studied. Similarly one of six HD-derived cell lines was also found to contain p53 point mutations.

Trümper and colleagues utilised the technique of single-cell-based PCR amplification to examine the status of the p53 gene in individual Hodgkin's cells and small lymphocytes (Trümper et al., 1993; see Trümper et al., this volume). In this technique single Hodgkin's cells were dropped into a lysis buffer, subjected to reverse transcription, and amplified using primers spanning p53 exons. In the one case reported so far, p53 mRNA was found in the majority of Hodgkin's cells but not in the small lymphocytes. A mutation in exon 7 was found in all clones sequenced from 5 of 7 cells. The remaining two cells had no detectable p53 mutations. These results, while preliminary, suggest that p53 mutation may occur in a subclone of neoplastic cells in HD.

There is a potential interaction between p53 and the EBV protein BZLF1. BZLF1 is an EBV immediate-early protein which is responsible for initiating the switch from latent to lytic infection. Both *in vivo* and *in vitro*, BZLF1 can interact with p53 (Zhang *et al.*, 1994). Wild-type p53 inhibits the ability of BZLF1 to mediate the switch from latent to productive viral infection. Thus, one might hypothesise a relationship between p53 protein expression and EBV infection in Hodgkin's cells. Niedobitek and colleagues examined the relationship between p53 protein expression by immunohistochemistry with EBV as assessed by EBER *in situ* hybridisation (Niedobitek *et al.*, 1993). They found p53 overexpression in 32% of cases and EBV positivity in 32% of cases. However, there was no correlation between p53 positivity and EBV infection of Hodgkin's cells.

RB

Study of retinoblastoma led to discovery of the RB gene, the first documented tumour suppressor gene. Similar to p53, the RB gene has a role in the regulation of the cell cycle (Weiss, 1994). The RB protein may be either phosphorylated or underphosphorylated, depending upon the stage of the cell cycle. RB protein in its underphosphorylated form may inhibit progression of the cell cycle from G1 to S phase, via binding to other transcription factors. It is thought that mutations occur at each allele of the RB gene. In congenital retinoblastoma, one of the defects is inherited and the other is acquired. In cancers occurring outside of the congenital syndrome, both defects are acquired.

Mutations of the RB gene usually manifest as deletions or point mutations that lead to abnormal or absent RB mRNA levels, resulting in protein absence. Although the RB protein is constitutively present in the nuclei of normal cells, it is lacking in many cells with mutations in the RB alleles. Therefore, detection of RB mutations may be indirectly detected by assessment of RB mRNA levels (e.g., by Northern blotting), or by immunohistochemical studies of RB protein. Polyclonal antibodies that react with RB protein in paraffin sections are available.

We have recently studied 22 cases of HD for expression of RB protein (Weiss, unpublished observations). The reliability of the staining procedure was confirmed by noting at least some expression of RB protein in the nuclei of non-neoplastic cells. In all 22 cases, there was detectable RB protein in the nuclei of Hodgkin's cells, inconsistent with the hypothesis that RB protein absence has a pathogenetic role in HD. In many cases, the level of protein expression was much higher than that of normal cells, raising the possibility that increased RB protein may be of significance in HD (Fig. 2). An alternative explanation is that increased RB protein expression in Hodgkin's cells may merely be a reflection of the increased cell cycling of these cells.

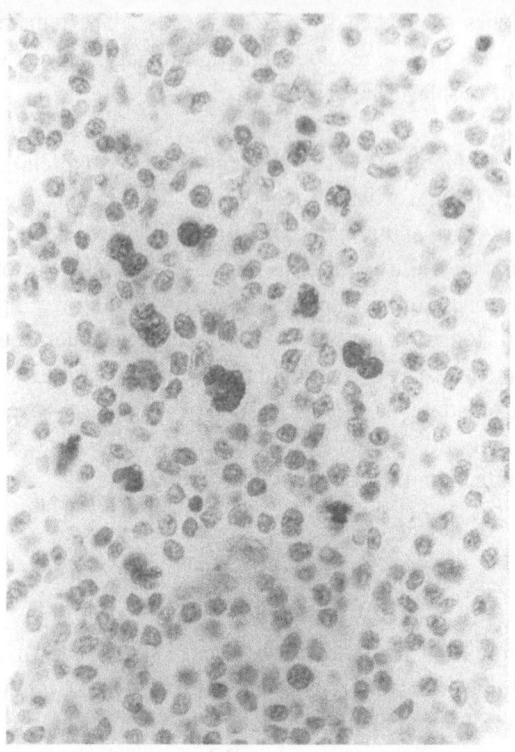

Fig. 2:   RB protein expression in a case of Hodgkin's disease. Nuclear staining of Hodgkin's cells is seen.

ROLE OF EPSTEIN-BARR VIRUS IN THE
PATHOGENESIS OF HODGKIN'S DISEASE

For many years, a relationship between EBV and HD had been hypothesised. In 1987, we reported the presence of EBV genomes in approximately 20% of cases of HD, using slot-blot and Southern blot hybridisation, in both initial diagnostic and recurrent specimens (Weiss *et al.*, 1987). All of the DNAs in the study were also screened for cytomegalovirus genomes and failed to show any hybridisation above background levels. In half of the patients with EBV-positive tumours, uninvolved tissues were available for study and showed no evidence of EBV genomes. Furthermore, using a probe directed against the EBV terminus, we showed that the EBV genome was present in a monoclonal population of cells. However, the specific cellular population harbouring the EBV could not be determined by these methods.

In order to answer the question of which cells were infected by EBV, *in situ* hybridisation studies were needed to determine the precise localisation of the EBV genomes. Using relatively insensitive *in situ* hybridisation methods, we showed that the EBV DNA was present within RS cells and variants, and not within the reactive lymphocytes or other cellular elements (Weiss *et al.*, 1989). Approximately 20 to 30% of the cases were positive, a percentage similar to that obtained in Southern blot studies; indeed, there was excellent correlation between the two methods.

Shortly after the early *in situ* hybridisation studies were published, studies utilising the PCR demonstrated a much higher incidence of EBV genomes in HD, ranging from 50 to 80% of cases. In our study, we found variable amounts of EBV by PCR in 52% of cases of HD (Shibata *et al.*, 1991). An explanation for the difference in the percentage of EBV-positive cases between the two techniques was not immediately evident. Did the PCR-amplifiable EBV DNA represent EBV exclusively from within RS cells and variants or was EBV from other cells also being detected? The latter possibility seemed probable, as these same PCR studies could also identify evidence of EBV in from 0 to 43% of "normal" lymph nodes. The detected EBV DNA in these latter cases was most likely present within latently infected B lymphocytes from EBV-seropositive individuals.

We and others have been able to address the issue of which cells of HD harbour the EBV by utilising a highly sensitive *in situ* hybridisation method that uses an oligonucleotide probe directed against EBER-1 RNA, a transcript copied up to 10^6-10^7 times in latently infected cells (Chang *et al.*, 1992). With this methodology, EBV present in rare lymphocytes of EBV-seropositive individuals can be detected. Applying this technology to HD, we identified EBV RNA in RS cells and variants in approximately 50% of cases of classical HD (Weiss *et al.*, 1991). Most significantly, in almost all of the positive cases, all or virtually all of the RS cells were positive. We have also been able to localise EBV RNA to occasional non-neoplastic lymphocytes in nearly 90% of cases (including cases in which the Hodgkin's cells are EBV-negative), indicating that these lymphocytes are the probable source of the high percentages of EBV-positivity obtained in the PCR studies. Double labelling *in situ* hybridisation/immunohistochemical studies have demonstrated that most of these rare EBV-positive lymphocytes are CD20-expressing (B lineage) lymphocytes, some cells are unlabelled by T and B lineage markers, and some cells are colabelled with CD3 or CD43, possibly representing T lymphocytes. In our studies, EBV was found in HD most frequently in the mixed cellularity subtype and less frequently in nodular sclerosing. We have not yet found a case of nodular, lymphocyte predominance in which the L&H cells are EBV-positive.

In recent months, we have analysed EBV in HD in patients with high stage disease at presentation for whom multiple tissue samples were available for study (Vasef *et al.*, submitted). Forty-three specimens from fourteen patients were studied. In 6 patients, all 18 sites showed evidence of EBV within Hodgkin's cells. In contrast, the other 25 sites from the other eight patients were all negative for EBV within the Hodgkin's cells. The unanimity of positivity or negativity in those studies provides additional evidence that the EBV infection occurs early in the pathogenesis of the neoplasm, and certainly prior to dissemination.

In another study, we found an extremely high prevalence of EBV in the RS cells of HD occurring in an underdeveloped area of Peru (Chang *et al.*, 1993). The characteristics of the patient population were typical of what has been described for "third-world" HD, with a median age of 9 years, a male:female ratio of 3.5:1, and a predominance of cases of mixed cellularity subtype. EBV RNA was identified in all or virtually all of the RS cells and variants in 30 of the 32 cases (94%). This rate of positivity was statistically significantly higher than that found in typical Western cases, even after controlling for age and histological subtype. In addition, we and others have found a very high incidence of EBV-positivity in HD occurring in HIV-infected patients, higher than the incidence of EBV-positivity in HIV-associated non-Hodgkin's lymphomas (Herndier *et al.*, 1993). Interestingly, the one EBV-negative case in our study of EBV in HIV-associated HD was of the nodular, lymphocyte predominance subtype.

Finally, we have studied 13 cases of the extremely rare entity of chronic lymphocytic leukaemia/small lymphocytic lymphoma with RS-like cells for evidence of EBV (Momose *et al.*, 1992). In all patients, there was firm immunological evidence that these patients indeed had a low-grade B-cell lymphoproliferative disorder. In addition, all cases showed occasional cells with nuclear features indistinguishable from Hodgkin's cells; only the lack of the typical polymorphous background cells distinguished these cases from HD. In five cases, the Hodgkin's-like cells had immunological features of B-cells, whereas six cases had cells with immunological features typical of Hodgkin's cells. In two cases, a transitional phenotype was seen. The small lymphocytes were EBV-negative in all cases. However, regardless of the phenotype of the Hodgkin's cells, they contained EBV in 12 of 13 cases. Three of the cases showed subsequent evidence of disseminated classical HD; in the one case for which study material was available, the RS cells in the subsequent HD were EBV-positive. We hypothesise that rare cases of HD may arise from chronic lymphocytic leukaemia or small lymphocytic lymphoma, and that this transformation is somehow mediated by EBV.

In summary, the studies performed to date by our group and others provide strong evidence for an association of EBV with HD in approximately 50% of cases. It is important to stress that in about 50% of cases of HD, the RS cells and variants have been shown not to contain EBV. None of the studies performed to date have provided compelling evidence for a specific role for EBV in either the aetiology or pathogenesis of HD. Certainly, additional work needs to be performed in EBV-positive cases, to further investigate the epidemiological features and to define its role. In EBV-negative cases, it will be important to identify other possible aetiological factors such as other infectious agents.

REFERENCES

Armstrong AA, Gallagher A, Krajewski AS, Jones DB, Wilkins BS, Onions DE, Jarrett RF (1992) The expression of the EBV latent membrane protein (LMP-1) is independent of CD23 and bcl-2 in Reed-Sternberg cells in Hodgkin's disease. *Histopathol* **21**:72

Aster J, Kobayashi Y, Shiota M, Sklar J (1992) Detection of the t(14;18) at similar frequencies in hyperplastic lymphoid tissues from American and Japanese patients. *Am J Pathol* **141**:291

Athan E, Chadburn A, Knowles DM (1992) The BCL-2 gene translocation is undetectable in Hodgkin's disease by Southern blot hybridisation and polymerase chain reaction. *Am J Pathol* **141**:193

Bhagat SKM, Medeiros LJ, Weiss LM, Wang J, Raffeld M, Stetler-Stevenson M (1993) bcl-2 expression in Hodgkin's disease: correlation with the t(14;18) translocation and Epstein-Barr virus. *Am J Clin Pathol* **99**:604

Chang KL, Albujar PF, Chen Y-Y, Johnson RM, Weiss LM (1993) High prevalence of Epstein-Barr virus in the Reed-Sternberg cells of Hodgkin's disease occurring in Peru. *Blood* **83**:496

Chang KL, Chen Y-Y, Shibata D, Weiss LM (1992) *In situ* hybridisation methodology for the detection of EBV EBER-1 RNA in paraffin-embedded tissues, as applied to normal and neoplastic tissues. *Diagn Molec Pathol* **1**:246

Doglioni C, Pelosio P, Mombello A, Scarpa A, Chilosi M (1991) Immunohistochemical evidence of abnormal expression of the antioncogene-encoded p53 phosphoprotein in Hodgkin's disease and CD30+ anaplastic lymphomas. *Hematol Pathol* **5**:67

Gupta RK, Whelan JS, Lister TA, Young BD, Bodmer JG (1992) Direct sequence analysis of the t(14;18) chromosomal translocation in Hodgkin's disease. *Blood* **79**:2084

Gupta RK, Norton AJ, Thompson IW, Lister TA, Bodmer JG (1992) p53 expression in Reed-Sternberg cells of Hodgkin's disease. *Br J Cancer* **66**:649

Gupta RK, Patel K, Bodmer WF, Bodmer JG (1993) Mutation of p53 in primary biopsy material and cell lines from Hodgkin disease. *Proc Natl Acad Sci USA* **90**:2817

Henderson S, Rowe M, Gregory C, Croom-Carter D, Wang F, Longnecker R, Kieff E, Rickinson A (1991) Induction of bcl-2 expression by Epstein-Barr virus latent membrane protein 1 protects infected B cells from programmed cell death. *Cell* **65**:117

Herndier B, Sanchez H, Chang KC, Chen Y-Y, Weiss LM (1993) High prevalence of detection of EBV RNA in the Reed-Sternberg cells of HIV-associated Hodgkin's disease. *Am J Pathol* **142**:1073

Khan G, Gupta RK, Coates PJ, Slavin G (1993) Epstein-Barr virus infection and bcl-2 proto-oncogene expression. Separate events in the pathogenesis of Hodgkin's disease? *Am J Pathol* **143**:1270

LeBrun DP, Ngan B-Y, Weiss LM, Huie P, Warnke RA, Cleary ML (1994) The *bcl*-2 oncogene in Hodgkin's disease arising in the setting of follicular non-Hodgkin's lymphoma. *Blood* **83**:223

Limpens J, de Jong D, van Krieken JHJM, Price CGA, Young BD, van Ommen G-JB, Kluin PM (1991) Bcl-2/JH rearrangements in benign lymphoid tissues with follicular hyperplasia. *Oncogene* **6**:2271

Louie DC, Kant JA, Brooks JJ, Reed JC (1991) Absence of t(14;18) major and minor breakpoints and of bcl-2 protein overproduction in Reed-Sternberg cells of Hodgkin's disease. *Am J Pathol* **139**:1231

Masih A, Sun J, Strobach S, Mitchell D, Wu K (1991) Detection of t(14;18) in Hodgkin's disease by the polymerase chain reaction: correlation with EBV genome and histologic subtype. *Lab Invest* **64**:77A, (abstr)

Momose H, Jaffe ES, Shin SS, Chen Y-Y, Weiss LM (1992) Chronic lymphocytic leukemia/small lymphocytic lymphoma with Reed-Sternberg-like cells and possible transformation to Hodgkin's disease: mediation by Epstein-Barr virus. *Am J Surg Pathol* **16**:859

Niedobitek G, Rowlands DC, Young LS, Herbst H, Williams A, Hall P, Padfield J, Rooney N, Jones EL (1993) Overexpression of p53 in Hodgkin's disease: lack of correlation with Epstein-Barr virus infection. *J Pathol* **169**:207

Poppema S, Kaleta J, Hepperle B (1992) Chromosomal abnormalities in patients with Hodgkin's disease: evidence for frequent involvement of the 14q chromosomal region but infrequent bcl-2 gene rearrangement in Reed-Sternberg cells. *J Natl Cancer Inst* **84**:1789

Reid A, Cunningham RE, Frizzera G, O'Leary TJ (1993) BCL-2 rearrangement in Hodgkin's disease. Results of polymerase chain reaction, flow cytometry and sequencing on formalin-fixed, paraffin-embedded tissue. *Am J Pathol* **142**:395

Said JW, Sassoon AF, Shintaku IP, Kurtin PJ, Pinkus GS (1991) Absence of bcl-2 major breakpoint region and JH gene rearrangement in lymphocyte predominance Hodgkin's disease. *Am J Pathol* **138**:261

Segal GH, Wittwer CT, Fishleder AJ, Stoler MH, Tubbs RR, Kjeldsberg CR (1992) Identification of monoclonal B-cell populations by rapid cycle polymerase chain reaction. A practical screening method for the detection of immunoglobulin gene rearrangements. *Am J Pathol* **141**:1291

Shibata DM, Hansmann M-L, Weiss LM, Nathwani BN (1991) Epstein-Barr virus infections and Hodgkin's disease: a study of fixed tissues using the polymerase chain reaction. *Hum Pathol* **22**:1262

Shibata D, Hu E, Weiss LM, Nathwani BN (1990) Detection of specific t(14;18) chromosomal translocations in fixed tissues. *Hum Pathol* **21**:199

Stetler-Stevenson M, Crush-Stanton S, Cossman J (1990) Involvement of the bcl-2 gene in Hodgkin's disease. *J Natl Cancer Inst* **82**:855

Trümper LH, Brady G, Bagg A, Gray D, Loke SL, Griesser H, Wagman R, Braziel R, Gascoyne RD, Vicini S, Iscove N, Cossman J, Mak TW (1993) Single-cell analysis of Hodgkin and Reed-Sternberg cells: molecular heterogeneity of gene expression and p53 mutations. *Blood* **81**:3097

Weiss LM, Chang KL (1993) The bcl-2 proto-oncogene. *Appl Immunohistochem* **1**:163

Weiss LM, Chang KL (1992) Molecular biologic studies of Hodgkin's disease. *Semin Diagn Pathol* **9**:272

Weiss LM, Warnke RA, Sklar J, Cleary ML (1987) Molecular analysis of the t(14;18) chromosomal translocation in malignant lymphomas. *N Engl J Med* **317**:1185

Weiss LM (1994) Retinoblastoma protein: another target for immunohistochemists. *Appl Immunohistochem* **2**:69

Weiss LM, Strickler JG, Warnke RA, Purtilo DT, Sklar J (1987) Epstein-Barr viral DNA in tissues of Hodgkin's disease. *Am J Pathol* **129**:86

Weiss LM, Warnke RA, Sklar J (1988) Clonal antigen receptor gene rearrangements and Epstein-Barr viral DNA in tissues of Hodgkin's disease. *Hematol Oncol* **6**:233

Weiss LM, Movahed LA, Warnke RA, Sklar J (1989) Detection of Epstein-Barr viral genomes in Reed-Sternberg cells of Hodgkin's disease. *N Engl J Med* **320**:502

Weiss LM, Chen Y-Y, Liu X-F, Shibata D (1991) Epstein-Barr virus and Hodgkin's disease. A correlative *in situ* hybridization and polymerase chain reaction study. *Am J Pathol* **139**:1259

Whitney K, Mouradian J, Chen Y-T (1992) Absence of bcl-2 translocation in Hodgkin's disease: analysis by polymerase chain reaction. *Mod Pathol* **5**:91A (abstr)

Zhang Q, Gutsch D, Kenney S (1994) Functional and physical interaction between p53 and BZLF1: implications for Epstein-Barr virus latency. *Molec Cell Biol* **14**:1929

TUMOUR SUPPRESSOR GENES IN HODGKIN'S DISEASE

Miguel A Piris, Juan C Martinez, Margarita Sanchez-Beato, Juan F Garcia, Carmen Bellas, Javier Menarguez, Raquel Villuendas, Emilia Lloret

Department of Pathology, Virgen de la Salud Hospital, Toledo; Ramon y Cajal Hospital, Madrid, Hospital Provincial Gregorio Marañon, Madrid, Spain

This work was made possible by grants 92/0721 and 93/0016, from the Fondo de Investigacion Sanitaria (FIS), Spain.

INTRODUCTION

Molecular studies of neoplastic transformation are uncovering a process involving an accumulation of genetic alterations in the initiation and progression of neoplasia. These include oncogene activation leading to growth promotion, and the inactivation of the so-called tumour suppressor genes (TSG) resulting in loss of the growth suppressor effect.

Although the precise mechanisms of action of oncogenes and TSG are not fully understood, an increasing number of studies are showing that some oncogenes and TSG play a fundamental role in the regulation of cell cycle progression. The equilibrium between the increase of cell numbers resulting from cell division and decrease due to programmed cell death (apoptosis) or cell cycle exit is emerging as a key issue in neoplastic transformation. This equilibrium is the result of the interaction of numerous oncogenic and TSG proteins, the dysregulation of which, in non-Hodgkin's lymphomas (NHLs), has been shown to possibly lead to:

a) Abnormal, prolonged survival resulting from either the altered expression of genes that inhibit apoptosis, such as bcl-2, or viral infection, such as Epstein-Barr virus (EBV) infection and subsequent expression of proteins like EBV latent membrane protein-1 (LMP-1).

b)   Alteration of DNA repair mechanisms due to p53 TSG inactivation, secondary either to p53 mutation or as result of MDM2 gene overexpression.

c)   Induction of cell transformation by activation of proto-oncogenes, such as the constitutive activation of the c-myc oncogene in Burkitt's lymphomas, due to translocation to the immunoglobulin locus.

d)   Absence of cell cycle control, due to p53 or Rb dysregulation or functional inactivation.

p53

p53 mutations represent the most common genetic alterations found in human tumours (Nigro *et al.*, 1989; Levine *et al.*, 1991). The p53 TSG is located in the short arm of chromosome 17 and encodes a nuclear phosphoprotein. Wild type (wt) p53 protein behaves as a negative regulator of cell proliferation inhibiting cell transformation (Finlay, 1989). However, point mutations can convert the gene into a dominant oncogene with transforming activities (Hinds *et al.*, 1989; Lane and Benchimol, 1990). Studies of DNA-damaged ultraviolet (UV) irradiated cells suggest that wt-p53 acts as a DNA damage response protein, which induces cell cycle block at the G1/S transition checkpoint to allow DNA damage repair, or alternatively as an inducer of apoptosis in cases of irreversibly damaged DNA (Fritsche *et al.*, 1993; Hall *et al.*, 1993).

Recent observations indicate that this action is not the direct result of wt-p53. Cell cycle progression block, induced by the p53 control checkpoint, results after p53-mediated transactivation of the WAF1 (for Wt-p53 Activated Fragment) gene which encodes a 21 kD protein (p21/WAF1) also called Cip1 (for Cdk-Interacting Protein). This is a potent inhibitor of cyclin-cyclin dependent kinase complexes with a negative regulatory effect on cell cycle progression (El-Diery *et al.*, 1993; Harper *et al.*, 1993; Xiong *et al.*, 1993). Furthermore, *in vitro* studies have shown that the p21/WAF1 protein also inhibits Rb protein phosphorylation by cyclin A-cdk2, cyclin E-cdk2 and cyclin D1-cdk4 complexes (Harper *et al.*, 1993). These alterations in the p53 gene will damage the cell autocontrol system, thereby allowing cells with DNA damage to complete the cell cycle and thus to contribute to the propagation of genetic alterations.

p53 gene mutations have been described in a wide variety of human haematopoietic tumours including acute myeloblastic leukaemia (AML) (Fenaux *et al.*, 1990; Slingerland *et al.*, 1991), blastic transformation of chronic myeloid leukaemia (CML) (Mashal *et al.*, 1990; Feinstein *et al.*, 1991) and acute lymphoblastic leukaemia (Fenaux *et al.*, 1992). In NHLs, p53 mutations have been described in Burkitt's lymphoma (Gaidano *et al.*, 1991; Ballerini *et al.*, 1993) and Burkitt's cell lines (Gaidano *et al.*, 1991; Sakashita *et al.*, 1992), chronic lymphocytic leukaemia (Gaidano *et al.*, 1991) and adult T-cell leukaemia lymphoma (Cesarman *et al.*, 1992; Sakashita *et al.*, 1992), high grade NHLs (Villuendas *et al.*, 1993; Ichikawa *et al.*, 1993; Cesarman *et al.*, 1993; Nakamura *et al.*, 1993) and transformation of follicular lymphoma (Sander *et al.*, 1993; LoCoco *et al.*, 1993). In spite of the existence of p53 mutations over a wide spectrum of NHLs, it appears that p53 expression is frequently independent of mutation in these tumours, since several studies have agreed in confirming the absence of p53 mutations in a significant percentage of cases with high levels of protein expression (Villuendas *et al.*, 1993; Cesarman *et al.*, 1993; Nakamura *et al.*, 1993; Pezzella *et al.*, 1993).

p53 expression in solid tumours has often been found to be a consequence of p53 mutations which modify the conformation of the protein, stabilising it. The majority of p53 mutations occur as missense changes grouped in the highly conserved regions of the coding sequence (exon 5 through to 8). These mutations have been associated with increased nuclear or cytoplasmic expression in human lung, breast, oesophageal, ovarian and colon carcinoma (Iggo *et al.*, 1990; Bartek *et al.*, 1990; Rodrigues *et al.*, 1990; Bennet *et al.*, 1991; Marks *et al.*, 1991; Thor *et al.*, 1992; Moll *et al.*, 1992). Higher levels of p53 protein seem to result from a longer half-life of the mutated protein, resulting from conformational changes which stabilise the protein (4-8 hours) (Finlay *et al.*, 1989; Halevy *et al.*, 1989; Gannon *et al.*, 1990). In contrast, the wt-p53 protein does not normally accumulate in amounts detectable by immunohistochemical techniques, its half-life being 20 minutes (Reich *et al.*, 1989).

However, several studies performed on NHLs have provided evidence that the nuclear or cytoplasmic stabilisation of p53 protein can also depend on other factors, such as interaction with other molecules (for instance MDM2), as has been described in soft tissue sarcomas (Oliner *et al.*, 1992). The presence of detectable p53, independent of mutation, has been described in normal human bone marrow blasts (Rivas *et al.*, 1992), AML (Zhang *et al.*, 1992) and in the normal cells of a member of a cancer family (Barnes *et al.*, 1992). Immunohistochemical findings have even shown p53 expression in reactive lymphocytes and epithelial basal cells in toxoplasmic lymphadenitis and reactive tonsils, as well as in PHA-stimulated lymphocytes (Villuendas *et al.*, 1992; Mateo *et al.*, 1994).

p53 inactivation independent of mutation has also been described due to association with viral oncoproteins such as the SV40 large T antigen, HPV E6 in cervical carcinomas (Scheffner *et al.*, 1990; Crook *et al.*, 1992), adenovirus E1B protein and cellular proteins such as the MDM2 oncoprotein in some human sarcomas (Oliner *et al.*, 1992). The stabilisation of p53 has also been described in cells infected by the lymphotropic virus HTLV-I; in this example co-precipitation of p53 and viral proteins was not observed (Reid *et al.*, 1993).

MDM2

The oncogene MDM2 (for Murine Double Minute) was identified in the spontaneously transformed BALB/c 3T3 murine cell line (Cahilly-Snyder *et al.*, 1987). MDM2 amplification and/or overexpression increases tumourigenic cell capacity (Fakharzaden *et al.*, 1991). The human MDM2 gene has recently been cloned and localised to chromosome 12q13-14. It has been found to be dysregulated in human sarcomas, gliomas, breast cancer and leukaemias (Ladanyi *et al.*, 1993; Reiferberger *et al.*, 1993; Sheikh *et al.*, 1993; Bueso-Ramos *et al.*, 1993).

The MDM2 protein is a nuclear protein with two zinc finger motifs and an acidic transactivator domain, suggesting that it is a DNA binding protein (Fakharzaden et al., 1991). MDM2 protein can bind both wt and mutant p53 (Barak and Oren, 1992; Hinds et al., 1990), and it has been shown that MDM2 overexpression inhibits p53 by concealing its transactivational domain (Momand et al., 1992; Oliner et al., 1993). This could be the transforming mechanism of action of MDM2 oncogenes. Besides its amplification in sarcomas and its overexpression in leukemias (Bueso-Ramos et al., 1993), it has very recently been reported that MDM2 activation takes place as a result of chromosomal translocation in the cell line SP2, derived from a murine plasmocytoma. This translocation juxtaposes the MDM2 gene with the immunoglobulin C_κ gene segment, in a process analogous to the c-myc activation in the 8;14 translocation (Barberich and Cole, 1994).

EBNA-LP

A potential role for other p53 binding proteins, such as EBNA-LP (also called EBNA-5), has also to be taken into consideration in the understanding of lymphoproliferative processes. EBNA-LP protein may bind and inactivate both p53 and Rb proteins. Inactivation of both proteins in EBV-infected cells is associated with an increase in EBV expression, and in the number of EBNA-2 and EBNA-LP positive cells. This suggests that EBV uses this inactivation mechanism in the induction of cell transformation (Szekely et al., 1993).

Rb

The retinoblastoma gene is the prototype tumour suppressor gene. It is located on chromosome 13 and encodes a 110 kD nuclear phosphoprotein, with multiple phosphorylation sites, associated with DNA-binding activity (Lee et al., 1987). It seems to act in a control checkpoint at the G1/S transition of the cell cycle.

Rb protein, under normal conditions, interacts with different nuclear proteins, such as the transcription factor E2F (Chellapan et al., 1991; Bandara et al., 1991). It blocks the expression of positive regulators of the cell cycle, such as c-myc (Rustgi et al., 1991). Several viral oncoproteins such as viral SV40 large T antigen, adenovirus E1A, human papillomavirus E7 and EBV EBNA-LP (also called EBNA-5) ((Whyte et al., 1988; Szekely et al., 1993) have Rb binding capacity, blocking its normal function. This Rb inactivation seems to be a required step in cell transformation by these oncogenic viruses.

Involvement in cell-cycle regulation was inferred from changes in Rb phosphorylation occurring during progression through the cell cycle (Buchovich et al., 1989; Mihara et al., 1989; Chen et al., 1989). More recent studies show that besides phosphorylation, the cellular concentration of Rb protein changes during the cell cycle, from low levels undetectable by immunohistochemistry in the G0 and middle G1 phase, to ten fold greater levels in the G2/M phase (Xu et al., 1991; Martinez et al., 1993).

The study of knockout mice with biallelic deletion of the Rb gene has shown early *in utero* death, involving massive apoptosis, around the 14th day of gestation. Histopathological study revealed the absence of haematopoietic and neural differentiation, suggesting that the Rb gene may play a role in the control of cell cycle by inducing cellular differentiation (Lee *et al.*, 1992; Jacks *et al.*, 1992; Clarke *et al.*, 1992). Moreover, other investigations have demonstrated that Rb protein expression is maximum in embryonic murine tissue during the differentiation phase, and that this protein is necessary for the MyoD induced muscular differentiation (Gu *et al.*, 1993). Our group has observed expression of Rb protein in epithelial and lymphoid cell compartments, such as germinal centres and the basal cells of tonsilar epithelium, where it is needed to stop proliferation in order to induce differentiation (Martinez *et al.*, 1993; Mateo *et al.*, 1994).

The deletion or inactivation of Rb alleles plays an essential role in the development of retinoblastomas and osteosarcomas in patients who have inherited a mutated gene (Friend *et al.*, 1986). Rb inactivation has also been observed in other sarcomas, pulmonary small cell cancer and epithelial tumours of breast bladder and prostate (Cance *et al.*, 1990; Lee *et al.*, 1988; Yokota *et al.*, 1988; T'ang *et al.*, 1988; Brookstein *et al.*, 1990). In NHLs, Rb protein is expressed mainly in high grade lymphomas, there being a general relationship between the cell growth fraction measured by the Ki-67 antibody and reactivity for Rb, similar to that found in reactive lymphoid tissue. However, there was a group with Rb dysregulation in which only a small amount of Rb protein was detected. Interestingly, in a significant fraction of these tumours, a higher level of Rb protein was found, far higher than that which would be expected following studies in reactive lymphoid tissue (Martinez *et al.*, 1993).

p53 EXPRESSION IN HODGKIN'S DISEASE

The application of immunohistological techniques to paraffin sections has led to the recognition that p53 expression is a very frequent finding in Hodgkin's disease (HD). Although initial studies performed on frozen sections suggested that p53 protein could be detected in Reed-Sternberg (RS) cells from only 35% of cases of HD (Doussis *et al.*, 1993), the use of microwave irradiating techniques on paraffin-embedded material has raised the level of detection to 90%. p53 protein appears to be expressed almost exclusively by RS cells and large mononuclear variants, and is only rarely detected in small lymphocytes (Fig. 1). p53 levels have been found in our material to be higher in cases of nodular sclerosis (HDNS) and lymphocytic depletion (HDLD) disease than they are in cases of mixed cellularity (HDMC) or in lymphocytic predominance (HDLP) disease (Fig. 2). Moreover, the level of p53 expression also appears to be related to the expression of EBV LMP-1. Thus, cases of HD with LMP-1 expression have lower p53 values than cases which are EBV-negative. This alternative expression of p53 and LMP-1 could suggest that either of these proteins may play a role in the genesis or progression of HD, while their simultaneous expression is not necessary (see Weiss, this volume). Levels of p53 protein have also been found to be related to levels of Bcl-2 protein. When cases of HD were stratified according to levels of Bcl-2 expression, a simultaneous expression of p53 and Bcl-2 protein was detected, p53 expression being higher in cases of HD with strong Bcl-2 expression. This simultaneous expression of p53 and Bcl-2 has been shown to constitute an adverse prognostic factor in NHLs in survival studies (Piris *et al.*, 1994). Its significance in HD therefore deserves further exploration.

Some data for HD suggest that p53 expression could be the consequence not only of p53 mutations, but also of the dysregulation of other proteins in the p53 pathway. One of these p53-associated proteins is MDM2.

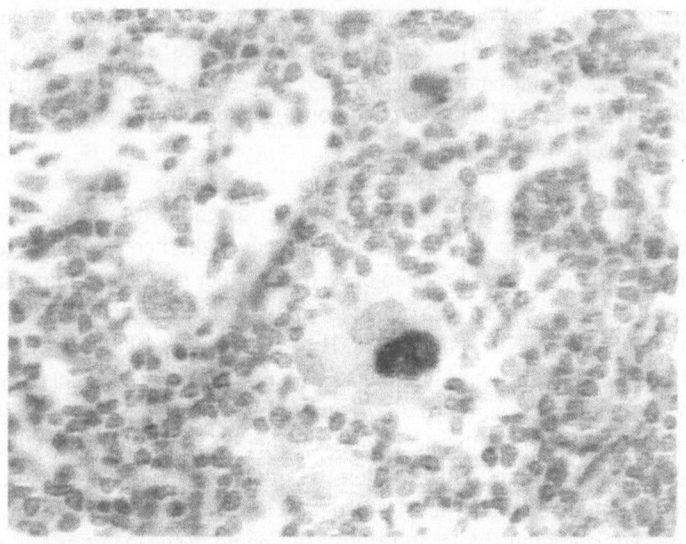

Fig. 1: p53 expression in a case of nodular sclerosis Hodgkin's disease. Staining of large mononuclear cells and Reed-Sternberg cells; staining is sometimes restricted to a single nucleus within a Reed-Sternberg cell.

P53, Rb, MIB1 AND EBV EXPRESSION

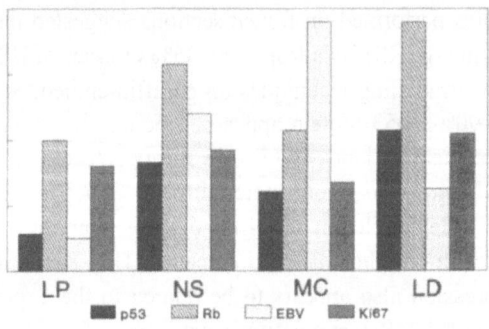

Fig. 2: p53, Rb, MIB1 and EBV expression in Hodgkin's disease. Distribution of p53, Rb, EBV LMP-1 and proliferative index (Ki-67) in the different histological subtypes of Hodgkin's disease, showing specific features for each histological type. LP, lymphocyte predominance Hodgkin's disease; NS, nodular sclerosis Hodgkin's disease; MC, mixed cellularity Hodgkin's disease; LD, lymphocyte depleted Hodgkin's disease.

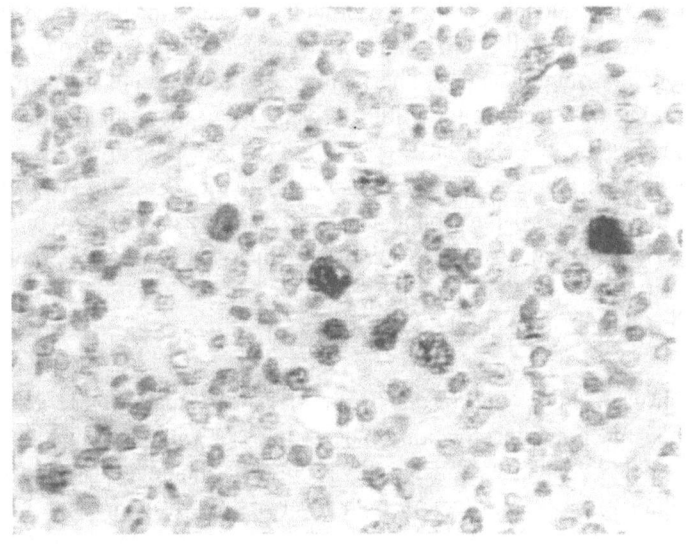

Fig. 3:   Rb staining of large cells was a rather constant feature in the cases of Hodgkin's disease studied.

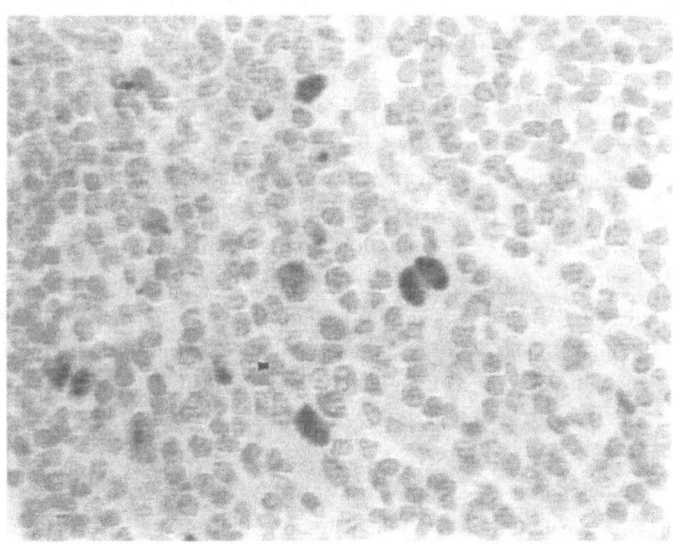

Fig. 4:   MDM2 staining paralleled p53 reactivity in most cases of Hodgkin's disease and many non-Hodgkin's lymphomas.

The results of a recent study (Martinez *et al.*, 1995) showed that in tonsils, stimulated lymphocytes, all cases of HD and the majority of high grade NHL specimens, MDM2 parallelled the pattern of p53 nuclear expression (Fig. 3). This was seen in a moderate number of large germinal centre cells but few interfollicular cells, some suprabasal epithelial cells and occasional prominent endothelial nuclei, Hodgkin's and RS cells, and also large neoplastic NHL cells. Immunoreactivity was mainly restricted to large nuclei in both non-tumour and tumour cells. The number of positive cells was greater, and the intensity of staining was stronger, for p53 than for MDM2. Interestingly, the most striking finding of the study was the constant expression of MDM2 protein by RS cells in all the cases of HD studied, irrespective of histological type. In these cases of HD, MDM2 protein was only identified in RS or large mononuclear cells, and was not present in small reactive lymphocytes. The simultaneous expression of p53 and MDM2 therefore probably reflects the balance between opposing biological signals which contribute to keeping cell proliferation under regulated control.

In the NHL group studied, our results seem to suggest that MDM2 expression may contribute to p53 functional inactivation. Indeed, in cases with p53 mutation, very low or null levels of MDM2 protein were detected. Nevertheless, a significant fraction of cases with wt configuration appear to have a high level of MDM2 expression. These results suggest a possible role for MDM2 dysregulation in lymphomagenesis, and are consistent with those previously found for soft tissue sarcomas and malignant gliomas (Leach et al., 1993; Reiferberger et al., 1993), in that simultaneous dysregulation of both p53 and MDM2 genes rarely seems to take place, suggesting that they are alternative mechanisms of inactivating p53.

This finding reinforces the probability of the existence of MDM2 overexpression independent of amplification, previously raised by the findings in acute leukaemia (Bueso-Ramos et al., 1993), where MDM2 overexpression has been found in 53% of acute leukemias to a level comparable with that found in sarcomas.

Another finding which suggests that p53 mutations may be an infrequent event in HD is their absence in most of the HD-derived cell lines studied. The only cell line in which a mutation has been found is CO, no mutation having been found in several other cell lines, such as DEV, L428, L540, HDLM-2, KM-H2 and HS445 (Gupta et al., 1993; Sanchez-Beato et al., submitted). This observation contrasts with the results obtained in NHLs, where p53 mutations appear to be a rather frequent finding in cell lines. It has been proposed that these mutations could give a selective growth advantage for the in vitro establishment of cell lines (Farrell et al., 1991).

p53 expression, independent of mutation, has already been described in cases of anaplastic large cell lymphoma, where it has been detected together with a wt-p53 in 73% of cases by Cesarman and colleagues (Cesarman et al., 1993). The opposite situation (p53 expression associated with mutation) was found in only 1/15 cases.

This and other findings suggest that the frequent expression of p53 by RS cells may be a consequence of p53 functional inactivation rather than p53 activation. However, two studies have provided some evidence in favour of p53 mutations in HD. One of these studies was performed with single RS cells isolated by a mechanical procedure (Trümper et al., 1993; see Trümper et al., this volume). Interestingly, p53 mutation was present in only some of the RS cells, wt-p53 being detected in others. The only other study describing p53 mutation in HD was performed by Gupta and co-workers (Gupta *et al.*, 1993), who showed p53 mutations in 2 patients, using a strategy of tumour cell enrichment by flow cytometry. Both patients were selected on the basis of a very high level of p53 protein expression.

RB EXPRESSION IN HODGKIN'S DISEASE

Rb protein expression is detectable in 95% of the HD cases studied (Fig. 4). Unlike the findings with p53, the expression of Rb is not restricted to RS and large mononuclear variants, as it is detectable in some small lymphocytes and endothelial cells. The level of Rb expression is greatest in cases of HDLD and HDNS, and lowest in HDLP and HDMC (Fig. 2). Analysis of the distribution of Rb according to proliferative index disclosed a strong relationship between Rb and Ki-67 in cases of HDLP, similar to that found in reactive lymphoid tissue. An association between Rb levels and p53 was mainly seen in cases with a high number of RS cells. In fact, both proteins appear to be quite constantly expressed by RS cells. An association between them has already been described, and the existence of multimolecular complexes containing a variable number of p53 and Rb molecules has been proposed in EBNA-LP expressing cells (Szekely *et al.*, 1993).

These results suggest that the absence of Rb expression is not a frequent lesion in HD. The most striking result obtained is the strong intensity of Rb expression. This high level of Rb protein has also been found in previous studies of NHL (Martinez *et al.*, 1993) and B-cell chronic lymphocytic leukaemia (Ginsberg *et al.*, 1992; Ahuja *et al.*, 1991). This raises the possibility that this high level of Rb could be a consequence of functional inactivation, as could occur if the Rb were sequestered and stabilised by another protein.

ACKNOWLEDGEMENTS

The authors would like to thank to Kevin Gatter and Ipatia Doussis, from the Department of Cellular Science, John Radcliffe Hospital, Oxford, UK, for the support they have given to this work, and Mrs D Gomez Donaire for her excellent technical assistance.

REFERENCES

Ahuja HG, Jat PS, Foti A, Bar-Eli M, Cline MJ (1991) Abnormalities of the retinoblastoma gene in the pathogenesis of acute leukemia. *Blood* **78**:3259

Ballerini P, Gaidano G, Gong JZ, Tassi V, Saglio G, Knowles DM, Dalla-Favera R (1993) Multiple genetic lesions in acquired immunodeficiency syndrome-related non-Hodgkin's lymphoma. *Blood* **81**:166

Bandara LR, Adamczewski JP, Hunt T, La Thangue NB (1991) Cyclin A and the retinoblastoma gene product complex with a common transcription factor. *Nature* **352**:249

Barak Y, Oren M (1992) Enhanced binding of a 95 kd protein to p53 in cells undergoing p53-mediated growth arrest. *EMBO J* **11**:2115

Barak Y, Juven T, Haffner R, Oren M (1993) MDM2 expression is induced by wild type p53 activity. *EMBO J* **12**:461

Berberich SM, Cole M (1994) The mdm-2 oncogene is translocated and overexpressed in a murine plasmocytoma cell line expressing wild-type p53. *Oncogene* **9**:1469

Barnes DM, Hanby AM, Gillett CE, Mohammed S, Hodgson S, Bobrow LG, Leigh IM, Purkis T, MacGeoch C, Spurr NK, Bartek J, Vojtesek B, Picksley SM, Lane DP (1992) Abnormal expression of wild type p53 protein in normal cells of a cancer family patient. *Lancet* **340**:259

Bartek J, Iggo R, Gannon J, Lane DP (1990) Genetic and immunohistochemical analysis of mutant p53 in human breast cancer cell lines. *Oncogene* **5**:893

Bennett WP, Hollstein MC, Zhu SM, Resau JH, Trump BF, Metcalf RA, Welsh JA, Midgley C, Lane DP, Harris CC (1991) Archival analysis of p53 genetic and protein alterations in Chinese esophageal cancer. *Oncogene* **6**:1779

Brookstein R, Shew JY, Chen PL, Scully P, Lee WH (1990) Suppression of tumourigenicity of human prostate carcinoma cells by replacing a mutated RB gene. *Science* **247**:712

Buchkovich K, Duffy LA, Harlow E (1989) The retinoblastoma protein is phosphorylated during specific phases of the cell cycle. *Cell* **58**:1097

Bueso-Ramos CE, Yang Y, de-Leon E, McCown P, Stass SA, Albitar M (1993) The human MDM2 oncogene is overexpressed in leukemias. *Blood* **82**:2617

Cahilly-Snyder L, Yang-Feng T, Francke U, George DL (1987) Molecular analysis and chromosomal mapping of amplified genes isolated from a transformed mouse 3T3 cell line. *Somat Cell Mol Genet* **13**:235

Cance WG, Brennan MF, Dudas ME, Huang CM, Cordon-Cardo C (1990) Altered expression of the retinoblastoma gene product in human sarcomas. *N Engl J Med* **323**:1457

Cesarman E, Inghirami G, Chadburn A, Knowles DM (1993) High levels of p53 protein expression do not correlate with p53 gene mutations in anaplastic large cell lymphoma. *Am J Pathol* **143**:845

Cesarman E, Chadburn A, Inghirami G, Gaidano G, Knowles DM (1992) Structural and functional analysis of oncogenes and tumour suppressor genes in adult T-cell leukemia/lymphoma shows frequent p53 mutations. *Blood* **80**:3205

Clarke AR, Maandag ER, van Roon M, van der Lugt NM, van der Valk M, Hooper ML, Berns A, te Riele H (1992) Requirement for a functional Rb-1 gene in murine development. *Nature* **359**:328

Crook T, Wrede D, Tidy JA, Mason WP, Evans DJ, Vousden KH (1992) Clonal p53 mutation in primary cervical cancer: association with human papillomavirus negative tumours. *Lancet* **339**:1070

Chellappan SP, Hiebert S, Mudryj M, Horowitz JM, Nevins JR (1991) The E2F transcription factor is a cellular target for the RB protein. *Cell* **65**:1053

Chen PL, Scully P, Shew JY, Wang JY, Lee WH (1989) Phosphorylation of the retinoblastoma gene product is modulated during the cell cycle and cellular differentiation. *Cell* **58**:1193

Doussis IA, Pezzella F, Lane DP, Gatter KC, Mason DY (1993) An immunocytochemical study of p53 and bcl-2 protein expression in Hodgkin's disease. *Am J Clin Pathol* **99**:663

El-Diery WS, Tokino T, Velculescu VE, Levy DB, Parsons R, Trent JM, Lin D, Mercer WE, Kinzler KW, Vogelstein B (1993) WAF1, a potential mediator of p53 tumour suppression. *Cell* **75**:817

Fakharzadeh SS, Trusko SP, George DL (1991) Tumourigenic potential associated with enhanced expression of a gene that is amplified in a mouse tumour cell line. *EMBO J* **10**:1565

Farrell PJ, Allan GJ, Shanahan F, Vousden KH, Crook T (1991) p53 is frequently mutated in Burkitt's lymphoma cell lines. *EMBO J* **10**:2879

Fearon ER, Vogelstein B (1990) A genetic model for colorectal tumourigenesis. *Cell* **61**:759

Feinstein E, Cimino G, Gale RP, Alimena G, Berthier R, Kishi K, Goldman J, Zaccaria A, Berrebi A, Canaani E (1991) p53 in chronic myelogenous leukemia in acute phase. *Proc Natl Acad Sci USA* **88**:6293

Fenaux P, Jonveaux P, Quiquandon I, Preudhomme C, Lai JL, Vanrumbeke M, Loucheux-lefebvre MH, Bauters F, Berger R, Kerckaert JP (1992) Mutations of the p53 gene in B-cell Lymphoblastic Acute Leukemia: a report on 60 cases. *Leukemia* **6**:42

Fenaux P, Collyn d`Hooghe M, Jonveaux P, Lai JL, Bauters F, Loucheux MH, Kerckaert JP (1990) Rearrangement and expression of the p53 gene in myelodysplastic syndrome and acute leukemia. *Nouv Rev Fr Hematol* **32**:341

Finlay CA (1993) The mdm-2 oncogene can overcome wild-type p53 suppression of transformed cell growth. *Mol Cell Biol* **13**:301

Finlay CA, Hinds PW, Levine AJ (1989) The p53 proto-oncogene can act as a suppressor of transformation. *Cell* **57**:1083

Friend SH, Bernardos R, Rogelj S, Weinberg RA, Rapaport JM, Albert DM, Dryja TP (1986) A human DNA segment with properties of the gene that predisposes to retinoblastoma and osteosarcoma. *Nature* **323**:643

Fritsche M, Haessler C, Brandner G (1993) Induction of nuclear accumulation of the tumour suppressor protein p53 by DNA-damaging agents. *Oncogene* **8**:307

Gaidano G, Ballerini P, Gong JZ, Inghirami G, Neri A, Newcomb EW, Magrath IT, Knowles DM, Dalla-Favera R (1991) p53 mutations in human lymphoid malignancies: association with Burkitt lymphoma and chronic lymphocytic leukemia. *Proc Natl Acad Sci USA* **88**:5413

Gannon JV, Greaves R, Iggo R, Lane DP (1990) Activating mutations in p53 produce a common conformational effect. A monoclonal antibody specific for the mutant form. *EMBO J* **9**:1595

Ginsberg AM, Raffeld M, Cossman J (1992) Mutations of the retinoblastoma gene in human lymphoid neoplasms. *Leuk Lymphoma* **7**:359

Gu W, Schneider JW, Condorelli G, Kaushal S, Mahdavi V, Nadal-Ginard B (1993) Interaction of myogenic factors and the retinoblastoma protein mediates muscle cell commitment and differentiation. *Cell* **72**:309

Gupta RK, Patel K, Bodmer WF, Bodmer JG (1993) Mutation of p53 in primary biopsy material and cell lines from Hodgkin disease. *Proc Natl Acad Sci USA* **90**:2817

Halevy O, Hall A, Oren M (1989) Stabilization of the p53 transformation-related protein in mouse fibrosarcoma cell lines: effects of protein sequence and intracellular environment. *Mol Cell Biol* **9**:3385

Hall PA, McKee PH, Menage HD, Dover R, Lane DP (1993) High levels of p53 protein in UV-irradiated normal human skin. *Oncogene* **8**:203

Harper JW, Adami GR, Wei N, Keyomarsi K, Elledge SJ (1993) The p21 Cdk- interacting protein Cip1 is a potent inhibitor of G1 cyclin-dependent kinases. *Cell* **75**:805

Hinds PW, Finlay CA, Quartin RS, Baker SJ, Fearon ER, Vogelstein B, Levine AJ (1990) Mutant p53 cDNAs from human colorectal carcinomas can cooperate with ras in transformation of primary rat cells. *Cell Growth Diff* **1**:571

Hinds P, Finlay C, Levine AJ (1989) Mutation is required to activate the p53 gene for cooperation with the ras oncogene and transformation. *J Virol* **63**:739

Ichikawa A, Hotta T, Takagi N, Tsushita K, Kinoshita T, Nagai H, Murakami Y, Hayashi K, Saito H (1992) Mutations of p53 gene and their relation to disease progression in B-cell lymphoma. *Blood* **79**:2701

Iggo R, Gatter K, Bartek J, Lane D, Harris AL (1990) Increased expression of mutant forms of p53 oncogene in primary lung cancer. *Lancet* **335**:675

Jacks T, Fazeli A, Schmitt EM, Bronson RT, Goodell MA, Weinberg RA (1992) Effects of an Rb mutation in the mouse. *Nature* **359**:295

Ladanyi M, Cha C, Lewis R, Jhanwar SC, Huvos AG, Healey JH (1993) MDM2 gene amplification in metastatic osteosarcoma. *Cancer Res* **53**:16

Lane DP and Benchimol S (1990) p53: oncogene or antioncogene. *Genes Dev* **4**:1

Leach FS, Tokino T, Meltzer P, Burrell M, Oliner JD, Smith S, Hill DE, Sidransky D, Kinzler KW, Vogelstein B (1993) p53 mutation and MDM2 amplification in human soft tissue sarcomas. *Cancer Res* **53**:2231

Lee EY, To H, Shew J-Y, Brookstein R, Scully P, Lee W-H (1988) Inactivation of the retinoblastoma susceptibility gene in human breast cancers. *Science* **241**:218

Lee WH, Shew JY, Hong FD, Sery TW, Donoso LA, Young LJ, Bookstein R, Lee EY (1987) The retinoblastoma susceptibility gene encodes a nuclear phosphoprotein associated with DNA binding activity. *Nature* **329**:642

Lee EY, Chang C-Y, Hu N, Wang Y-CJ, Lai C-C, Herrup K, Lee W-H, Bradley A (1992) Mice deficient for Rb are nonviable and show defects in neurogenesis and haematopoiesis. *Nature* **359**:288

Levine AJ, Momand J, Finlay CA (1991) The p53 tumour suppressor gene. *Nature* **351**:453

Lo Coco F, Gaidano G, Louie DC, Offit K, Chaganti RS, Dalla-Favera R (1993) p53 mutations are associated with histologic transformation of follicular lymphoma. *Blood* **82**:2289

Marks JR, Davidoff AM, Kerns BJ, Humphrey PA, Pence JC, Dodge RK, Clarke-Pearson DL, Iglehart JD, Bast RC, Berchuck A (1991) Overexpression and mutation of p53 in epithelial ovarian cancer. *Cancer Res* **51**:2979

Martínez JC, Piris MA, Sánchez-Beato M, Villuendas R, Orradre JL, Algara P, Sánchez-Verde L, Martínez P (1993) Retinoblastoma (Rb) gene product expression in lymphomas. Correlation with Ki67 growth fraction. *J Pathol* **169**:405

Martinez JC, Mateo M, Sanchez-Beato M, Villuendas R, Orradre JL, Algara P, Sanchez-Verde L, Garcia P, Lopez C, Martinez P, Piris MA (1995) MDM2 expression in lymphoid cells and reactive and neoplastic lymphoid tissue. Comparative study with p53 expression. *J Pathol* (in press)

Mashal R, Shtalrid M, Talpaz M, Kantarjian H, Smith L, Beran M, Cork A, Trujillo J, Gutterman J, Deisseroth A (1990) Rearrangement and expression of p53 in the chronic phase and blast crisis of chronic myelogenous leukemia. *Blood* **75**:180

Mateo MS, Sanchez-Beato M, Martinez JC, Orfao A, Orradre JL, Piris MA (1995) p53, bcl2 and Rb expression along the cell cycle. A study in PHA-stimulated lymphocytes and microwave irradiated lymphoid tissue sections. *J Clin Pathol* **48**:151

Mihara K, Cao XR, Yen A, Chandler S, Driscoll B, Murphree AL, T'Ang A, Fung YK (1989) Cell cycle-dependent regulation of phosphorylation of the human retinoblastoma gene product. *Science* **246**:1300

Moll UM, Riou G, Levine AJ (1992) Two distinct mechanisms alter p53 in breast cancer: mutation and nuclear exclusion. *Proc Natl Acad Sci USA* **89**:7262

Momand J, Zambetti GP, Olson DC, George D, Levine AJ (1992) The MDM2 oncogene product forms a complex with the p53 protein and inhibits p53-mediated transactivation. *Cell* **69**:1237

Nakamura H, Said JW, Miller CW, Koeffler HP (1993) Mutation and protein expression of p53 in acquired immunodeficiency syndrome-related lymphoma. *Blood* **82**:920

Nigro JM, Baker SJ, Preisinger AC, Jessup JM, Hostetter R, Cleary K, Bigner SH, Davidson N, Baylin S, Devilee P, Glover T, Collins FS, Weston A, Modali R, Harris CC, Vogelstein B (1989) Mutations in the p53 gene occur in diverse human tumour types. *Nature* **342**:705

Oliner JD, Kinzler KW, Meltzer PS, George DL, Vogelstein B (1992) Amplification of a gene encoding a p53-associated protein in human sarcomas. *Nature* **358**:80

Oliner JD, Pietenpol JA, Thiagalingam S, Gyuris J, Kinzler KW, Vogelstein B (1993) Oncoprotein MDM2 conceals the activation domain of tumour suppressor p53. *Nature* **362**:857

Pezzella F, Morrison H, Jones M, Gatter KC, Lane D, Harris AL, Mason DY (1993) Immunohistochemical detection of p53 and bcl-2 proteins in non- Hodgkin's lymphoma. *Histopathol* **22**:39

Piris MA, Pezzella F, Martinez-Montero JC, Orradre JL, Villuendas R, Sanchez-Beato M, Cuena R, Cruz MA, Martinez B, Garrido MC, Gatter K, Aiello A, Delia D, Giardini R, Rilke F (1994) p53 and bcl-2 expression in high-grade B-cell lymphomas: correlation with survival time. *Br J Cancer* **69**:337

Reid RL, Lindholm PF, Mireskandari A, Dittmer J, Brady JN (1993) Stabilization of wild-type p53 in human T-lymphocytes transformed by HTLV-1. *Oncogene* **8**:3029

Reifenberger G, Liu L, Ichimura K, Schmidt EE, Collins VP (1993) Amplification and overexpression of the MDM2 gene in a subset of human malignant gliomas without p53 mutations. *Cancer Res* **53**:2736

Rivas CI, Wisniewski D, Strife A, Perez A, Lambek C, Bruno S, Darzynkiewicz Z, Clarkson B (1992) Constitutive expression of p53 protein in enriched normal human marrow blast cell populations. *Blood* **79**:1982

Rodrigues NR, Rowan A, Smith ME, Kerr IB, Bodmer WF, Gannon JV, Lane DP (1990) p53 mutations in colorectal cancer. *Proc Natl Acad Sci USA* **87**:7555

Rustgi AK, Dyson N, Bernards R (1991) Amino-terminal domains of c-myc and N-myc proteins mediate binding to the retinoblastoma gene product. *Nature* **352**:541

Sakashita A, Hattori T, Miller CW, Suzushima H, Asou N, Takatsuki K, Koeffler HP (1992) Mutations of the p53 gene in adult T-cell leukemia. *Blood* **79**:477

Sander CA, Yano T, Clark HM, Harris C, Longo DL, Jaffe ES, Raffeld M (1993) p53 mutation is associated with progression in follicular lymphomas. *Blood* **82**:1994

Scheffner M, Werness BA, Huibregtse JM, Levine AJ, Howly PM (1990) The E6 oncoprotein encoded by human papillomavirus types 16 and 18 promoted the degradation of p53. *Cell* **63**:1129

Sheikh MS, Shao ZM, Hussain A, Fontana JA (1993) The p53-binding protein MDM2 gene is differentially expressed in human breast carcinoma. *Cancer Res* **53**:3226

Slingerland JM, Minden MD, Benchimol S (1991) Mutation of the p53 gene in human acute myelogenous leukemia. *Blood* **77**:1500

Szekely L, Selivanova G, Magnusson KP, Klein G, Wiman KG (1993) EBNA-5, an Epstein-Barr virus encoded nuclear antigen, binds to the retinoblastoma and p53 proteins. *Proc Natl Acad Sci USA* **90**:5455

T'ang A, Varley JM, Chakraborty S, Murphree AL, Fung Y-K (1988) Structural rearrangements of the retinoblastoma gene in human breast carcinoma. *Science* **242**:263

Thor AD, Moore DH, Edgerton SM, Kawasaki ES, Reihsaus E, Lynch HT, Marcus JN, Schwartz L, Chen LC, Mayall BH, Smith H (1992) Accumulation of p53 tumour suppressor gene protein: an independent marker of prognosis in breast cancers. *J Natl Cancer Inst* **84**:845

Trümper LH, Brady G, Bagg A, Gray D, Loke SL, Griesser H, Wagman R, Braziel R, Gascoyne RD, Vicini S, Iscove SS, Cossman J, Mak TW (1993) Single-cell analysis of Hodgkin and Reed-Sternberg cells: molecular heterogenity of gene expression and p53 mutations. *Blood* **81**:3097

Villuendas R, Piris MA, Algara P, Sanchez-Beato M, Sanchez-Verde L, Martinez JC, Orradre JL, Garcia P, Lopez C, Martinez P (1993) The expression of p53 protein in non-Hodgkin's lymphoma is not always dependent on p53 gene mutations. *Blood* **82**:3151

Villuendas R, Piris MA, Orradre JL, Mollejo M, Algara P, Sanchez L, Martinez JC, Martinez P (1992) p53 protein expression in lymphomas and reactive lymphoid tissue. *J Pathol* **166**:235

Whyte P, Buchkovich K, Horowitz JM, Friend SH, Raybuck RA, Weinberg RA, Harlow E (1988) Association between an oncogene and an anti-oncogene: the adenovirus E1A proteins bind to the retinoblastoma gene product. *Nature* **334**:124

Weinberg RA, Harlow E (1988) Association between an oncogene and an anti-oncogene: the adenovirus E1A proteins bind to the retinoblastoma gene product. *Nature* **334**:124

Wu X, Bayle JH, Olson D, Levine AJ (1993) The p53-mdm-2 autoregulatory feedback loop. *Genes Dev* **7**:1126

Xiong P, Hannon G, Zhang H, Casso D, Kobayashi R, Beach D (1993) p21 is a universal inhibitor of cyclin kinases. *Nature* **366**:701

Xu HJ, Hu SX, Benedict WF (1991) Lack of nuclear RB protein staining in G0/middle G1 cells: correlation to changes in total RB protein level. *Oncogene* **6**:1139

Yokota J, Akiyama T, Fung Y-K, Benedict WF, Namba Y, Hanaoka M, Wada M, Terasaki T, Shismosato Y, Sugimura T, Terada M (1988) Altered expression of the retinoblastoma (RB) gene in small-cell carcinoma of the lung. *Oncogene* **3**:471

Zhang W, Hu G, Estey J, Deisseroth A (1992) Altered conformation of the p53 protein in myeloid leukemia cells and mitogen-stimulated normal blood cells. *Oncogene* **7**:1645

INDEX

Main entries are indicated in bold; figures and tables are indicated in italics.

223

—D—

Dendritic reticulum cells	169
Disease	
Hodgkin's	*See Hodgkin's disease*
Kawasaki's	40, 104
Rosai-Dorfman	101
Dustbin	
disappearance of EBV-positive Hodgkin's derived	
lines into	175

—E—

EBV	*See Epstein-Barr virus*
Epidemiology of Hodgkin's disease	1
age-incidence	1, *2*, 76
childhood Hodgkin's disease	3
clustering	*See Clustering*
delayed exposure hypothesis	
	See Late-host-response model
developing countries	76, 95
family studies	*See Family studies*
gender and	2
histological subtypes, distribution of	2, 76
hormonal factors and	6
infectious mononucleosis	
	See Infectious mononucleosis
late-host-response model	
	See Late-host-response model
MacMahon's hypothesis	
	See Multiple aetiology hypothesis
multiple aetiology hypothesis	
	See Multiple aetiology hypothesis
polio model	*See Polio model*
risk factors	3, 4, 28
seroepidemiology	*See Seroepidemiology*
socio-economic status	3
	See also Epidemiology, developing countries
time trends in disease incidence	2
two-disease hypothesis	*See Two-disease hypothesis*
young adult Hodgkin's disease	3, 4, 82
Epithelial membrane antigen	67, 68, 69
Epstein-Barr virus	33, 53, 66
See also Hodgkin's disease and Epstein-Barr virus	
adhesion molecules and	*See Adhesion molecules*
AIDS-related lymphoma and	
	See AIDS-related lymphoma
anaplastic large cell lymphoma and	
	See Anaplastic large cell lymphoma
angiocentric T-cell lymphoma and	
	See Angiocentric T-cell lymphoma
BamHI A fragment of	58, 66
B-cells, infection of	33, 41
BCRF1 of	67, 68
BHRF1 of	66, 67, 68
Burkitt's lymphoma	*See Burkitt's lymphoma*
BZLF1 of	67, 68, 203
CD23, modulation by	59, 69
CD30, modulation by	69, 110
cell-mediated immunity against	
	See Epstein-Barr virus, cytotoxic T-cells
cellular genes, modulation of	66, 67
cytotoxic T-cells	33, 44, 162

Epstein-Barr virus, continued	
EBER *in situ* hybridisation	41, 55, *56*, 75, 77, 93,
	163, 174, 180, 189, 200, 203, 205
EBER RNAs of	*35*, *36*, 38, 41, 66, 187, 190, 193
	See also Epstein-Barr virus,
	EBER in situ hybridisation
	See Hodgkin's disease and Epstein-Barr virus,
	EBER RNAs
EBNA of	35
EBNA-1 of	35, 39, 44, 45, 58
EBNA-2 of	**35**, 37, 39, 67, 162, 180
	187, 189, 190, 193, 212
EBNA-3A of	**35**, 37, 162
EBNA-3B of	**35**, 37, 162
EBNA-3C of	**35**, 37, 162
EBNA-5 of	*See EBNA-LP*
EBNA-LP of	**35**, *36*, 37, *67*, 68, 212, 217
epithelial cells, infection of	34, 41
genome	35
genomes, clonality of	40, 54, 189, 204
healthy carrier and	33
Hodgkin's disease and	
	See Hodgkin's disease and Epstein-Barr virus
immunocompetent host and	39
immunocompromised host and	33, 38
	See also Hodgkin's disease, HIV
immunotherapy of	44
latent gene expression, pattern of	**34**, *37*, 58, 66
latent infection	**34**
	See also Epstein-Barr virus,
	latent gene expression, pattern of
latent proteins	35
lethal midline granuloma and	
	See Lethal midline granuloma
LMP-1 of	**35**, 37, 39, 44, 45, 59, 66, *67*, 180,
	187, 189, 190, 193, 200, 209
LMP-2 of	**35**, *36*, 44, 45, 66, *67*, 180
lymphoblastoid cell line	33, 35
lymphoma and	**33**
nasopharyngeal carcinoma	
	See Nasopharyngeal carcinoma
non-Hodgkin's lymphoma and	**33**, 39
oral hairy leukoplakia	*See Oral hairy leukoplakia*
pattern of latent gene expression	**34**, *37*, 58, 66
peripheral T-cell lymphomas	40
persistent infection	34, **41**
post-transplant lymphoproliferative disease	
	See Post-transplant lymphoproliferative disorders
primary infection by	34, 41
promoter usage	35, **45**
SCID mice and	180, 188
	See also SCID mice
subtypes of	37, 58
subtypes, biological differences of	38
subtypes, geographical distribution of	37
T-cells, infection of	40
transcription	35
transmission	33, 41
Zebra protein of	*See Epstein-Barr virus, BZLF1 of*

—F—

Family studies	**27**
EBV and	29